AN INFORMAL INTRODUCTION TO TURBULENCE

FLUID MECHANICS AND ITS APPLICATIONS
Volume 63

Series Editor: R. MOREAU
MADYLAM
Ecole Nationale Supérieure d'Hydraulique de Grenoble
Boîte Postale 95
38402 Saint Martin d'Hères Cedex, France

Aims and Scope of the Series

The purpose of this series is to focus on subjects in which fluid mechanics plays a fundamental role.

As well as the more traditional applications of aeronautics, hydraulics, heat and mass transfer etc., books will be published dealing with topics which are currently in a state of rapid development, such as turbulence, suspensions and multiphase fluids, super and hypersonic flows and numerical modelling techniques.

It is a widely held view that it is the interdisciplinary subjects that will receive intense scientific attention, bringing them to the forefront of technological advancement. Fluids have the ability to transport matter and its properties as well as transmit force, therefore fluid mechanics is a subject that is particulary open to cross fertilisation with other sciences and disciplines of engineering. The subject of fluid mechanics will be highly relevant in domains such as chemical, metallurgical, biological and ecological engineering. This series is particularly open to such new multidisciplinary domains.

The median level of presentation is the first year graduate student. Some texts are monographs defining the current state of a field; others are accessible to final year undergraduates; but essentially the emphasis is on readability and clarity.

For a list of related mechanics titles, see final pages.

An Informal Introduction to Turbulence

by

ARKADY TSINOBER

Department of Fluid Mechanics,
Faculty of Engineering, Tel Aviv University,
Tel Aviv, Israel

KLUWER ACADEMIC PUBLISHERS

DORDRECHT / BOSTON / LONDON

A C.I.P. Catalogue record for this book is available from the Library of Congress.

ISBN 1-4020-0110-X (HB)
ISBN 1-4020-0166-5 (PB)

Published by Kluwer Academic Publishers,
P.O. Box 17, 3300 AA Dordrecht, The Netherlands.

Sold and distributed in North, Central and South America
by Kluwer Academic Publishers,
101 Philip Drive, Norwell, MA 02061, U.S.A.

In all other countries, sold and distributed
by Kluwer Academic Publishers,
P.O. Box 322, 3300 AH Dordrecht, The Netherlands.

Printed on acid-free paper

TO MY WIFE

TABLE OF CONTENTS

PREFACE

Over the last decade a number of books on turbulence have appeared. To mention a few: Biswas and Eswaran (2000), Bohr *et al.* (1998), Chorin (1994), Durbin and Pettersson (2001), Frisch (1995), Holmes *et al.* (1996), Lesieur (1997), Libby (1996), Mankbadi (1994), Mathiew and Scott (2000), McComb (1990), Piquet (1999), Pope (2000), Scott and Mathiew (2000). So why one more book on the subject? Does it make any difference?

The key words are *informal introduction*. The book, which is *not a textbook*, is essentially an *introduction,* and it is an *informal* introduction that, as far as possible, presents its material in a qualitative form. There are several reasons why an introduction to turbulence should be as informal (in several meanings) as possible.

First, there seems to be little chance in the foreseeable future of creating a pure, formal theory of turbulence - even for its simplest cases. To quote A. N. Kolmogorov (1985): *I soon understood that there was little hope of developing a pure, closed theory, and because of absence of such a theory the investigation must be based on hypotheses obtained on processing experimental data,* see Tikhomirov, 1991, p. 487. This refers – in the words of E. Spiegel - to the big T-problem, i.e. the true dynamical problem (most probably) described by the Navier-Stokes equations, which is the main focus of this book; surrounding the big T-problem there are several 'little t-problems' such as turbulent diffusion or more generally behaviour of passive objects in real turbulent or some artificial random velocity field ('scalar turbulence'). Indeed the heaviest and the most ambitious armoury from theoretical physics and mathematics was tried for more than fifty years, but without much success - turbulence, as a physical and mathematical problem remains unsolved. There is even no consensus on what is (are) the problem(s) of turbulence[1], neither is there an agreement on what are/should be the aims/goals of turbulence research/theories and what would constitute its (their) solution. Therefore lots of formalisms are avoided, since the methods mostly brought in from linear analysis (such as

---

[1]*One of the things that I always found troubling in the study of the problem of turbulence is that I am not quite sure what the theoretical turbulence problem actually is... One reason I think we have so much difficulty in solving it, is that we are not really sure what it is* (Saffman, 1991).

various decompositions, perturbation methods, etc.) failed, and genuinely nonlinear analytic methods applicable to turbulence mostly do not exist.

Second, the existing theoretical (mathematical, physico-theoretical and traditional fluid mechanical) material is rather complicated and extremely large in scope. The same is true of the experimental information (laboratory, field and numerical). Many existing books are overloaded with technical details, a notable exception being the most successful course by Tennekes and Lumley (1972) and also the book by Pope (2000). The unwary reader is totally lost in the enormous ocean of existing references. Therefore highly technical information has, for the most part, not been included in this volume. Instead references are given in the text, where appropriate, to the above mentioned books and other sources.

Third, the subject as no other is of intimate vital interest and importance for scientists in a really enormous variety of fields.[2]

Finally, many of the existing texts tend to avoid most of the controversies, contradicting views, unresolved questions (in turbulence there are more of them than 'solutions'); they attempt to 'smooth the angles' as much as possible, in ways that are, to my view, both inappropriate and misleading. Consequently some interpretations/views expressed in this book may appear to some as flagrant/egregious.

For these reasons all the subject(s) will be discussed in an essentially informal form/style, maximally avoiding complicated formalisms, which are useful to a very limited extent only - if at all. Following the advice of Leonardo da Vinci *Remember, when discoursing about water, to induce first experience, then reason* the emphasis is on the physical aspects based in the first place on observations and empirical facts as distinct from intuitive conjectures/hypotheses. This does not mean that the latter are ignored and that the presentation is oversimplified or even easy.

The informal nature of the book allows also to make it 'thin'. The very nature of the problem of turbulence and turbulent flows (absence of a systematic theory, extreme difficulty and enormous scope of the subject) left its mark/imprint on this small book - there are no simple analytical solutions (with one exception – the 4/5 law of Kolmogorov – they do not exist in the field of turbulence), etc. On the contrary, it contains as much questions and similar things (or even more) as answers, (hopefully) unbiased discussion of the unresolved issues, controversies and major problems. In

---

[2]Mathematics including Applied and Computational, Physics, Engineering (Aeronautical, Naval/Marine, Hydraulic, Civil/Environmental, Chemical/Petroleum, Material Processing,...), Geo-Astrophysical Sciences (Atmospheric, Meteorology, Oceanography, Fluid Dynamics of Earth Interior, Astrophysics, Cosmology), Bio-Medical Fluid Dynamics. ... *turbulence undoubtedly represents a central principle for many parts of physics, and a thorough understanding of its properties must be expected to lead to important advances in many fields* (Neumann, 1949).

particular there are no lapses into brevity at difficult places.[3]

Due to the above mentioned nature of the problem/subject the book has to some extent a character of an updated guide to major sources dealing in more detail with various aspects of the problem.

For the same reasons visual material is used and/or referred wherever possible and useful. It is supplied by extended figure captions and in some cases boxes.

The book is based in part on the graduate course delivered by the author in the Department of Fluid Mechanics, Tel Aviv University, and in part on lectures delivered by the author in Delft Technical University, Swiss Federal Institute of Technology – Lausanne and Zurich, Ecole Normale Supérieure, Université Paris VII and in Laboratoire Modélisation en Mécanique, Université Paris VI, and in part on its revision and updating on the basis of the latest work during the period of the Programme on Turbulence held in the Isaac Newton Institute, Cambridge, January 6 - July 2, 1999 and the research Program on Physics of Hydrodynamics of Turbulence held in the Institute of Theoretical Physics, Santa Barbara, January 31 - June 30, 2000.

It should be emphasized that this is an *informal introduction* only, it is *not a textbook*, but an introduction to turbulence as a physical phenomenon. Those who want to go deep(er) into the field are warned not to underestimate the numerous difficulties, disappointments and even frustration awaiting them. This book alone gives only an impression of what is turbulence. It can serve as an introduction to the research and the literature in the field in conjunction with a self-contained text, such as the book written by Pope (2000).

The scope of the book is mostly limited by purely basic aspects on turbulent flows of incompressible fluids. The prerequisites include a basic course in fluid dynamics (including turbulence) and standard knowledge of physics and mathematics at mid-graduate level. Therefore no systematic introduction to general fluid dynamics is included, neither is any material on probabilistic tools.[4] Both are usually included in books on turbulence, and the latter sometimes at a quite elaborate level However, its use in these books is very limited with most of the high probability and stochastic tools remaining unused. In order to aid the reader glossaries of some terms and

---

[3]*It is much easier to present nice rational linear analysis than it is to wade into the morass that is our understanding of turbulence dynamics. With the analysis, professor and students feel more comfortable; even the reputation of turbulence may be improved, since the students will find it not as bad as they had expected. A discussion of turbulence dynamics would create only anxiety and a perception that the field is put together out of folklore and arm waving* (Lumley, 1987).

[4]The recently published book by Pope (2000) contains well balanced information on all these and other useful tools for treating turbulence.

brief discussions of basic fluid mechanics are given in the appendices as well as some other useful information. More details about the scope of the book are given at the end of the first Chapter. This is an introduction in which the most important points are mentioned, and discussed in more detail in the subsequent Chapters along with some additional material. The list of references is limited to major sources: books/monographs, collections of essays, review papers and selected specific papers with some emphasis on the most recent ones, overlooking essentially all the rest, which are many indeed. This book is biased unavoidably by my views of what matters, and in this sense the book is to a large extent a personal view of the author on the subject. In selecting the references I used the (genuinely small) parameter introduced by Saffman (1978), which he called information density, ϵ_I, and defined as the ratio, S/N, in the literature, with $S =$ signal (*understanding*), and $N =$ noise (*mountains of publications*). In order to increase the value of ϵ_I I did all my best to concentrate on the numerator, S, and to reduce the denominator, N, to the best of my knowledge, ability and judgement/understanding. However, absence of references does not necessarily mean that - in my view - they belong to N, but is due to my ignorance and/or the lack of space needed to discuss them here.

The book is intended for as broad a readership as possible with the aim of makimg it interesting and useful *both* to graduate students and scientists in all the above mentioned fields. It is hoped that this aim is relatively realistic due to the *informal* nature of this book, with its emphasis on turbulence as a physical phenomenon, observations, misconceptions and unresolved issues rather than on conventional formalistic elements and models.

The list of acknowledgements is too long to be reproduced here. I am grateful to all those who responded to my queries and requests and to my hosts in the places mentioned above for their hospitality.

Tel Aviv, Israel   A. Tsinober

March 2001.

INTRODUCTION

About The Main Features of Turbulent Flows and The Main Problems

This introductory chapter comprises the basis and to a large extent is a guide for the rest of the book. It starts with a brief discussion of the history of the subject, which is followed by a number of various representative examples. The principle aim of bringing these examples is to demonstrate the variety of situations in which turbulent flows occur and the diversity of their manifestations. Where appropriate the examples are supplied by comments emphasizing the differences between turbulent and laminar flows and some specific properties of the former. Instead of a definition of turbulence a subsection is devoted to major qualitative universal properties of turbulent flows with cross references to the previous subsection on the representative examples. This is followed by a subsection attempting to give an idea as to why turbulence is such an extremely difficult problem. The last subsection contains an overview of the contents of the following chapters and the rest of the book.

1.1. Brief history

> *The Rise and Fall of Ideas in Turbulence.*
> Liepmann (1979).

Only a brief outline of some major milestones in turbulence research is given below. Full appreciation of these comes only after reading this book. More details are given in Monin and Yaglom (1971, 1992) and Loitsyanskiı̆ (1966); see also Frisch (1995) for interesting historical digressions/excursus and Lumley and Yaglom (2000) for additional comments on the developments during the last century.

> 1500 Recognition of two states of fluid motion by Leonardo da Vinci and use of the term *la turbolenza*.
> 1839 'Rediscovery' of two states of fluid motion by G. Hagen.
> 1883 Osborne Reynolds' experiments on pipe flow. Concept of critical Reynolds number - transition from laminar to turbulent flow regime.
> 1887 Introduction of the term 'turbulence' by Lord Kelvin.
> 1895 Reynolds decomposition.
> 1909 D. Riabuchinsky invents the constant-current hot-wire anemometer.
> 1912 J.T. Morris invents the constant-temperature hot-wire anemometer.
> 1921, 1935 Statistical approach by G.I. Taylor.
> 1922 L.F. Richardson's hierarchy of eddies.
> 1924 L.V. Keller and A.A. Friedman formulate the hierarchy of moments.
> 1938 G.I. Taylor discovers the prevalence of vortex stretching.
> 1941 A.N. Kolmogorov local isotropy, 2/3 and 4/5 laws.
> 1943 S. Corrsin establishes the existence of the sharp laminar/turbulent interface in shear flows.
> 1949 Discovery of intrinsic intermittence by G. Batchelor and A. Townsend
> 1951 Turbulent spot of H.W. Emmons
> 1952 E. Hopf functional equation.
> 1967 Bursting phenomenon by S.J. Kline *et al.*
> 1976 Recapitulation of large scale coherent structures by A. Roshko.

Most of the developments in turbulence research have occurred since Osborne Reynolds undertook his experiments. During this period the research in turbulence was conducted almost exclusively by the engineering community and in some other practical fields, such as in atmospheric and ocean sciences, and astrophysics. The last three decades have been marked by an increasing involvement of physicists and applied mathematicians.

1.2. Nature and major qualitative universal features of turbulent flows

1.2.1. REPRESENTATIVE EXAMPLES OF TURBULENT FLOWS

Unlike other complicated phenomena turbulence is easily observed, but is extremely difficult to understand or explain.

There exist a number of beautiful collections of images of turbulent flows. To mention some: Corrsin (1961), Fantazy of Flow (1993), Atlas of Visualization (1997), van Dyke (1982) and Werlé (1987). Precise and penetrating in capturing the essential aspects of turbulent flows, the drawings by Leonardo da Vinci deserve a special mention; see Pedretti (1982), Popham (1994) and Richter (1939).

What follows is a selection of such pictures attempting to illustrate the diversity of circumstances in which turbulent flows occur and the variety of their manifestations.

FLOWS IN PIPES

A qualitative repetition of the Reynolds experiment on his original facility in Manchester is shown in figure 1.1

In a particular example shown in figure 1.2 the flow becomes turbulent at the value of Reynolds number, $Re \sim 2700$, though with appropriate measures it can be kept laminar for Re up to 10^5 and, in principle, at much higher values of Reynolds number, since this flow is stable to small enough (infinitesimal) disturbances. The only other flow known to possess a similar property is the Couette flow in a plane channel. Both become turbulent at relatively low Reynolds numbers with disturbances of finite amplitude.

In the laminar regime the pressure difference at distance, l, along the pipe is proportional to the mean velocity, U, whereas in the turbulent regime it is much larger and is approximately proportional to $U^{7/4}$ in pipes with smooth walls and to U^2 in pipes with rough walls, see, for instance, figure 20.18 in Schlichting (1979). The latter means that the rate of energy losses, $\Delta p d^2 U$, i.e. rate of energy dissipation, in the turbulent regime in pipes with rough walls, is proportional to U^3/d and is independent of viscosity. In other words the nondimensional rate of energy dissipation per unit mass $\epsilon \equiv \frac{\Delta p d^2 U}{\rho l d^2 (U^3/d)}$ or simply dissipation, is Reynolds independent. This appears to be true of most turbulent flows at large enough Reynolds numbers, Idelchik (1986).

The visual observations shown in the above and subsequent figures were made by using some dye, i.e. via observing a *passive scalar*[1], which is not a dynamical quantity such as velocity or vorticity. In order to be able to 'see' some dynamical variable one has to use more elaborate methods such as tracking small enough neutral (i.e. of the same density as the fluid) particles or use the data from direct numerical simulations (DNS) of the Navier-Stokes equations (NSE), see colour plate 1. One can see the difference between the quiet laminar and the restless turbulent flow regimes looking at the water jet from a tap, see figure 1 in Corrsin (1961) or figure 4.1 in Mullin (1993). Note that this is more 'pipe turbulence' rather than 'jet turbulence'.

---

[1]Most visualizations are made in such a way. It has to be noted that what one sees looking at a pattern of a passive scalar in a turbulent flow may have nothing to do with the behaviour of dynamical variables such as velocity and vorticity. We return to this issue in chapter 4.

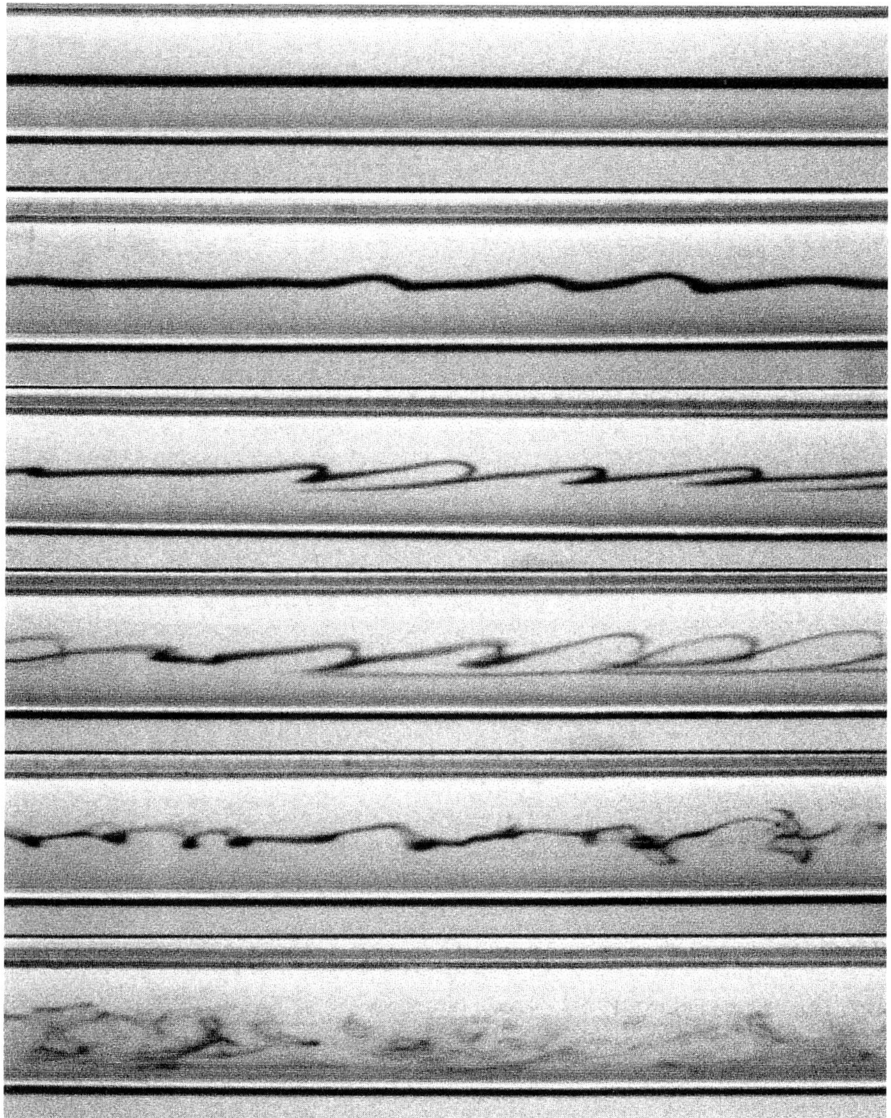

Figure 1.1.   Reynolds experiment in a circular pipe. The smallest mean velocity corresponds to the upper frame where the flow regime is laminar, the largest mean velocity corresponds to the lower frame where the flow regime is turbulent. Courtesy of Professor J.D. Jackson, School of Engineering, University of Manchester.

BOUNDARY LAYERS

Boundary layer flows belong to the same category as the flows in pipe, channels and other *wall bounded* flows.

The flows shown in figures 1.1-1.4 have many common features and

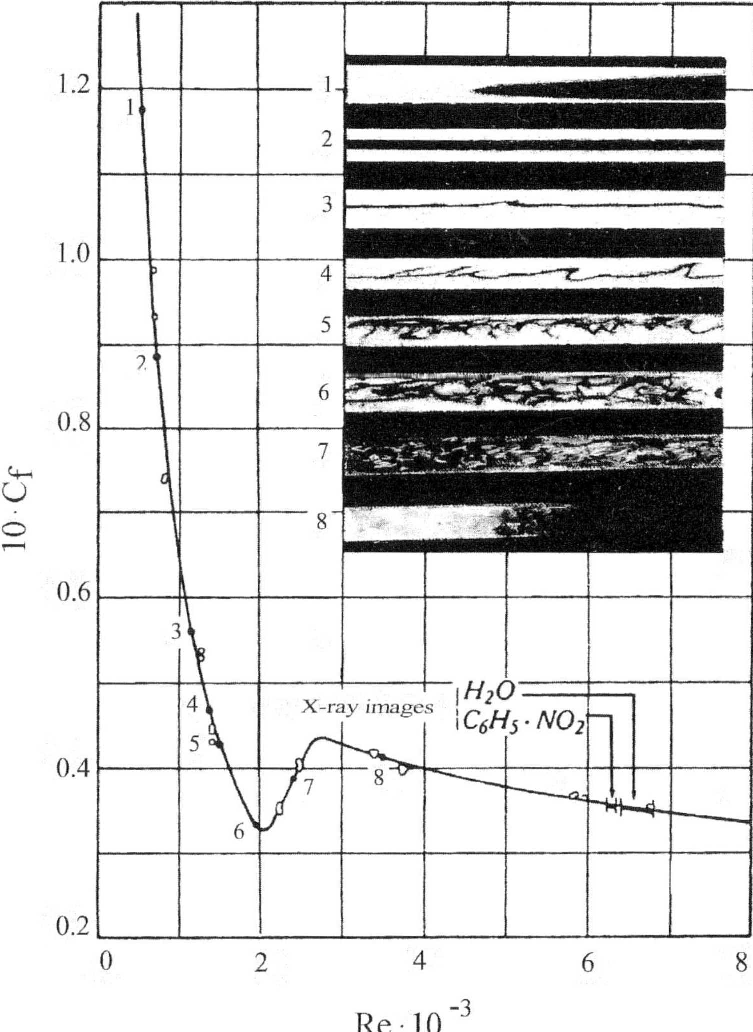

Figure 1.2. Reynolds number dependence of the friction factor, C_f, of flow with mean velocity, U, in a circular pipe of diameter, d, with corresponding flow visualization at particular values of Reynolds number $Re = Ud/\nu$, where ν is the kinematic viscosity of the fluid. The friction factor is defined as $C_f = (\Delta p d)/(l \frac{1}{2}\rho U^2)$, where Δp is the pressure drop on the distance, l, along the pipe, and ρ is the fluid density. Note that there is a range of values of Reynolds number in which the friction factor, C_f, follows the laminar law, $C_f = 64/Re$, but the flow pattern (pictures 4-6) is far from looking as purely laminar. Adapted from Dubs (1939).

properties, such as their near-wall behaviour. However, turbulent boundary layers possess specific essential features.

Figure 1.3.   Side view of a turbulent boundary layer visualized by (top) smoke traces, courtesy of Professor H. Nagib, and (bottom) by hydrogen bubbles, Courtesy of Professor A.J. Grass.

First, these flows are *partly* turbulent[2] in the sense that the turbulent regime coexists with the laminar one (which is usually close to irrotational), as well as with the transitional states between laminar and turbulent regimes (see figure 1.4). Second, the 'boundary' between the laminar and turbulent flow regions is strongly corrugated with many space-time

____
[2]This term comes from Scorer (1978).

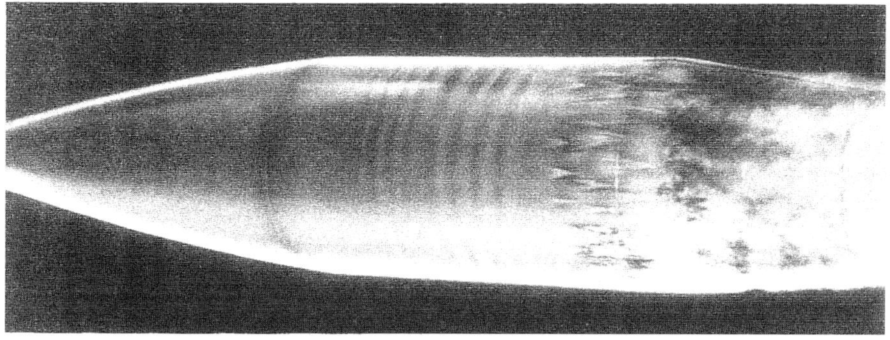

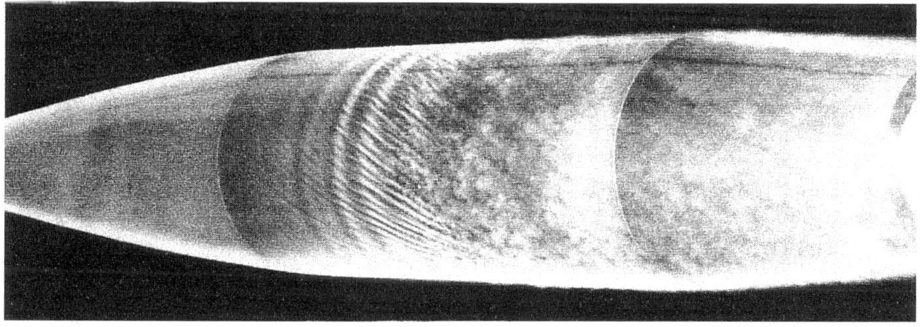

Figure 1.4. Coexistence of different flow regimes - laminar, transitional and turbulent - in a boundary layer on an axisymmetric body visualized in a smoke photograph. The upper picture corresponds to a nonspinning body, the lower picture corresponds to a spinning body. In the former regular waves (Tollmien-Schlichting waves) are seen at the early stage of transition. In the latter the T-S waves coexist with the so-called cross-flow vortices. Courtesy of Prof. T.J. Mueller, (see Mueller et al. 1981).

scales involved - it is 'fractal'-like, but nonetheless distinct. The largest scales are such that the 'laminar' fluid is found quite close to the wall. Third, the fluid from the laminar region is continuously entrained into the turbulent region through the boundary between the two. That is the entrainment process is such that laminar fluid becomes turbulent in the proximity of the laminar/turbulent boundary. This is one of the basic processes of transition of the flow state from laminar to turbulent. An important overall characteristics of this process is the entrainment rate.

The three features mentioned above are also observed in the so-called *free* (of rigid or other boundaries) turbulent shear flows: jets and plumes, wakes, mixing layers and flows in separation regions as well in more complicated situations in geophysical and astrophysical contexts. The simplest example is the entrainment process at the bottom of the turbulent surface (warmer) layer of the ocean. These flows all are also partially turbulent with corrugated boundary between the laminar and turbulent regions, and entrainment of fluid from the former into the latter. It seems that in free shear flows the approximately irrotational fluid can be found deep in the region occupied by the turbulent fluid [3] just as in the turbulent boundary layer, though both observations are mainly based on images of passive scalars. However, frequently (but not always, see chapter 4) they are in agreement, at least qualitatively, with observations made by other methods such as optical (shadowgraph, schlieren and interferometric). A somewhat dissimilar example of the coexistence of different flow regimes is shown in figure 1.5.

TURBULENT JETS, PLUMES

Turbulent jet like flows are ubiquitous in engineering and nature, e.g. jets from aircraft and rocket engines and discharges from smoke stacks, volcanoes and other geologic nozzle eruptions, see figure 1.6.

TURBULENT WAKES PAST BODIES AND MIXING LAYERS

The flow in the wake past a sphere undergoes a number of changes from laminar to fully turbulent[4] as the Reynolds number grows. At $Re < 1$ the flow is time independent without separation; at $Re \sim 1 \div 10$ the flow remains steady and symmetric but separates at the back of the sphere in the form of a vortex ring, which becomes larger with increasing Reynolds number; at $Re \sim 10 \div 10^2$ the flow loses its symmetries: it becomes time dependent but (approximately) periodic, and the vortex ring is deformed into a helical vortex, which rotates around the (former) axis of the flow. The separation line is no more a circle and assumes a complicated form, which is changing in time. At $Re \sim 10^2 \div 10^5$ the wake becomes (apparently) random and aperiodic but with large scale structures, which are more complicated than the destroyed helical vortex. The boundary layer (before the line of separation) on the sphere remains laminar. At $Re >\sim 10^5$ the boundary layer also becomes turbulent. It is noteworthy that the separation zone past the sphere (plate 2) for $Re = 3 \cdot 10^5$ is much narrower due to the onset of turbulence in the boundary layer of the sphere. This causes a strong *decrease*, by a factor of 6, of its drag, e.g., see figure 1.5 in Schlichting (1979). This is an example when transition to turbulence is reducing the losses instead of increasing them, as in most cases. The reason is that the delayed separation leads to a much narrower wake and hence smaller losses. This feature is also exhibited in changes of pressure distribution over the surface of the sphere. No such phenomena are observed in flows past

---

[3]See figures below and figures 109, 117, 134, 151, 158, 166, 167, 174, 176, 177 in van Dyke (1982). Note the figure 151, showing a turbulent wake behind a projectile at supersonic speed with a remarkably sharp boundary between the turbulent flow in the wake and the ambient irrotational flow.

[4]The term *fully developed turbulence (voll ausgebildet turbulenz)* is most probably derived from L. Prandtl.

Figure 1.5. Coexistence of different flow regimes - laminar, transitional and turbulent - in a fountain in the center of Washington, photo by the author (1985). Here the process is reverse due to the decrease of Reynolds number in each subsequent step: the turbulent flow regime in the upper part is replaced at the next step by the transitional one with sporadic outbursts of turbulent activity. This in turn is replaced by regular wavelike motions similar to those observed in the initial transitional stage in a boundary layer flow shown in figure 1.4. Note the difference in the meaning of the term 'coexistence' in this example and the one shown in figure 1.4. However, in both cases the changes in the flow regime are related to the changes of the value of the 'local' Reynolds number. Hence the similarity between the two examples.

bodies with sharp edges, e.g. past a circular disc with its plane normal to the direction of the undisturbed flow. In such a case the normalized drag $C_D = \frac{F_{drag}}{\rho U^2 d^2}$ (and consequently the energy dissipation) remains independent of the Reynolds number up to the highest achievable Reynolds numbers of

Figure 1.6.　Coexistence of different flow regimes - laminar, transitional and turbulent - in a circular jet. Courtesy of Professor H. Nagib.

Figure 1.7.　Left: a turbulent jet from testing a Lockheed rocket engine in the Los Angeles hills, courtesy of Professor P.E. Dimotakis. Right: Eruption of Mount St. Helen volcano on 18 May 1980, US Geological Survey.

the order 10^6, figure 1.8. Other numerous examples of this kind can be found in Idelchik (1996).

Wakes past bodies exhibit very large scale fluctuations (undulations) in the far field in which the wake undulates essentially as a whole (figure 1.9).

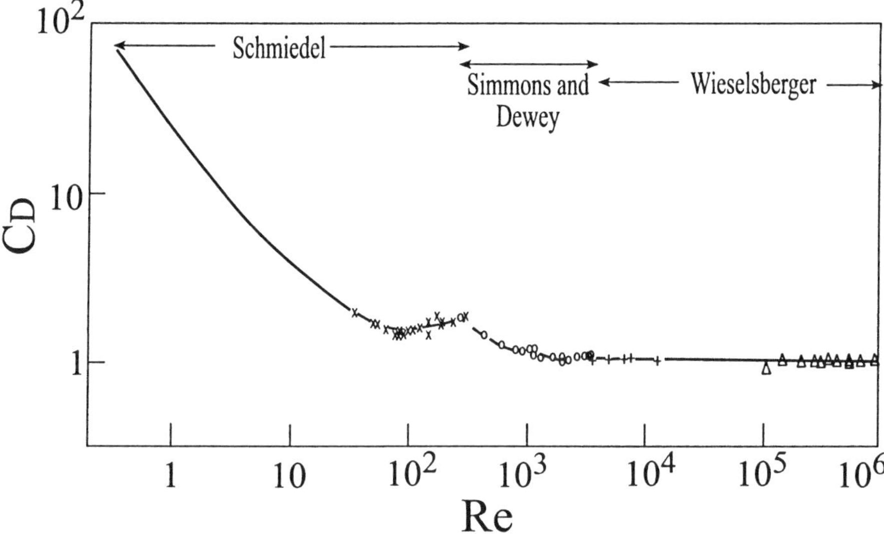

Figure 1.8.   Reynolds number dependence of the resistance coefficient, C_D, of a circular disc with its plane normal to the direction of the undisturbed flow. Adapted from Schiller (1932); see also Muttray (1932).

This is an indication that these large scale undulations are the result of a large scale instability not directly related to the turbulent nature of the flow within the wake. This is probably the main reason for the similarity of the large scale features between the two flows shown in figure 1.10, in spite of a large difference in the value of their Reynolds numbers. Similar large scale features are observed in other free shear flows such as mixing layers (figure 1.10) and plumes (plate 3).

QUASI-ISOTROPIC AND HOMOGENEOUS FLOWS

Quasi-isotropic and homogeneous flows are realized in laboratory in a flow past a grid or a grating such as those shown in figure 1.11 and plate 4. Another way to produce such flows is via direct numerical simulations of the Navier-Stokes equations with appropriate forcing at the right hand in cubic box with periodic boundary conditions. An example of some results in such a computation is shown in plate 5.

The two last flows (figure 1.11 and plates 4 and 5) are considered by many as of purely academic interest, since they represent an approximation to an idealized situation which is called homogeneous isotropic turbulent flow. Nevertheless these kinds of flows are of special interest for several reasons. Initially this interest was due to the relative 'simplicity' and analytical convenience of such flows. However, there is one more aspect which makes the flows shown in the last three examples of special interest and impor-

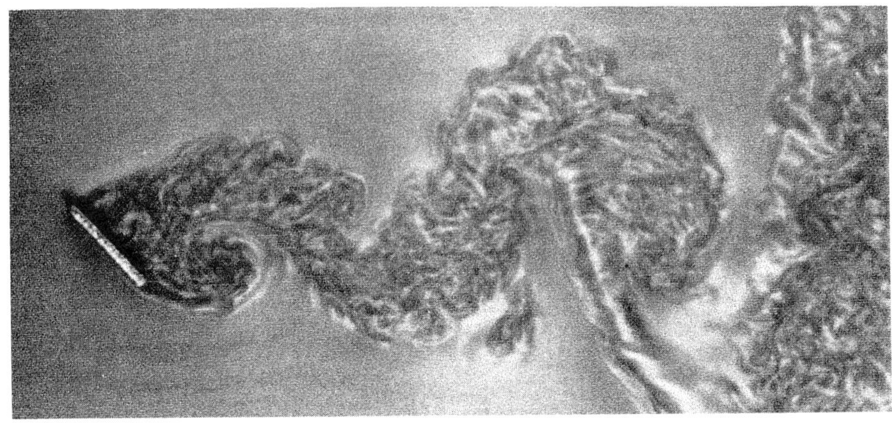

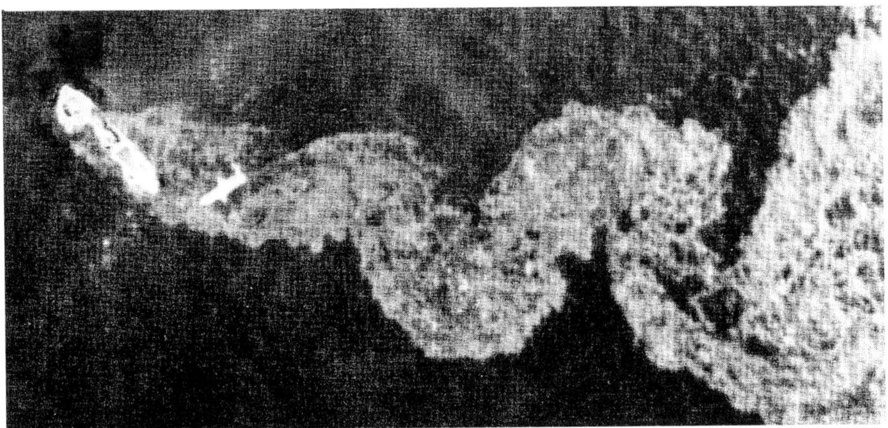

Figure 1.9.   Top: a wake past an inclined plate at a Reynolds number 4300. Bottom: a wake of leaking oil past a grounded tanker at a Reynolds number $\sim 10^7$. Courtesy of Professor B. Cantwell.

tance. Namely, it appears that there exist many universal features at the level of basic physical processes of turbulent flows, which are manifested in this idealized situation, regardless of the origin of a specific turbulent flow, at least qualitatively. The particular significance of these kinds of flows follows from the fact that (quasi-) homogeneous and isotropic flows are free from external influences, like mean shear, centrifugal forces (rotation), buoyancy, magnetic field, etc., which usually act as an organizing factor, favouring the formation of the so-called (large scale) coherent structures of different kinds (quasi-two-dimensional, helical, etc.). These external influences in many cases have usually a strong (linear) masking effect on the intrinsic nonlinear turbulent processes in such situations as those described by rapid distortion theory, in which nonlinear interactions between turbu-

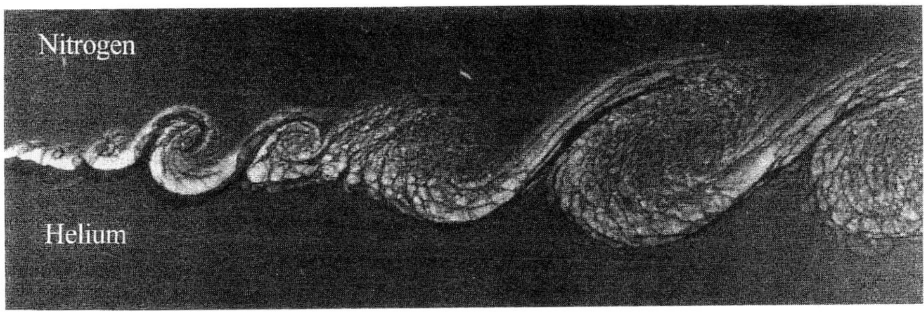

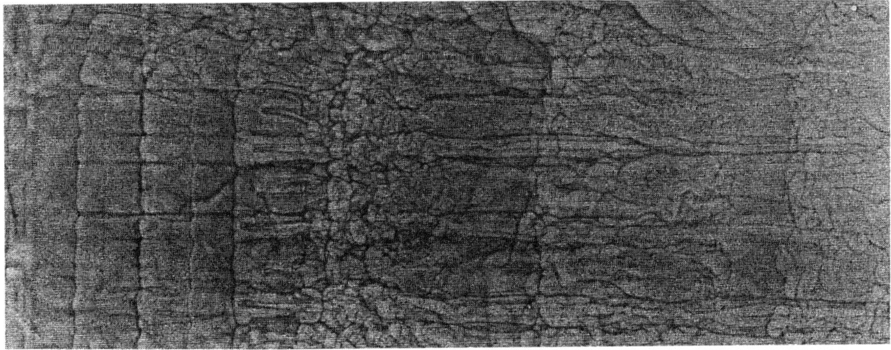

Figure 1.10. An example of a turbulent mixing layer. Top - side view, bottom - (half) plan view, both shadowgraph. The undisturbed velocity of the upper part of the flow (nitrogen) is $10m/s$ and the undisturbed velocity of the lower part (mixture of helium and air) is $3.8m/s$. From Konrad (1976).

lent fluctuations are neglected for the duration of the distortion (Savill, 1987; Hunt and Carruthers, 1990). In other words the nonlinear nature of turbulent flows is manifested more distinctly in (quasi-) homogeneous and isotropic turbulence.

It is argued here that most of the intrinsic properties of turbulent flows, that is their physics, are seen in the cleanest and relatively 'simple' way in quasi-homogeneous isotropic flows. This is the reason for the emphasis given to such flows.

As mentioned there are many factors and influences which cause a real turbulent flow in nature and technology to deviate from this idealized state, sometimes strongly. In the latter case turbulent flows may loose most of their resemblance to the three-dimensional homogeneous isotropic flow, but can be quite similar to the (quasi) two-dimensional one. The aspect of various influences on turbulence is addressed in chapter 8.

Before finishing this section it is important to note that all the above examples show the *spatial* intricacy of turbulent flows. The *temporal* behaviour obviously cannot be seen from a single snapshot. An example of time recordings of several quantities is shown in plate 6. All the quantities

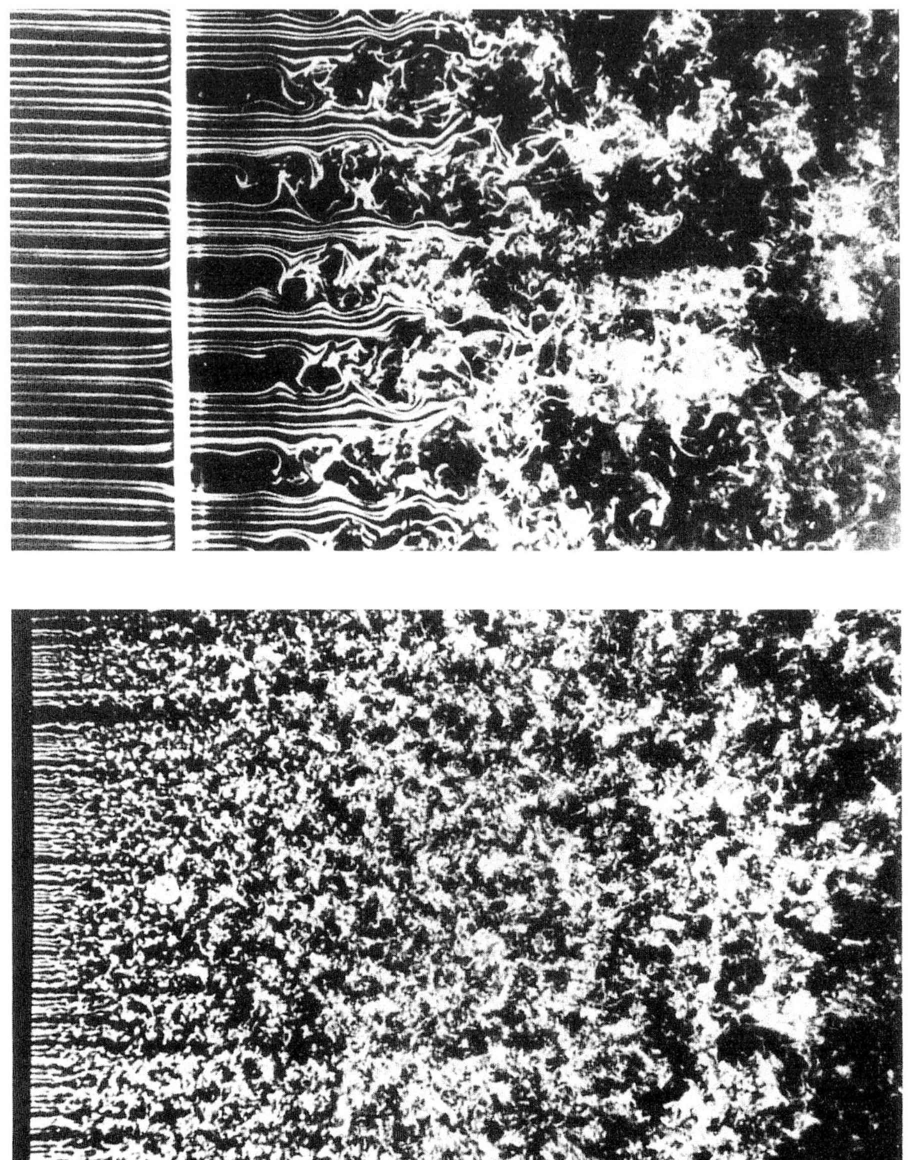

Figure 1.11.   Examples of a turbulent flow past a grid. Top - in the proximity of the grid. Bottom - in the far field. Courtesy Prof. H. Nagib.

exhibit apparently random temporal patterns, though quite different. Similar differences occur in the spatial variations of these quantities. We will return to these differences and many other related issues in the subsequent chapters.

The above examples of flows show that one can easily observe at least some manifestations of turbulence and even can *describe* some them qualitatively, i.e. without mathematics. It is naturally to use the *qualitative* manifestations of turbulent flows as a first step to 'define' what is turbulence.

1.2.2. IN LIEU OF DEFINITION: MAJOR QUALITATIVE UNIVERSAL FEATURES OF TURBULENT FLOWS

> *Turbulence is a phenomenon which sets in in a viscous fluid for small values of the viscosity coefficient ν, ... hence its purest, limiting form may be interpreted as the asymptotic, limiting behavior of a viscous fluid for $\nu \to 0$ (Neumann, 1949).*

Already from the observation of the examples given in the previous section it is easy to arrive at a conclusion that turbulence is an extremely intricate and complicated phenomenon. Therefore, it would be naïve and hopeless to attempt its definition in few sentences[5], see the collection of citations in the appendix A. *What is turbulence?* showing that this is really not easy if not impossible. Indeed, let us try to give such a 'definition': *Turbulence is the manifestation of the spatio-temporal chaotic behaviour of fluid flows at large Reynolds numbers, i.e. of a strongly nonlinear disspative system with extremely large number of degrees of freedom* (most probably) *described by the Navier-Stokes equations.* Obviously, it is practically useless and absolutely not sufficient to say this or something similar. So instead of futile attempts to define what is turbulence, one starts by description of its main qualitative features, as did Tennekes and Lumley (1972). The following is an updated list of such features with cross references to the examples given in the previous section.

- • - Intrinsic spatio-temporal randomness, irregularity. Turbulence is definitely chaos. However, vice versa, generally, is not true: many chaotic flow regimes are not turbulent, e.g. Lagrangian/kinematic chaos or 'Lagrangian turbulence', laminar 'turbulent' flows.

One of the most important aspects is that the stochastic/random nature of turbulent flows is its *intrinsic* property (self-stochastization or self-randomization). There is no necessity for external *random* forcing either in the interior of the fluid flow or at its boundaries, nor one needs to start the turbulent flow with some random initial conditions *provided* that the

---

[5]Though there are many attempts to do so. Even if such a definition would be possible it is likely to come *after* the basic mechanisms of turbulence as a physical phenomenon are fully understood. In any case there is considerable 'turbulence' in the attempts to define what is turbulence.

Reynolds number is large enough[6]. The fascinating question is how such a behaviour can arise from such purely deterministic equations as the Navier-Stokes equations are, deterministic forcing along with smooth initial and boundary conditions. The answer is its extreme sensitivity to disturbances whatever small (initial conditions, boundary conditions, external noise). Turbulence is both a strongly non-linear (stochastic) oscillator and an amplifier with an (almost) 'infinite' gain.

The feature described here is seen in all the examples given in the previous section.

• - EXTREMELY WIDE RANGE OF STRONGLY INTERACTING SCALES. Turbulent flows are large systems. In atmospheric flows, relevant scales range from hundreds of *kms* to parts of a *mm*, i.e. there exist $\sim 10^{29}$ excited degrees of freedom[7], many of which are strongly interacting. Hence extreme complexity of turbulence (along with its intrinsic randomness) means that description of turbulent flows should be of statistical nature. We emphasize that statistical *description* is not synonymous to statistical *theorization*.

The interaction between the many degrees of freedom, resulting from the *nonlinearity* of turbulent flows is essential (linear systems can have arbitrarily many degrees of freedom as well, but they do not interact, and each degree of freedom lives its own life without knowing anything about other degrees of freedom), but not sufficient: the nonlinear interaction may lead to strongly organized regular behaviour, e.g. solitons in the systems described by the Korteveg de Vries and Shrödinger equations, and shocks in the Burgers equation and .

Again most of the examples given in the previous section clearly show the multi-scale nature of turbulent flows in space. Turbulent flows exhibit also quite complex behaviour in time as well, see plate 6.

• - LOSS OF PREDICTABILITY. Two initially nearly (but not precisely) identical turbulent flows become unrecognizably different on the time scale of dynamical interest. The details of any realization are strongly different from any other realization. This is because any individual realization is extremely sensitive to small perturbations/disturbances. However, *different realizations of the same turbulent flow have the same statistical properties*, such as drag of a sphere or rate of mixing of some contaminant. That is the statistical properties (not only some means, but almost *all* statistical properties) of turbulent flows are insensitive to disturbances – *turbulent flows*

---

[6]In this sense the random behaviour of nonlinear systems as a response to random forcing is not necessarily turbulence. One of the most popular examples is the Burgers equation: without external random forcing it does not exhibit any chaotic behaviour.

[7]This includes all the scales of atmosphere. Not all of them are considered as turbulent in the atmospheric community. For example, Orlanski (1975) considers only the first several hundred meters as turbulent. With this approach the number of excited degrees of freedom is still enormous $\sim 10^{18}$.

are statistically stable, they possess statistically stable properties. This *insensitivity* to disturbances is only *statistical*. In other words turbulent flows possess both predictable and unpredictable features. The well known problem of predictability from meteorology is essentially about the dynamics and statistics of an initial error, which is the measure of the differences between some two realizations of a turbulent flow under *almost* (hence the error) the same conditions. An important aspect is that the error also possesses stable statistical properties in the sense that errors corresponding to different pairs of realizations have the same statistical properties.

• - TURBULENT FLOWS ARE HIGHLY DISSIPATIVE. A source of energy is required to maintain turbulence (gradients of mean velocity, buoyancy, or other external forces). The energy supply is mostly at large scales, its dissipation is at small ones[8]. Statistical irreversibility is involved, i.e. the processes in turbulent flows are 'one way' in time.

The immediate examples that come to mind are the resistance in pipes and the drag of bluff bodies, which in the turbulent regime are orders of magnitude larger than their laminar counterparts at the same Reynolds number.

• - TURBULENT FLOWS ARE THREE-DIMENSIONAL AND ROTATIONAL. They are 'random' fields of vorticity, $\boldsymbol{\omega} \equiv curl\ \mathbf{u}$, with predominant vortex stretching, i.e. continuous positive net production of enstrophy, ω^2, by inertial nonlinear processes, which is 'dissipated' by viscosity. An important concomitant process is a positive net production of strain, $s_{ij} = 1/2(\partial u_i/\partial x_k + \partial u_k/\partial x_i)$. Both are the result of *self*-amplification of the field of velocity derivatives in/by turbulent flows, and comprise one of the most basic specific dynamical properties of turbulence. The latter, i.e. production of strain, is directly related to the dissipative nature of turbulent flows, whereas the former, i.e. amplification of vorticity, is related to dissipation only in an indirect manner. Random potential flows are not turbulence.

It is not difficult to observe the three-dimensional nature of turbulent flows, but their rotational nature can be 'seen' from the direct numerical simulations of the Navier-Stokes equations, as shown in the colour plate 1.

There is no consensus whether two-dimensional chaotic flows even with many degrees of freedom should be qualified as turbulence. The main reason is that such flows lack the mechanism of vorticity and strain amplification. This issue is discussed in chapter 8.

---

[8]The common view of turbulence dynamics involves the Richardson-Kolmogorov cascade of energy (the famous poem by Richardson). However, there are numerous examples in which turbulence develops from small scales into larger ones, e.g. in all spatially developing turbulent flows. All the examples given in the previous section on the *partly* turbulent flows are such, both free and wall-bounded. The important point is that the so-called 'cascade' takes place *not* in the physical space. We will return to this point and to the notion of *scale* later in chapter 5 (see also appendix C).

• - STRONGLY DIFFUSIVE (random waves are not). Turbulent flows exhibit strongly enhanced transport processes of momentum, energy, passive objects (scalars, e.g. heat, salt, moisture, particles; vectors, e.g. material lines, gradients of passive scalars, magnetic field). It should be emphasized that in respect with passive objects *only* this property is true of a much broader class of systems. Namely, *any* random velocity field and even laminar flows, which are Lagrangian chaotic, exhibit enhanced transport of passive objects.

This aspect is manifested in the large resistance occurring in pipe flows, the enlarged drag of various bodies, and the enhanced heat and mass transfer and rate of mixing in turbulent flows as compared to their laminar counterparts. The latter process is illustrated in plate 4. It shows also how two separate passive scalars are mixed – an aspect important for chemical reactions and combustion. Using the terms 'turbulent viscosity' and 'turbulent diffusivity', one can say that they are orders of magnitude larger in turbulent flows than their molecular counterpart. It is the right place to note that turbulence is a property of fluid flows not fluids. For instance, molecular viscosity and diffusivity are properties of fluids (liquids, gases) and are independent of the flow. On the contrary the so-called 'turbulent viscosity' and 'turbulent diffusivity' are properties of fluid flows and depend on the fluid flow in question. The difference is more than essential.

These mostly wide known qualitative features of all turbulent flows are essentially the same, i.e. it is meaningful to speak about *qualitative universality* of turbulent flows. The concept of *qualitative universality* is not just a fuzzy idea: in chapter 9 it will be given a number quantitative attributes. Indeed these *qualitative* features of turbulent flows are *universal* for all turbulent flows arising in qualitatively different ways and circumstances and generally characterize turbulent flows as a whole. In addition to these general *qualitative* features, there are universal *quantitative* features which are more specific for turbulent flows. They are described later in this book, since they require more specific information and terminology. Many (but not all) *quantitative* properties may vary largely with the range of scales of interest. The *large scale* properties of turbulent flows depend on *particular* mechanisms generating turbulence and, generally, are *quantitatively* not universal, though as mentioned they are *qualitatively* universal. It is the *small scale* turbulence which, since Kolmogorov, is believed to possess a number of universal properties independent of the large scale flow structure. This point of view is not accepted universally: ... *perhaps there is no 'real turbulence problem', but a large number of turbulent flows and our problem is the self imposed and possibly impossible task of fitting many phenomena into the Procrustean bed of a universal turbulence theory* (Saffman, 1978; see also Hunt et al.,1994). The issue is one of several continuously debated controversies in the problem of turbulence. This includes the *meaning* of the term 'universality'. For example, one issue involves the invariance of Reynolds number of (some) properties of a particular turbulent flow at large enough Reynolds numbers. Another issue is concerned with the universality of scaling properties of small scale turbulence, which has remained for more than fifty years one of the most active fields of inquiry. Derivation of scaling properties of fully developed turbulent flows directly from the Navier-Stokes equations analytically is one of the most popular illusive goals of theoretical research. This (scaling) and other *phenomenological* aspects are extensively reviewed in Monin and Yaglom (1971, 1975, 1992, 1996), Frisch (1995), and Sreenivansan and Antonia (1997). For this reason and some more specific reasons explained later these issues are discussed only briefly in this volume. Basic information on scales and scaling and related matters in turbulence can be found in appendix C.

Two notes are in order.

First, the most accepted division of turbulent flows on large and small scales is to a large extent artificial and in some sense even unphysical due to strong coupling between the two and due to an ambiguity of the very term 'scale' (see chapter 5 and appendix C).

Second, it should be stressed that some of the (possibly universal) properties of turbulent flows (such as scaling) are characteristic of a much broader class of nonlinear systems, others are *specific* to turbulent flows. Both are addressed later in the book with the emphasis on the properties of fluid dynamical turbulence.

1.3. Why turbulence is so impossibly difficult[9]? The three N's

> *Turbulent flow constitutes an unusual and difficult problem of statistical mechanics, characterized by extreme statistical disequilibrium, by anomalous transport processes, by strong dynamical nonlinearity, and by perplexing interplay of chaos and order (Kraichnan, 1972).*
>
> *The experience of 100 years should suggest, if nothing else, that turbulence is a difficult problem, that is unlikely to suddenly succumb to our efforts. We should not await sudden breakthroughs and miraculous solutions (Lumley, 1996).*

1.3.1. ON THE NAVIER-STOKES EQUATIONS

The basic equations and some other essential information are given in the appendix C. Here we provide some general notes regarding the Navier-

---

[9]See also the collection of citations in appendix B.

Stokes equations (NSE) and related matters.

Though there exist a set of deterministic differential equations (NSE) probably containing (almost) all of turbulence, most of our knowledge about turbulence comes from observations and experiments (laboratory, field and later numerical)[10]. This was understood long ago by A.N. Kolmogorov:
I soon understood that there was little hope of developing a pure, closed theory, and because of absence of such a theory the investigation must be based on hypotheses obtained on processing experimental data (see Tikhomirov, 1991)[11]. Much later he wrote that *the observational material is so large, that it allows us to foresee rather subtle mathematical results, which would be very interesting to prove* (Kolmogorov, 1978).

However, the conclusion that NSE are useful as only an experimental tool would be incorrect. It is true that there is little substantial theoretical use of NSE in turbulence, since there is almost no way to use them explicitly in theoretical approaches, e.g. by solving them by 'hand'. However, there are several ways to do this implicitly, i.e. by indirect use of NSE and their consequences. For example, looking at the NSE and their consequences themselves enables us to recognize the dynamically important quantities and physical processes involved. In other words, NSE and their consequences tell us what quantities and relations should be studied. So far this can be done mostly experimentally, but this kind of 'guiding' should also be useful theoretically. The most elegant exception is a set of theoretical results on the *a priory* upper bounds[12] of long time averages of dissipation and global transport of mass, momentum and heat (for a recent review see Doering, 1999; Kerswell, 1999; and also Foias, 1997). In this sense the NSE tell us more than any estimates of the dimensions of attractors and similar things.

Though the NSE have (at best) a limited kinetic foundation (for gases only), they are commonly believed to be adequate in the sense that their solutions correspond to real fluid flows. This is not obvious, since the NSE are a gradient expansion. So in principle, higher order terms *may* become dominant in regions with large velocity gradients. Also, NSE are the result of coarse graining over the stochastic (molecular) effects. However, one can take the standpoint of continuum mechanics at the very outset. In the latter case one encounters the problem of the relation between the stress tensor and the rate of strain tensor in the fluid flow. The Newtonian fluid is the one in which this relation is linear (see, e.g. Serrin, 1959). There exists

---

[10]This is the main reason that this book is biased experimentally.

[11]Therefore, the importance of experimental research in turbulence goes far beyond the view of those who think of an experimentalist as a superior kind of professional fixer, knowing how to turn nuts and bolts into a confirmation of their theories.

[12]Unfortunately, there are no results on the lower bounds, except of trivial values corresponding to the laminar flows.

large empirical evidence that NSE are adequate, at least, at all accessible Reynolds numbers, so we will take the standpoint of continuum mechanics. This does not exclude the possibility that in very 'hot spots', where the strain rate is extremely large, the Newtonian fluids become non-Newtonian (see chapter 9). So far there is no direct evidence of this.

1.3.2. ON THE NATURE OF THE PROBLEM

Formally, the problem is to solve the Navier-Stokes equations subject to initial and boundary conditions. At present, it is possible to obtain fully resolved solutions at moderate Reynolds numbers via direct numerical simulations of the Navier-Stokes equations. However, the important point is that looking at the behaviour of a *particular* solution does not solve the problem, since any particular solution (which is not in an analytical form) may not contribute much to the *understanding* of the basic physics of turbulent flows[13]. Ideally, one needs for this a *method of understanding the qualitative content of equations* (Feynmann, 1963). In other words, *nothing less than a thorough understanding of the* [global behaviour of the] *system of all their* [NSE] *solutions would seem to be adequate to elucidate the phenomenon of turbulence* (Neumann, 1949). That is in order to understand the dynamics, or the main characteristics of the dynamics, it is necessary to understand a significant portion of the phase flow (see glossary of terms), especially the unstable solutions[14]. However, at present (if ever) it is impossible due to very high dimension and complicated structure of the underlying attractors (assumed to exist): *one may never be able to realistically determine the fine-scale structure and dynamical details of attractors of even moderate dimension. ... The theoretical tools that characterize attractors of moderate or large dimensions in terms of the modest amounts of information gleaned from trajectories* [i.e. particular solutions]... *do not exist ...they are more likely to be probabilistic than geometric in nature* (Guckenheimer, 1986).

However, the situation is not that bad, since in many cases an individual realization of a turbulent flow in a large enough space/time domain allows us to obtain important information of rather general nature. The reason is the property of ergodicity, which is believed to be true in turbulence (and

---

[13]There is no consensus on what is (are) the problem(s) of turbulence and what would constitute its (their) solution. Neither is there agreement on what constitutes understanding. It is definitely not the proof of the uniqueness of the smooth solution of the NSE on a three-dimensional domain for all time or anything similar. The discussion of this matter is postponed to the last chapter.

[14]In dynamical systems the unstable solutions give the key to understanding of the global behaviour in some significant parts of the phase space.

NSE) and has a considerable empirical support[15].

1.3.3. NONLINEARITY

The nonlinearity of turbulence is the most frequently fingered as the main 'guilty party'. There are several 'howevers'.

First, there are nonlinear problems that are completely integrable. The well known example, are systems displayng solitons or solitary waves. In these systems the many degrees of freedom are so strongly coupled that they do not display any chaotic/irregular behaviour. Instead they are entirely organized and regular (see Zakharov, 1990 and references therein). By a quite questionable analogy it is thought that the so-called coherent structures in turbulent flows may be treated/viewed in a similar way.

Second, nonlinearity is frequently blamed for the difficulties in the so called closure problem which is associated with some form of decomposition, such as the Reynolds decomposition of the flow field into the mean and the fluctuations, or similar decompositions into resolved and unresolved scales associated with large eddy simulations. The essence of the problem is that the equations for the mean field (resolved scales) contain moments of the fluctuations (unresolved scales) due to the nonlinearity of the NSE. However, a similar problem exists for the so-called advection-diffusion equation describing the behaviour of a passive scalar in some flow field. But this equation is linear. The problem arises due to the multiplicative nature of the velocity field, since velocity enters this equation as its coefficients.

1.3.4. NONINTEGRABILITY

In 1788 Lagrange wrote: *One owes to Euler the first general formulas for fluid motion ... presented in the simple and luminous notation of partial differences... By this discovery, all fluid mechanics was reduced to a single point analysis, and if the equations involved were integrable, one could determine completely, in all cases the motion of a fluid moved by any forces.(Mécanique analitique,* Paris, 1788, Sec X. p. 271).

The *if* in the above citation is crucial: the Navier-Stokes equations are not integrable.

Integrable systems, such as those having a solution 'in closed form' exhibit regular organized behaviour, even those having (formally) an infinite number of strongly coupled degrees of freedom. A prominent example is provided by the solitons in the systems described the Koreteveg de Vries and Shrödinger equations. Another example is the Burgers equation, which

---

[15]But no theoretical/mathematical support: *The problem with this ergodicity assumption is that nobody has ever even come close to proving it for the Navier-Stokes equations* (Foias, 1997).

is an integrable equation, and exhibits random behaviour only under random forcing, otherwise its solutions are not random[16]. That is, these examples represent the response of nonlinear systems to random forcing and which otherwise are not random, and should be distinguished from problems involving genuine turbulence. Navier-Stokes equations at sufficiently large Reynolds number have the property of intrinsic stochasticity in the sense that they possess mechanisms of self-randomization (most probably at all scales) which are not fully understood.

1.3.5. NONLOCALITY

This is probably one of the main reasons the problem of turbulence is so difficult.

Formally, nonlocality is due to the fact that the Navier-Stokes equations are integro-differential for the velocity field, and hence the velocity field is nonlocal in physical or any other space. Physically, it is because of the presence of long range forces due to pressure. The property of nonlocality of NSE is two-fold. On the one hand, it is due to pressure ('dynamic' nonlocality), since $\nabla^2 p = \rho \frac{\partial^2 u_i u_k}{\partial x_i \partial x_j}$, and therefore pressure is nonlocal due to nonlocality of the operator ∇^{-2}: the pressure is defined in each space point by the velocity in the whole flow field. This aspect of nonlocality is strongly associated with the essentially non-Lagrangian nature of pressure, which is related to the 'memory' of turbulence, i.e. nonlocality in time.

One cannot get rid of nonlocality by taking the *rot* of the NSE, thereby getting rid of the pressure gradient ∇p, and looking at the resulting equation for vorticity $\boldsymbol{\omega}$. The reason is that the equation for vorticity is nonlocal in vorticity, since it contains the rate of strain tensor $s_{ij} = \frac{\partial u_i}{\partial x_j} + \frac{\partial u_j}{\partial x_i}$, $\mathbf{u} = rot^{-1}\boldsymbol{\omega}$ due to the nonlocality of the operator rot^{-1} ('kinematic' nonlocality). In other words, since (for incompressible fluids) $\nabla^2 \mathbf{u} = -rot\boldsymbol{\omega}$, the whole flow field is defined in each space point by the vorticity in the whole flow field and boundary conditions on velocity[17]. This in turn means that the large scales as represented by the velocity field and the small scales as represented by vorticity (and strain) should be strongly coupled, as indeed is the case. Note that this coupling is *bidirectional*, i.e. the small scales cannot be seen as passive or as 'slaved' to the large scales – the small scales

[16]There is no consensus on the meaning of the term integrability, but it is agreed mostly that integrable systems behave nicely and are globally 'regular', whereas the nonintegrable systems are not 'solvable exactly' and exhibit chaotic behaviour. See Zakharov (1990) for more examples and discussion on what is integrability.

[17]Note that, just like in the case of vorticity, the whole flow field is determined entirely by the field of strain. This is seen from the equation $\nabla^2 u_i = 2\partial s_{ik}/\partial x_k$, which together with the boundary conditions uniquely defines the velocity field.

react back in a nonlocal manner. There is no such relation in the case of a passive contaminant (or any other passive object) in a turbulent flow. The relation in this latter case, though also nonlocal, is one-directional: the fluid flow does not 'know' anything about the presence of the passive object.

The nonlocality due to the coupling between large and small scales is also manifested in problems related to various decompositions of turbulent flows and in the so-called closure problem. For example, in the Reynolds decomposition of the flow field into the mean and the fluctuations and in similar decompositions associated with large eddy simulations (LES), the relation between the fluctuations and the mean flow (or resolved and unre-solved scales in LES, etc.) is not pointwise in space/time, it is a functional. That is the field of fluctuations at each time/space point depends on the mean (resolved) field in the whole time/space domain. Vice versa, the mean (resolved) flow at each time/space point depends on the field of fluctuations (unresolved scales) in the whole time/space domain. This is because the equations for the fluctuations (unresolved scales) contain as coefficients the mean (resolved) field. This means that in turbulent flows point-wise flow independent 'constitutive' relations analogous to real material constitutive relations for fluids (such as stress/strain relations) can not exist, though the 'eddy viscosity' and 'eddy diffusivity' are frequently used[18] as a crude approximation for taking into account the reaction back of fluctuations (un-resolved scales) on the mean flow (resolved scales). The fact that the 'eddy viscosity' and 'eddy diffusivity' are flow (and space/time) dependent is just another expression of the strong coupling between the large and the small scales. More details on the issue of nonlocality are given in chapter 6.

1.3.6. ON PHYSICS OF TURBULENCE

The difficulties described above are mostly of a formal/technical nature. There is another difficulty of a more general nature – lack of knowledge about the basic physical processes of turbulence and its generation and origin, and poor understanding of the processes which are already known. For example, the underlying mechanisms of predominant vortex stretching, that is why in turbulent flows vorticity is stretched more than compressed, are (at best) poorly understood and essentially not known (see chapter 6).

1.3.7. ON STATISTICAL THEORIES

Just like in statistical physics, the statistical approach should be adopted in turbulence theories from the outset/start due to the extreme complexity

---

[18]But Scorer (1978) is quite critical about *the doubtful meaning of an eddy transfer coefficient.*

1.3.7. ON STATISTICAL THEORIES

Just like in statistical physics, the statistical approach should be adopted in turbulence theories from the outset/start due to the extreme complexity of turbulence phenomenon(a). In both cases certain statistical hypotheses are made. But the former was quite successful in making a number of important predictions, whereas the latter, with few exceptions, such as the Kolmogorov four fifths law (Kolmogorov, 1941b), was unable to produce genuine predictions based on the first principles. All the rest – in the words of P.G. Saffman – are postdictions. Apart from the above-mentioned reasons for such a failure it should be mentioned that, unlike statistical physics, in turbulence neither 'simple objects' (such that a collection of these objects would adequately represent turbulent flows) 'to do statistical mechanics' with them, nor 'right' statistical hypotheses have so far been found. The question about the very existence of both remains open.

1.4. Outline of the following material

Our main emphasis is on the basic properties of turbulent flows. Therefore origins of turbulence and ways of its creation are discussed only briefly in chapter 2, together with some points concerning the transition to chaos. Instability and transition to turbulence and chaos comprise several separate disciplines, in which many thousands of publications already exist.

chapter 3 is devoted to the problem of describing turbulent flows, with the stress on the principal points. Some technical issues regarding the methods of studying of turbulent flows are briefly discussed in the appendix E.

Selected kinematic issues are discussed in chapter 4. Namely, we concentrate on those issues of the behaviour of passive objects in random flows and kinematic (Lagrangian) chaos which are relevant for comparison with the dynamical aspects of turbulent flows.

The primary aspects of phenomenology are discussed in chapter 5, with the stress placed on the concept of cascade, decompositions and related matters. This and the next two chapters mainly deal with the case of homogeneous (and mostly isotropic) turbulent flows.

Chapters 6 and 7 are devoted to the dynamics of turbulence and its structure, with the emphasis placed on the dynamics of the field of velocity derivatives: vorticity and strain, and their interaction. Special attention is given to nonlocality, non-Gaussianity and geometrical statistics, intermittency and its relation to structure(s), and qualitative difference between vortex stretching and stretching of material lines.

An overview of turbulent flows under various influences and physical circumstances, some of which serve also as sources of energy for sustaining the turbulence, is given in chapter 8. These include shear, buoyancy, rotation, (electro) magnetic field, compressibility and additives. For obvious reasons the material of this chapter is limited by only the most important essential features and changes in turbulent flows under the various influences.

Chapters 2 to 7 are all concluded by a brief summary.

Chapter 9 is devoted to the recapitulation of some main points with somewhat different emphasis, and to the discussion of issues of the general nature not addressed in previous chapters.

This last chapter is followed by a list of references and appendices. The latter contain glossaries of some terms, and discussion of basic fluid mechanics, research methods in turbulence, and listing some misconceptions. These are followed by subject and author indices.

1.5. In lieu of summary

MAJOR QUALITATIVE UNIVERSAL PROPERTIES OF TURBULENCE

– Intrinsic spatio-temporal randomness, irregularity. Turbulence is chaos (but not necessarily vice versa); its intrinsic property is self-stochastization or self-randomization.

– Loss of predictability, but stable statistical properties.

– Extremely wide range of strongly and nonlocally interacting degrees of freedom ('scales' in time and space).

– Highly dissipative, statistically irreversible.

– Turbulent flows are three-dimensional and rotational with continuous self-production of vorticity and strain.

– Strongly diffusive with enhanced transport of momentum, energy, and passive objects.

– Strongly nonlinear, nonintegrable, nonlocal and non-Gaussian.

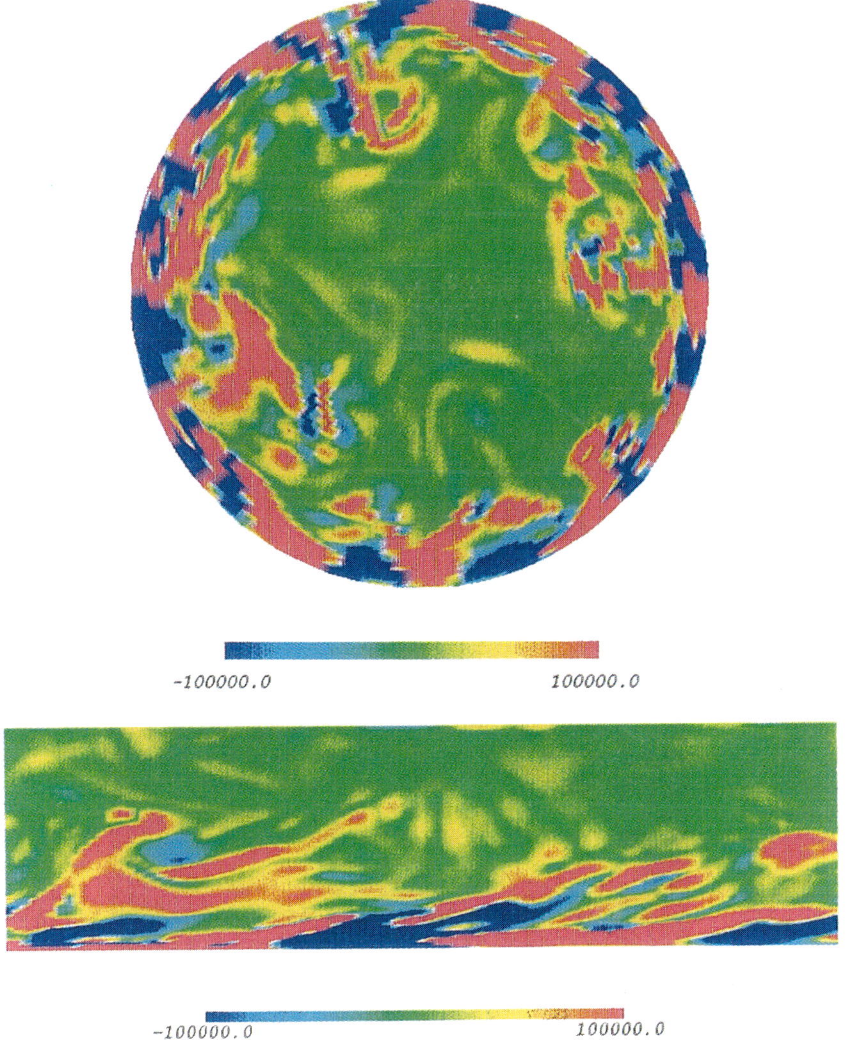

Plate 1. Visualization of the field of the enstrophy production $\omega_i \omega_k s_{ik}$ from the data in DNS of NSE in circular pipe flow at $Re \approx 7000$ performed by Eggels *et al.* (1994). One of the prominent features of *all* turbulent flows is that in the mean the enstrophy production is always positive, $\langle \omega_i \omega_k s_{ik} \rangle > 0$. Courtesy of Prof. F.T.M. Niewstadt and Dr. J.M.J. den Toonder.

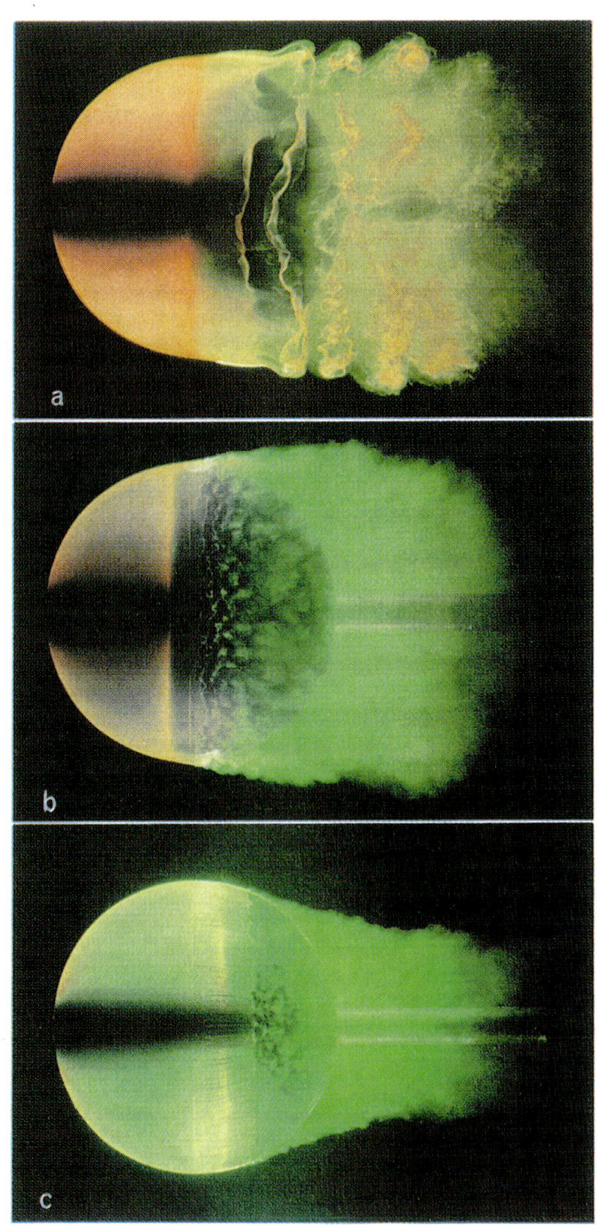

Plate 2. Flow in the near wake region past a sphere at three values of the Reynlolds number, based on the free stream velocity and the sphere diameter: (a) - $2 \cdot 10^4$, (b) - $2 \cdot 10^5$, (a) - $3 \cdot 10^5$. From Werlé (1987), by permission ONERA.

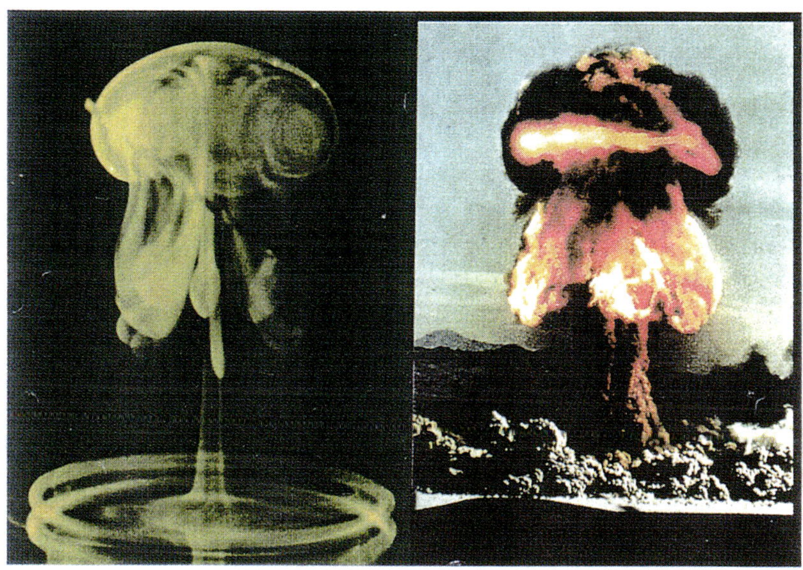

Plate 3. Similarity in large scale flow patterns of a drop dyed with fluorescin into clear water (inverted) at Reynolds number $Re \sim 10^2$ and of a nuclear test in Nevada at Reynolds number $Re \sim 10^9$, Sigurdson (1997), by permission CRC Press LLC.

Plate 4. Flow past a grating of five circular cylinders visualized by two colours, by Tetsuo SAGA, The University of Tokyo, Fantazy of Flow (1993), by permission Ohmsha, Ltd.

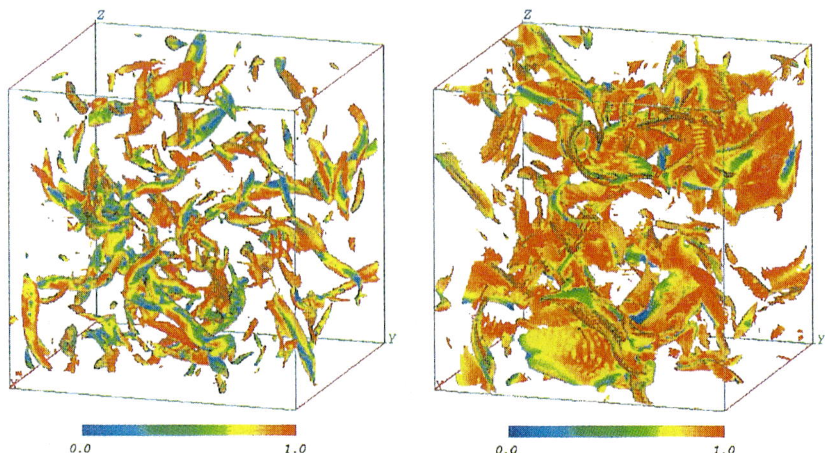

Plate 5. Visualizations of the magnitude of (left) the vorticity, ω, and (right) the scalar gradient, G ($G_i \equiv \partial\theta/\partial x_i$), in a direct numerical simulation of the Navier-Stokes equations in a cubic box with periodic boundary conditions at $Re_\lambda = 58$. The statistically stationary state is achieved by adding an appropriate forcing in the RHS of NSE. The threshold values for both ω and G are larger than 2.5 of their rms value. Contour colouring corresponds to the $cos(\boldsymbol{\omega}, \boldsymbol{\lambda}_2)$ and $cos(\mathbf{G}, \boldsymbol{\lambda}_3)$ alingments. The vectors $\boldsymbol{\lambda}_i$ form the eigen frame of the rate of strain tensor, s_{ij}, coressponding to its eigenvalues Λ_i ($\Lambda_1 > \Lambda_2 > \Lambda_3$; $\Lambda_1 > 0, \Lambda_3 < 0$). Red indicates strong alignment. From Ph. D. Thesis by Flohr (1999).

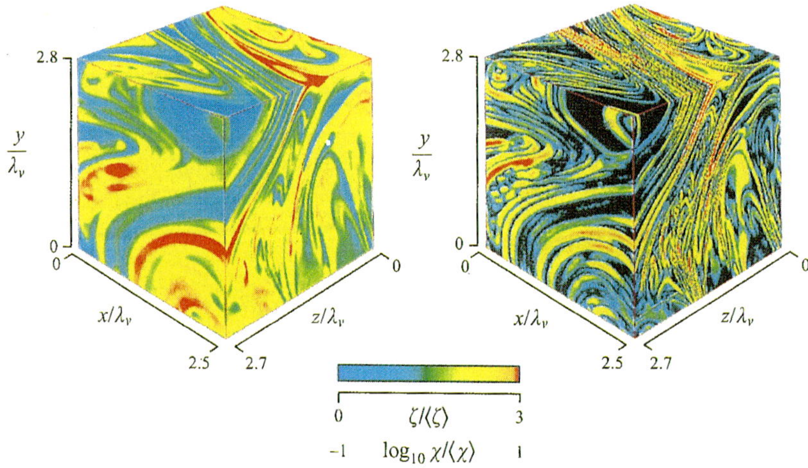

Plate 8. Fully resolved three-dimensional data volume of a) the scaler field, θ, and b) its 'disipation', $(\nabla\theta)^2$; Frederiksen, et al. (1997).

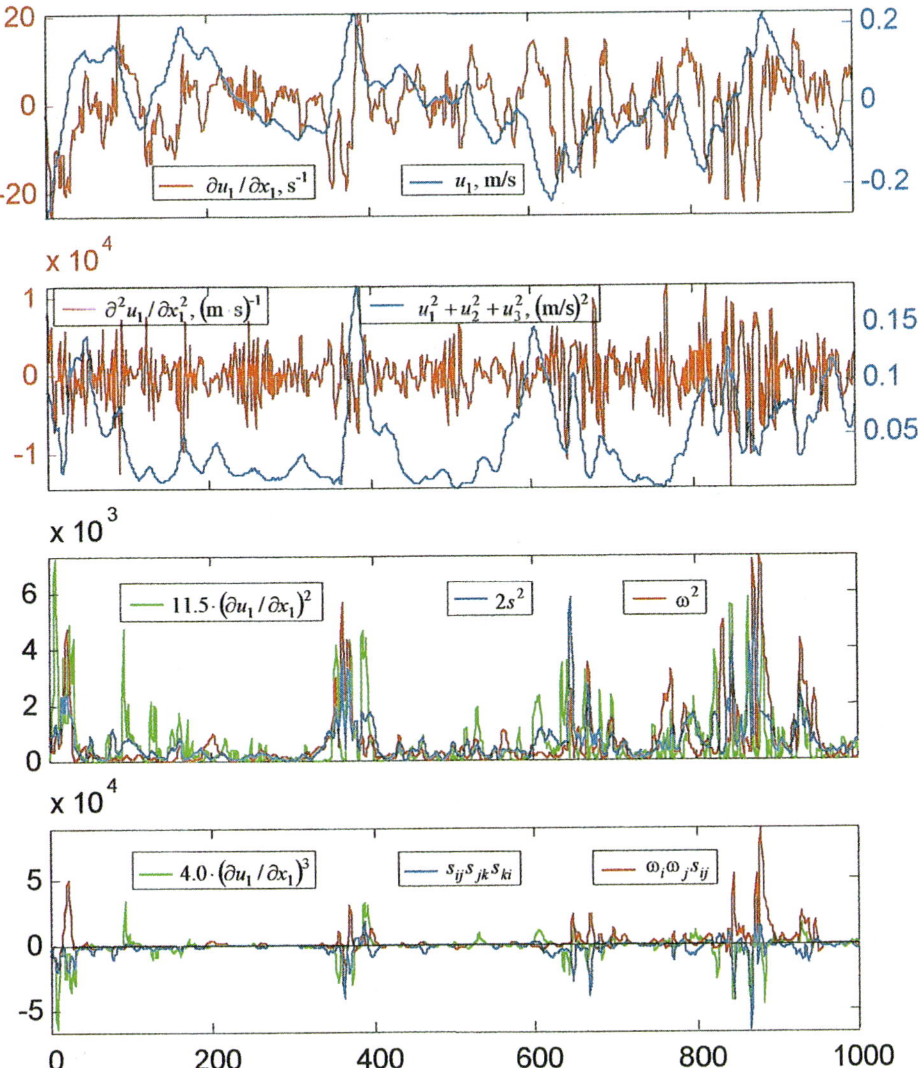

Plate 6. An example of time recording of various quantities in a field experiment at $Re_\lambda = 10^4$ in the atmospheric surface layer in a point at the hight $10m$ from the surface. The measurements in this experiment included all the three velocity components in five neighbouring points. This allowed to evaluate all the nine velocity derivatives $\frac{\partial u_i}{\partial x_k}$, and to obtain such quantities as instantaneous enstrophy, ω^2, total strain, $s^2 \equiv s_{ik}s_{ik}$, enstrophy production, $\omega_i\omega_k sik$ and many others, Kholmyansky and Tsinober (2000). Note the intermittent nature of the signals associated with velocity derivatives: peaks which are hundreds times larger than the mean and very low values are not rare.

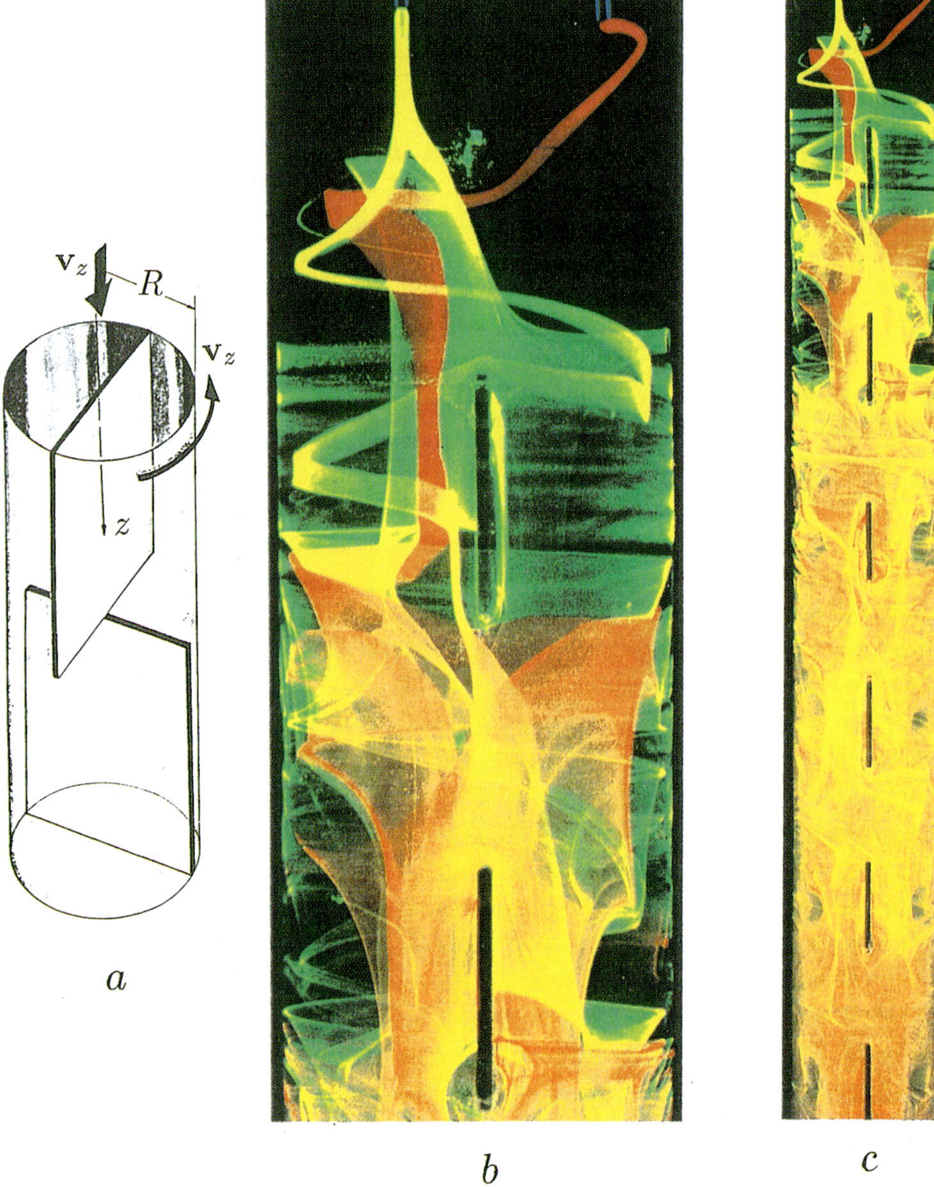

Plate 7. Mixing in PPM - partitioned-pipe mixer at very low Reynolds number. $\mathrm{Re}_{PPM:axial} = \langle v_z \rangle R/\nu = 0.3$ and $\mathrm{Re}_{PPM:cs} = v_R R/\nu = 1.8$; here $\langle v_z \rangle$ − average axial velocity and $v_R = \frac{1}{2}(|v_{1|\max} + |v_{1|\min})$ − characteristic cross-sectional velocity. $0 < \mathrm{Re}_{PPM:axial} < 0.8$ and $0 < \mathrm{Re}_{PPM:cs} < 0.8 < 8$. a) schematic of the PPM, b) is a close up of the upper part of c). From Kusch and Ottino (1992). For other examples see Acrivos (1991), Kim and Stringer (1992), Aref and El Nashie (1994) and references therein.

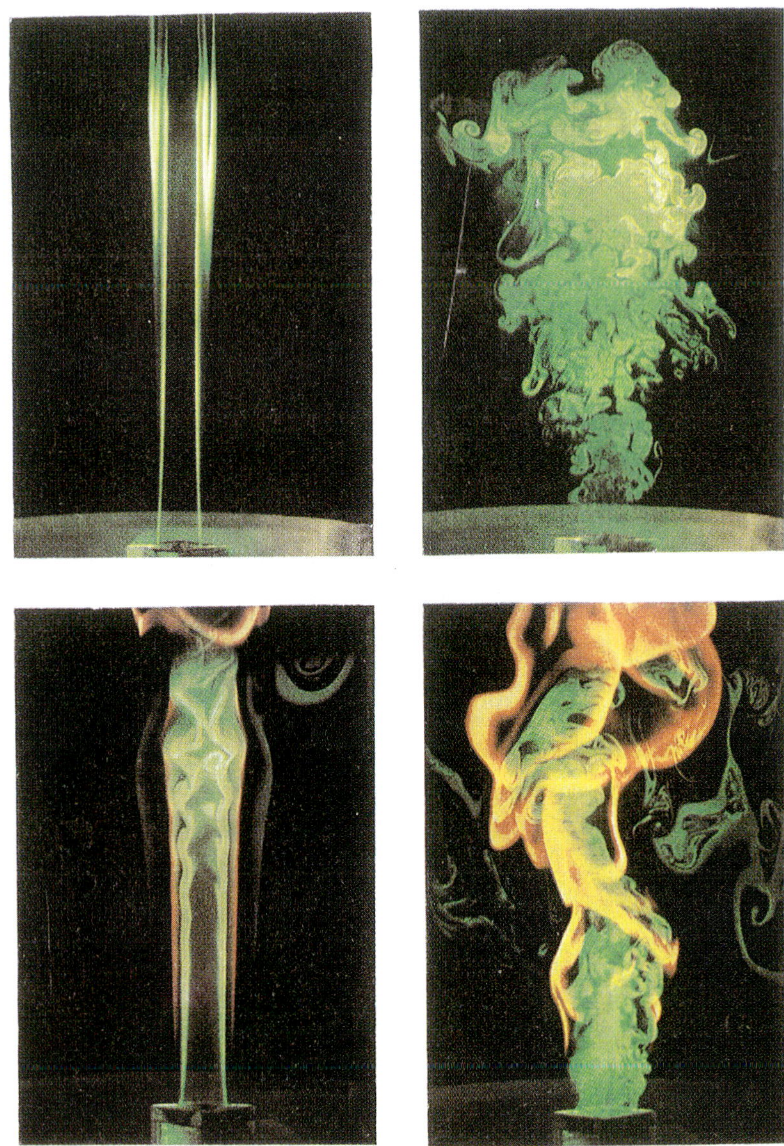

Plate 9. Flow patterns of a square jet of cold flow (nitrogen, top) and combusting gas (propane, bottom) exhibiting strong dependence on forcing (4 Hz, left column) at the jet exit by piezoelectric actuators. Courtesy of Professor A. Glezer, Phys. Fluids, **A4**, 1877.

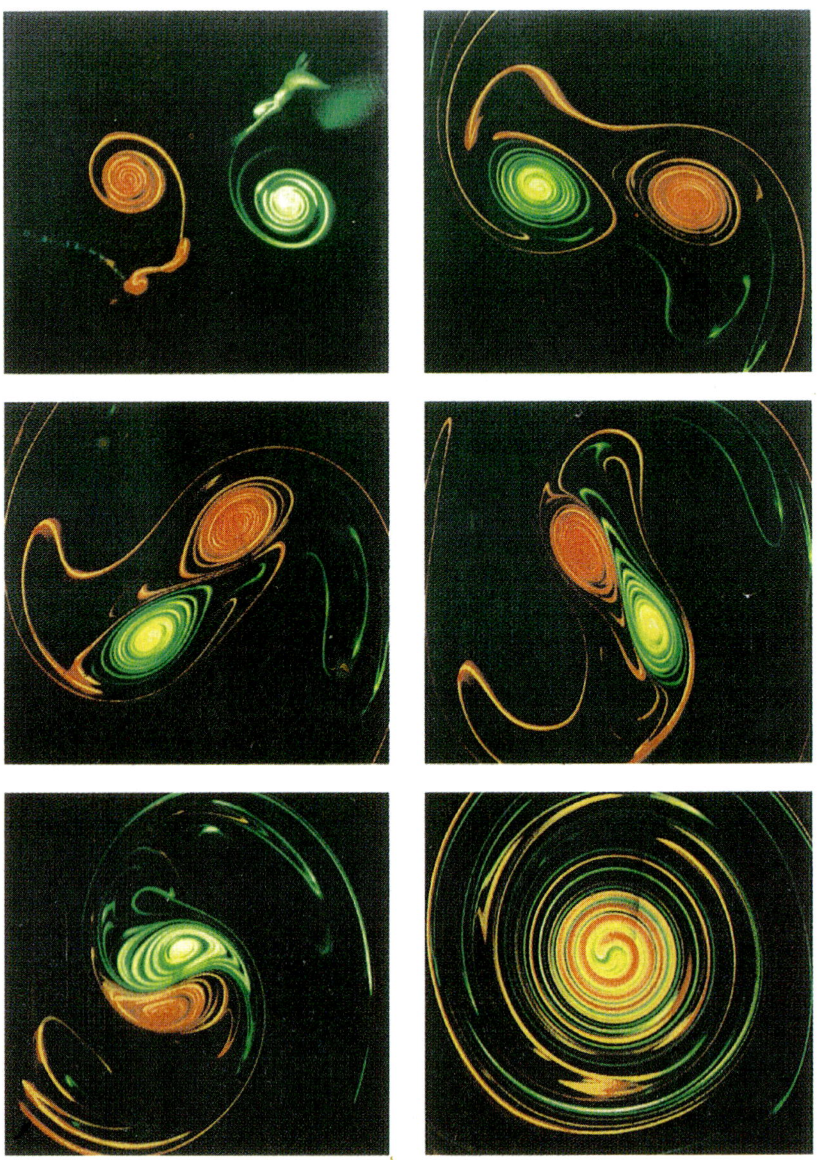

Plate 10. Dye visualization of two-corotating vortices with the vorticity of the same sign. Time is increasing from left to right and from top to bottom. Note the resulting rather complicated pattern of dyes, whereas the coresponding fluid flow consists of just one 'simple' vortex. Coutesy of Dr. T. Leweke, see Meunier and Leweke (2000).

CHAPTER 2

ORIGINS OF TURBULENCE

Overview of instability, transition and chaos

2.1. Instability

> *Yet not every solution of the equation of motion, even if it is exact, can actually occur in Nature. The flow that occur in Nature must not only obey the equations of fluid dynamics but also be stable (Landau and Lifshits, 1959).*
>
> *Kolmogorov's scenario was based on the complexity of the dynamics along the attractor rather than on its stability (Arnold, 1991).*
>
> *To the flows observed in the long run after the influence of the initial conditions has died down there correspond certain solutions of the Navier-Stokes equations. These solutions constitute a certain manifold $\mathcal{M} = \mathcal{M}(\mu)$ (or $\mathcal{M} = \mathcal{M}(Re)$) in phase space invariant under phase flow... The notion of stability here refers to the whole manifold and not to the single motions contained in it. (Hopf, 1948).*

It is a common view that the origin of turbulence is in the instability of some basic laminar flow(s). This is understood in the sense that any flow is started at some moment in time from rest, and as long as the Reynolds number (or a similar parameter) is small, the flow remains laminar. As the Reynolds number increases, some instability sets in, which is followed by further (secondary, tertiary..) instabilities (bifurcations), transition and a fully developed turbulent state[1]. Such sequences of events occur not only throughout the whole flow field, but also at successive downstream locations of a single flow, such as the spatially developing flows as shown in the figures 1.4 and 1.5 of chapter 1. However, it is important to stress that transition to turbulent regime may be quite sudden. For example, this may

[1] The literature on fluid flow instabilities and transition is vast. Comprehensive reviews and lists of references can be found in Drazin and Reid (1981), Huerre and Rossi (1998), and Monin and Yaglom (1997, 1998), and also Monin (1986). For recent developments, see Fasel and Saric (2000).

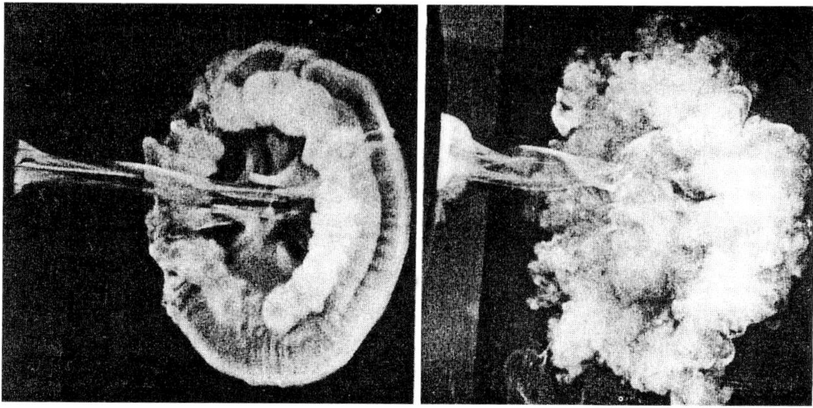

Figure 2.1. Sudden transition of a laminar vortex ring to a turbulent state (Schultz-Grunov, 1980).

happen in pipe flows under certain conditions[2] or in the process involving the impingement of a laminar vortex ring upon a rigid wall (see figure 2.1).

From the mathematical point of view the transitions from one flow regime to another with increasing Reynolds number – as we observe them in *physical* space – are believed to be a manifestation of generic structural changes of the mathematical objects called phase flow and attractors in the *phase* space through bifurcations in a given flow geometry (Hopf, 1948). However, partly turbulent flows (a special feature of these flows is the co-existence of regions with laminar and turbulent states of flow) are not easily 'fit' in this picture. Note that in partly turbulent flows there is a continuous transition of laminar flow into turbulent as result of the entrainment process occurring across the boundary between the two (see section 8.2).

Whereas the processes by which flows become turbulent are quite diverse, all known quantitative properties of many (but not all) turbulent flows do not depend either on the initial conditions or on the history and particular way of their creation, e.g. whether the flows were started from rest or from some other flow and/or how fast the Reynolds number was changed. The qualitative properties of all turbulent flows are the same.

The diversity of the processes by which flows become turbulent is in part due to the sensitivity of the instability and transition phenomena to various details characterizing the basic flow and its environment. For example, the Orr-Sommerfeld equation governing the linear(ized) (in)stability contains the second derivative of the basic velocity profile. Many flows (some of the so-called open flows, such as flows in pipes, boundary layers, jets, wakes,

---

[2]For example in a pipe flow which is held laminar at rather large Reynolds number by special precautions, say at $Re \sim 3 \cdot 10^4$, and then subject to disturbance of finite amplitude.

mixing layers) are very sensitive to external noise and excitation. There are essential differences in the instability features of turbulent shear flows of different kinds (wall bounded - pipes/channels, boundary layers, and free - jets, wakes and mixing layers), thermal, multidiffusive and compositional convection, vortex breakdown, breaking of surface and internal waves and many others[3]. It is important that such differences occur also for the *same* flow geometry, which display in words of M.V. Morkovin *bewildering variety of transitional behaviour*. The specific route may depend on initial conditions, level of external disturbances (receptivity), forcing, time history and other details in most of the flows mentioned above (see plate 7 for an example of such sensitivity). This diversity is especially distinct for the very initial stage - the (quasi)linear(ized) instability. Later nonlinear stages are less sensitive to such details. Hence there is a tendency to universality in strongly nonlinear regimes, such as developed turbulence.[4] We shall discuss this tendency in more detail in later chapters.

One of the important common features of processes resulting in turbulence is that all of them tend to enhance the rotational and dissipative properties of the flow in the process of transition to turbulence. The first property is associated with the production of vorticity, whereas the second property is due to the production of strain (see chapters 5, 6 and 8).

2.2. Transition to turbulence versus routes to chaos

One of the main achievements of modern developments in deterministic chaos is the recognition that chaotic behaviour is an intrinsic fundamental property of a wide class of nonlinear physical systems (including turbulence) and not a result of external random forcing or errors in the input of the numerical simulation on the computer or the physical realization in the laboratory. The nonlinear systems and the equations describing them produce an apparently random output 'on their own', 'out of nothing' – it is their very nature. However, there is a variety of qualitatively different systems exhibiting such a behaviour just like there is a large diversity of such behaviours.

The qualification of turbulence as a phenomenon characterized by a large number of strongly interacting degrees of freedom enables us to make a clear distinction between transition to turbulence and transition to chaotic behaviour.

The main point of distinction is as follows.

---

[3]For a description of particular examples and references see Tritton (1988, chapters 17, 18, 22-24), Sherman (1990, chapter 13) and Huerre and Rossi (1998).

[4]By tendency, it is meant that unversality occurs on the qualitative, but not necessarily on the quantitative level.

Low dimenstional chaotic systems like the famous Lorenz (1963) system
or the spherical pendulum studied by Miles (1984) change their behaviour
from simple regular (as periodic) to distinctly chaotic as some parameter of
the system changes. However, obviously the number of degrees of freedom
of all such systems remains the same, only the character of the interaction
of these degrees of freedom changes.

At early stages of transition to turbulence some of fluid dynamical systems (but not all)[5] exhibit the same behaviour as found in the so called
'routes to chaos' in low dimensional dynamical systems. Namely, they have
only few excited degrees of freedom and a fixed number of them, which are
strongly correlated over the whole flow domain. Hence their dynamics is
essentially temporal: it is chaotic but rather 'simple' - the chaos is temporal only, the spatial structure of the flow is not changing. However, at more
advanced stages of transition, the number of excited degrees of freedom in
fluid flows increases rapidly with the Reynolds number (or similar parameter as the Rayleigh number in thermal convection) and in the developed
stage, it is $\sim Re^{9/4}$ (see appendix 3 and chapter 5). This steep increase in
the number of excited degrees of freedom results in a qualitative change in
the behaviour of the flow - it is chaotic as well, but qualitativey different,
much more complicated kind of chaos - it is both temporal and spatial and
high dimensional: 'more is different' (Anderson, 1972, 1991, 1995). The idea
that the essential feature of transition to turbulence is an increase of the
number of excited degrees of freedom dates back to Landau (1944) and
Hopf (1948) and is correct, though the details of their scenario appeared
to be not precise (see Monin, 1986). However, *Kolmogorov's ideas on the
experimentalist's difficulties in distinguishing between quasi-periodic systems with many basic frequencies and genuinely chaotic systems have not
yet been formalized* (Arnold, 1991). In other words it is very difficult if not
impossible to make such a distinction in practice.

Here is the right place to note that there is an important difference between the number of degrees of freedom rougly proportional to the number
of ordinary differential equations necessary to adequately represent a system described by partial differential equations (NSE) and the dimension
of the attractor of the system (if such exists). In a particular dynamical
system, the former is obviously fixed and is independent of the parameters
of the system, whereas the latter is changing with the parameters but is
bounded. In turbulence both are essentially increasing with the Reynolds
number and become very large at large Reynolds number.

---

[5]The so called closed systems, like small aspect ratio Raylegh-Bènard convection and
Taylor-Couette flow, see e.g. Aref and Gollub (1996), Mullin (1993) and Tritton (1988,
section 24.7) and references therein.

2.3. Many ways of creating turbulent flows

Any turbulent flow is maintained by an external source of energy produced by one or more mechanisms. The mechanisms maintaning/sustaining turbulence, at least some of them, are believed to be closely related (but are not the same) to those by which laminar and transitional flows become turbulent. We address this issue in chapter 8. Here we note that, apart from 'natural' ways resulting from instabilities, turbulent flows can be produced by 'brute force', i.e. by applying external forcing of various kinds both in real physical systems and in computations by adding some forcing in the right hand side of the Navier-Stokes equations. For example, one of the simplest kinds of turbulent flow - quasi-homogeneous and isotropic - can be established by moving a grid through a quiescent fluid or placing such a grid in a wind tunnel, or oscillating such grids in a water tank, or forcing turbulent motion by electromagnetic forces, e.g. in electrolytes, liquid metals or plasma. A similar kind of turbulent flow can be produced numerically by adding a force (random or deterministic) to the right hand side of the Navier-Stokes equations.[6]

An important point is that the nature of forcing (deterministic, random or whatever) is secondary in establishing and sustaining a turbulent flow, provided that the Reynolds number is large enough and the forcing is mostly in the large scales. Another important point is that the forcing does not have to be random. Even if a turbulent flow is produced by random forcing, the primary role of such forcing (as of any forcing) is to supply energy to the flow and to trigger the intrinsic mechanisms of self-stochastization or self-randomization of turbulent flow, i.e. creation of randomness out of 'nothing'. This is the reason why both kinds of turbulent flows - those arising 'naturally', e.g. by a simple (time independent and smooth in space) deterministic forcing, or produced by some external random source, are the same qualitatively and in many essential respects quantitatively. At small enough Reynolds numbers, the flow produced by deterministic forcing is not random, it is laminar, but the flow produced by random forcing, though random, is in many respects trivial, e.g. there is no interaction between its degrees of freedom/modes.

Thus a turbulent flow originates not necessarily out of a laminar flow with the same geometry. It can arise from any initial state including a 'turbulent' one, such as random initial conditions in direct numerical simulations of the Navier-Stokes equations. That is, the transition from laminar to turbulent regime is not the only causal relation. This problem is related

---

[6]The random force method is due to Novikov (1964). It is now a common practice in direct numerical simuations of the Navier-Stokes equations (see Avelius, 1999; Machiels and Deville, 1998; Nie and Tanveer, 1999; Overholt and Pope, 1998; Sain et al., 1998; and references therein).

to a somewhat 'philosophical' question on whether flows become or whether they just are turbulent, and to the unknown properties of the phase flow, attractors and related matters, which are far beyond the scope of this book.

2.4. Summary

There is a great variety of ways/routes in which a laminar flow becomes turbulent, just like there are many ways to establish the *same* turbulent flow. In other words, the view that turbulent flows *always* develop from the laminar ones is too narrow.

Once a flow becomes turbulent, it seems impossible to find out its origin. The reason is due to the chaotic nature and the irreversibility of turbulent flows.

The main difference between the transition to chaos and to turbulence is that in the former the number of degrees of freedom remains fixed, whereas in the latter the number of degrees of freedom increases strongly with increases in the Reynolds number and/or other similar parameters.

CHAPTER 3

METHODS OF DESCRIBING TURBULENT FLOWS

Deterministic, structural or statistical?

In contrast to this experimental cornucopia, theory can offer only a few crumbs (Siggia, 1994).

One of the major problems in describing turbulence stems from its extremely intricate effectively/apparently random behaviour along with huge number of strongly and nonlocally interacting degrees of freedom. Other reasons why the turbulence problem is so impossibly difficult have been mentioned in chapter 1. One more consideration involves the fact that adequate tools to handle both the problem and the phenomenon of turbulence are not developed enough. In this respect the state of matters is not very much different from the one depicted by von Neumann (1949):

The entire experience with the subject indicates that the purely analytical approach is beset with difficulties, which at this moment are still prohibitive. The reason for this is probably as was indicated above: That our intuitive relationship to the subject is still too loose – not having succeeded at anything like deep mathematical penetration in any part of the subject, we are still quite disoriented as to the relevant factors, and as to the proper analytical machinery to be used.

That is there is no way and tools so far, if ever, to treat turbulence analytically – turbulence is beyond analytics (TBA). Unfortunately, this is true of other theoretical approaches such as attempts to construct statistical and/or other theories.

Another matter the technical tools, which are of purely experimental and observational nature. Unlike the theoretical issues/problems much essential progress has occurred in developing numerical, laboratory and field experimental approaches to turbulent flows (see Appendix E). It is noteworthy that in view of the state of the theoretical field, the experimental research in turbulence was and remains the main source of knowledge of turbulent flows. Therefore, the role of experiments in turbulence goes far beyond the view of those who think of experimentalists as a superior kind of professional fixers knowing how to turn nuts and bolts into a confirmation of other people's theories. From the basic point there is almost nothing to be confirmed so far. On the contrary *the essential mathematical complications of the subject were only disclosed by actual experience with the physical counterparts of these equations* (Neumann, 1949), *and the observational material is so large, that it allows to foresee rather subtle mathematical results, which would be very interesting to prove* (Kolomogorov, 1978). These statements remain as valid today as they were earlier. The experiment remains a major exploratory tool in elucidating the properties of turbulence as a physical phenomenon. We remind here that this does not mean that the Navier-Stokes equations are useful as an experimental tool only (see chapter 1, subsection 1.3.1.)

This chapter is devoted to some matters of principle regarding the methods of describing turbulent flows and related issues.

3.1. Deterministic versus random/stochastic or how 'statistical' is turbulence?

> *Turbulence flow constitutes an unusual and difficult problem of statistical mechanics, characterized by extreme statistical disequilibrium, by anomalous transport processes, by strong dynamical nonlinearity, and by perplexing interplay of chaos and order.* (Kraichnan, 1972).
>
> The "statistical" *community ... strongly disputes the possibility of any coherence or order associated to turbulence.* (Lesieur, 1997).
>
> *The transition from laminar to turbulent flow is a nonequilibrium phase transition to a more organized motion.* (Klimontovich, 1996).

It is quite common to contrapose the 'traditional' statistical and the deterministic/structural approaches in turbulence research. However, contrasting the terms 'deterministic' and 'random' has lost most (but not all) of its meaning with the developments in 'deterministic chaos': it is well established that even simple systems governed by purely deterministic nonlinear sets of equations, such as those described by only three nonlinear ordinary differential equations, as a rule exhibit irregular apparently random/stochastic behaviour. In fact, in respect with turbulence this was known long before the 'discovery of chaos'. However, since Leray (1934) one was not sure about the (theoretical but not observational) possibility that turbulence is a manifestation of breakdown of the Navier-Stokes equations.

The early justifications the necessity of statistical approaches to turbulence were usually based on the extreme complexity of the individual realizations of turbulent flows:

In ... turbulent motion, an enormous number of degrees of freedom are always excited, and hence the variation with time (and space) of any physical value will be described ...by functions of extremely complicated nature.

The theory of turbulence by its very nature cannot be other than statistical, i.e. an individual description of the fields of velocity, pressure, temperature, and other characteristics of turbulent flow is in principle impossible. Moreover, such description would not be useful even if possible, since the extremely complicated and irregular nature of all the fields eliminates the possibility of using exact values of them in any practical problems...

In the present-day statistical fluid mechanics, it is always implied that the fluid mechanical fields of a turbulent flow are random fields in the sense used in probability theory. (Monin and Yaglom, 1971, pp. 3-4, 7).

From the very beginning it was clear that the theory of random functions of many variables (random fields), whose development only started at that time, must be the underlying mathematical technique. (A. N. Kolmogorov, 1985 in notes preceding the papers on turbulence in the first volume of his selected papers, English translation, Tikhomirov, 1991, p. 487).

Later the very large dimension (see Table 3.1) and complicated (stochastic?) structure of the underlying attractors, assumed to be in existence, was invoked in the justification of the unavoidable necessity of statistical methods of *description* (and 'theories') of turbulent flows: *one may never be able to realistically determine the fine-scale structure and dynamical details of attractors of even moderate dimension. ... The theoretical tools that characterize attractors of moderate or large dimensions in terms of the modest amounts of information gleaned from trajectories* [i.e. paricular solutions]... *do not exist ...they are more likely to be probabilistic than geometric in nature* (Guckenheimer, 1986).

Apart from the extreme intricacy of turbulent flows and their stochastic nature, there is another advantage in favour of the statistical methods of description (not necessarily 'theories'). Namely, all the experience accumulated during the period of a century of studying turbulent flows shows that turbulent flows possess stable statistical properties (SSP)[1]. It is important to stress that stable statistical properties means not only means (averages) and other simple things, but much more, presumably all statistical properties together with those which are related to what can be called the structure of turbulence, which is not the same as what is called structures of turbulence, e.g. its instantaneous structure-like features (see chapter 7). The existence of SSP seems to be an indication of the existence of what mathematicians call attractors. But again the matters are more complicated than that. Many statistical properties of nonstationary

---

[1] *The existence of an asymptotic statistical state is strongly suggested experimentally, in the sense that reproducible statistical results are obtained. However, physical plausibility aside, it is embarrassing that such an important feature of turbulence as its statistical stability should remain mathematically unresolved, but such is the nature of the subject* (Orszag, 1977).

TABLE 3.1. Dimensions of "attractors". Here D_L - is the so-called Lyapunov dimension, D_{KL} - is roughly the number of modes to account for 90% of the energy, and D_l - is the well known Landau estimate, which is roughly proportional to the number of ODE's needed to adequately represent the flow (it was removed from the last Russian and subsequent English editions). This is an updated table based on the one by Sirovich (1997).

	D_L	D_{KL}	$D_l \cdot 10^{-6}$
Isotropic			
Landau			$Re^{9/4}$
Pipe			
Huang & Huang (1989)	> 11		
Sieber (1987)	> 10		
Plane Channel			
Keefe et al (1992)	800	400	26
Sirovich et al (1991)	1500	4200	100
Webber et al (1997)	300	650	15

(time-dependent in the statistical sense) turbulent flows are quite similar to those of statistically stationary ones as long as the Reynolds number of the former is not too small at the particular time moment of interest. For example, in a decaying turbulent flow past a grid, almost everything is very similar qualitatively and in many respects quantitatively to what is happening in decaying statistically stationary forced turbulence in a cubic box with periodic boundary conditions (see de Bruyn Kops and Riley, 1998; Galanti and Tsinober, 2000; Tsinober et al., 1997; and references therein).

It should be emphasized that our concern here is not with statistical *theories* all of which are using various *ad hoc* assumptions on the nature and properties mainly of the small scale structure and its relation with the rest of the flow, and/or attempting to represent turbulence as a collection of more or less *simple* objects. Our concern is much less ambitious – the focus is on statistical methods of *description and interpretation* of the data from laboratory, field and numerical experiments on turbulent flows via appropriate *processing* of the data. The latter is likely to be a prerequisite for any worthy 'theory' of turbulence. Quoting A.N. Komogorov: *...I soon understood that there was little hope of developing a pure, closed theory, and because of absence of such a theory the investigation must be based on hypotheses obtained on processing experimental data* (Tikhomirov, 1991, p. 487). This view goes back to Leonardo da Vinci: *Remember, when discoursing about water, to induce first experience, then reason.*

It should be stressed that even the simplest nonlinear systems exhibiting chaotic behaviour are analyzed via various statistical means. Also, the so called 'coherent structures' in turbulent flows are hunted using essentially statistical methods, such as conditional statistics though with limited success (Bonnet, 1996). Finally, methods of dimensional analysis, similarity and symmetries (group theoretical methods) and phenomenological arguments are applied exclusively to quantities expressing the statistical properties of turbulent flows.

3.2. On statistical theories, reduced (low dimensional) representations and related matters

As mentioned statistical *methods of describing* turbulent flows should not be confused with statistical *theories* of turbulent flows. The latter are outside the scope of this book for the reasons explained above, so only a few brief remarks are given below.

The natural tendency to simplify the problem is manifested in numerous searches for a reduced description of turbulent flows[2]:

One of the most basic questions in turbulence "theory" (which usually is not asked) whether there exists a closed representation that is simple enough to be tractable and insightful, but powerful enough to be faithful to the essential dynamics (Kraichnan and Chen, 1989).

An early goal of the statistical theory of turbulence was to obtain a finite, closed set of equations for average quantities, including the mean velocity and energy spectrum. That goal now is viewed to be unrealistic. The goal is now to reduce to a manageable number the many degrees of freedom necessary to describe the flow, to determine the equations governing the dynamics of the reduced degrees of freedom, and to solve those equations analytically or numerically to calculate fundamental quantities that characterize the flow. (Frisch and Orszag, 1990).

If we assume as a basic starting point in every theory of turbulence its representation in terms of spectral coefficients, statistical or physical averages, or more generally simple objects conditionally extracted by a weak background, turbulence modelling could be defined reductively, as the art of writing the equations that produce directly such quantities. (Germano, 1999).

These citations represent the most popular view/hope/belief/assumption and an implicit claim that such a reduction is possible and results in an adequate description of the remaining degrees of freedom, which presumably include some of the so-called 'coherent structures'.[3] Such a belief goes

---

[2]The whole issue is closely related to the problem of decomposition/representation of turbulent flows (next section) and their structure (chapter 7).

[3]This is related to a more general assumption that perhaps(!) large systems actually boil down to a much smaller number of degrees of freedom than actually excited, because many of them are strongly correlated within a group representing a 'coherent structure'.

back to the early forties :..*it is necessary to separate random processes from the nonrandom processes* (Dryden, 1948), and, in fact, is the essence of the concept of 'eddy viscosity' (Boussinesq, 1877; Kraichnan, 1989). The implication is that such a separation is possible. But it is not obvious at all that such a reduction is possible, as it is seen from the futility of enormous efforts to do so throughout the whole history of turbulence research. The difficulty is a nontrivial one. For example, one even does not know how to separate random gravity-wave motion and genuine turbulence in a stably stratified fluid (Stewart, 1959).

There is, however, a less popular view too:

Most problems in classical stochastic processes are reduced to solubility by statistical independence, or the assumption of a normal distribution (which is equivalent) or some other stochastic model; because of the governing differential equations, the turbulent velocity at two space-time points is, in principle, never independent - in fact, the entire dynamical behaviour is involved in the departure from statistical independence. The equations, in fact, preclude the assumption of any ad hoc model, although this is often done in the absence of a better idea (Lumley, 1970).

Perhaps the biggest fallacy about turbulence is that it can be reliably described (statistically) by a system of equations which is far easier to solve than the full time-dependent three-dimensional Navier-Stokes equations (Bradshaw, 1994).

Theoretical estimates of the dimension of the attractor for channel turbulence appear to preclude truly low-dimensional description (Omurtag and Sirovich, 1999).

In spite of these warnings there is a general belief that an adequate reduced (low dimensional) description is possible, e.g. via reduction of the huge number of degrees of freedom by retaining the so-called *relevant/important* ones, though the meaning of what are the relevant/important modes/degrees of freedom is quite problematic (Holmes et al., 1997; Kraichnan, 1988). Most frequently it is argued that these are 'modes' containing most of the energy, but - at least from the physical point of view - the 'modes', e.g. carrying most of the energy dissipation and vorticity are not less relevant/important in some sense. 'Mixed modes' related to both small and large scales such as eigenfunctions of $\langle u_i(\mathbf{x})\omega_j(\mathbf{x}+\mathbf{r})\rangle$ may appear even more relevant/important[4]. Even in such a case it is not clear whether it is possible to obtain a low dimensional approximation representing adequately the flow field[5]. For instance, such a 'simple' turbulent flow as the flow in

---

[4]Note that two most important quantities: Lamb vector $\boldsymbol{\omega} \times \mathbf{u} \equiv \epsilon_{ijk}\omega_j u_k$ and the vortex stretching vector $W_i = \partial(u_i\omega_j)/\partial x_j$ are precisely of this kind and are closely related to the tensor $u_i(\mathbf{x})\omega_j(\mathbf{x}+\mathbf{r})$.

[5]Inadequate (too) low dimensional approximations may lead to spurious chaotic behaviour, which disappears when the number of the basic functions becomes large enough

a plane channel at rather low Reynolds number $Re = 3300$, which attractor dimension is estimated to be of the order 10^3 (see table 3.1). Possible exceptions are when the flow, though turbulent, at the outset is strongly dominated by some 'low dimensional subsystem'/coherent structures (e.g. Holmes et al., 1996, 1997; Lesieur and Metais, 1996; O'Neil and Meneveau, 1997). As mentioned in chapter 1 it is quite possible that such large scale structures are the result of a large scale instability of the flow as a whole not related directly to the turbulent nature at least of free shear flows. The situation is more complicated in wall-bounded turbulent flows. These matters are discussed in chapter 8. We add that in a recent attempt Farge et al. (2000) used orthogonal wavelets to represent what they call the coherent part of velocity and vorticity in a three-dimensional flow in a periodic box. The claim is that out of total $1.4 \cdot 10^7$ degrees of freedom $4 \cdot 10^5$ (2.8% of the total) are sufficient in order to adequately represent the 'coherent' part of the flow. It is still quite a lot. However, the advantage of their approach is that their 2.8% include both large and small scales, so that it seems that these 2.8% represent reasonably the whole flow field rather than its 'coherent' part only.

It is natural to seek a closed representation that is mathematically *simple enough to be tractable and insightful, but powerful enough to be faithful to the essential dynamics* (Kraichnan and Chen, 1989). But this does not justify oversimplified treatment of small scales via methods like eddy viscosity, because the small scales contain a great deal of essential physics of turbulent flows, much of which is not known or poorly understood, and which are intimately and bidirectionally related to the large scales, (see chapter 6, section 6.6).

Various models of turbulent flows - a really huge number of them - are all statistical theories in the sense mentioned above. They differ from, e.g. 'physical' theories only by different methods of 'closure' of the resulting equations for the chosen statistical variables. All of them have in common some *ad hoc* assumptions of unknown validity and obscured physical and mathematical justification. In this sense all the statistical theories are not rigorous.

Hans Liepmann wrote in 1979: *Turbulent modelling is still on the rise owing to rapid development of computers coupled with the industrial need for management of turbulent flows. I am convinced that much of this huge effort will be of passing interest only. Except for rare critical appraisals...much of this work is never subjected to any kind of critical or comparative judgement. The only encouraging prospect is that current progress in understanding turbulence will restrict the freedom of such modelling and guide these efforts toward a more reliable discipline* (Liepmann, 1979).

and adequate resolution is used (Curry et al., 1984; see also Rempfer, 1999,2000).

Liepman's criticism was directed at (already at that time) the great number of publications which used a variety of assumptions, most of them very remote from any physical basis, to say nothing of any rigorous mathematical basis. This state of affairs seems to be changed due to two recent developments. First, it was rigorously proved (Fursikov and Emanuilov, 1995 and references therein) that the Keller-Fridman (1925) chain of equations for the moments (and consequently the Hopf equation) has a unique solution for initial conditions in an appropriately chosen functional space[6]. In other words a positive answer was given to the question whether the closure problem has a solution, and an estimate of convergence of approximations for the closure of the infinite chain of equations for moments was given. Thus, it became clear - at least in principle - that turbulence modelling can be put on a rigorous foundation. Second, some developments in LES seem to move in the same direction (Meneveu and Katz, 2000). However, just like direct numerical simulation by itself does not bring understanding, neither does modelling of whatever sophistication.

3.3. Turbulence versus deterministic chaos

Another rather recent and popular view emerged from the developments in the 'deterministic chaos' already mentioned in chapter 2 , i.e. comparatively simple nonlinear systems exhibiting chaotic behaviour. This, however, did not 'solve the problem of turbulence' neither 'the right path was found' as was claimed quite frequently, e.g. *There have been important changes in our understanding of the mechanism whereby turbulence occurs. Although a consistent theory of turbulence is still a thing in the future, there is reason to suppose that the right path has finally been found.* (M. I. Rabinovich, in: L. D. Landau and E. M. Lifshitz, Fluid Mechanics, 2nd edition, Pergamon Press 1987).

Today it seems that the application of dynamical systems methods and results to turbulence in fluids is hardly appropriate[7]. Methods of dynamical systems theory, after an initial period of euphoria and even claims that the problem of turbulence was solved, have proved to be ineffective/irrelevant for the theory of fully developed turbulence. Quoting G.K. Batchelor (1989) : *'... considerations of the properties of fully developed turbulence require rather different ideas, ...'* (*J. Fluid Mech.*, **205**, 593).

It is now recognized that *despite the considerable successes of the present studies of the application of modern ideas on chaos to well-controlled fluid flows, they appear to have little relevance when applied to the more general*

---

[6]Provided that the corresponding three-dimensional problem for the Navier-Stokes equations does have a unique solution.

[7]Though some authors hold an opposite opinion, Bohr et al. (1998), Ruelle (1990).

problem of fluid turbulence (Mullin, 1993, p. 93; see also Tritton, 1988, p. 410).

So it is quite plausible that any fluid flow which is adequately represented by a low dimensional system is not turbulent – a kind of definition of 'non-turbulence'. The immediate examples are low dimensional chaotic fluid flows.

3.4. Statistical methods of looking at the data only? Or what kind of statistics one needs?

Whatever the origins of real turbulent flows[8], turbulent flow states are so complicated that the use of statistical tools is unavoidable. The question is about what kind of statistics one has to use. It is directly related to the most difficult question on *what are the problems,* i.e. to the skill/art to ask the right and correctly posed questions, which is quite a problem in turbulence research.

At early stages, the interest was in relatively simple quantities like means[9], correlations, spectra, and probability density functions (PDFs) of various quantities. With the digital methods of data acquisition and processing conditional statistics became a powerful tool of data analysis (Van Atta, 1974; Antonia, 1981).

Standard statistical tools like means and correlations smooth out some important qualitative features of (typical) individual realizations. The 'mean fields', e.g. large scale averages of velocity or concentration of some species or particles, are smooth whereas the individual realizations are not. They are corrugated, highly intermittent and contain clusters/regions of high level of some quantity/ies (enstrophy, dissipation, passive tracer, reacting species, particles, etc.) surrounded by low level 'voids' of this quantity. In other words, 'standard' 'traditional' statistical methods to a large extent ignore the structure(s) of turbulent flows, which was the main reason for numerous objections against statistical methods often understood as averaging only. More subtle statistical properties of turbulent flows associated with their structure(s) both in small and large scales are important in many applications. For instance, special information on small scale structure(s) is needed in problems concerning, e.g. combustion, turbulent flows with chemical reactions, some environmental problems and propagation of sound and light in turbulent environments. In such problems, not only special statistical properties are of importance like those describing the behaviour of

---

[8]Whether turbulence *is a priory* random/stochastic because the Nature is such or the intricacy of turbulent flows arises out of deterministic equations like NSE or any other unknown reason.

[9]*The true aim of turbulence theory is to predict the mean properties and their dependence on boundary conditions* (Saffmann, 1960).

smallest scales of turbulence, but also actual 'nonstatistical' features like maximal concentrations in such systems as an explosive gas which should be held below the ignition threshold, some species in chemical reactions, concentrations of a gas with strong dependence of its molecular weight on concentration (such as hydrogen fluoride used in various industries e.g. in production of unleaded petrol) and toxic gases. Similarly, problems such as the manipulation (and possibly control) of turbulence and turbulence induced noise require information on large scale structure(s) of turbulent flows far beyond such simple statistical characteristics as averages, correlations, spectra and PDFs.

In other words statistical methods have their limitations, so that in many cases one has to look not only at the properties of turbulent flows *en masse*, but also at some specific properties of individual realizations like those involved in weather forecasting. After all one does not need ensemble averaging to be sure that the coffee will be well mixed via only one and pretty short realization. Likewise, not much can be done statistical-wise to cope with a destructive hurricane or a tropical cyclone[10]

Each particular statistical tool has its own limitations; being useful in one context/respect, it may say nothing in many others. A typical example is correlation, widely used in many aspects of turbulence research. Usually if a correlation between two quantities is not small, it reflects some important relation. However, if the correlation is small, it is not necessarily insignificant. For instance, let us have a look at the famous Reynolds stress $\langle u_1 u_2 \rangle$ – the correlation between the velocity fluctuations in the direction of the mean flow (x_1) and those normal to the wall (x_2) in a wall-bounded turbulent flow. The typical value of the corresponding correlation coefficient is $\langle u_1 u_2 \rangle / u_1' u_2' \sim 0.4$. However, the real quantity entering the equation for the mean flow (RANS, see Appendix C) is the derivative $d \langle u_1 u_2 \rangle / dx_2$. In a developed turbulent flow with its mean properties independent of the streamwise coordinate, x_1, (flat channel, pipe), $d \langle u_1 u_2 \rangle / dx_2 = \langle (\boldsymbol{\omega} \times \mathbf{u})_1 \rangle \equiv \langle \omega_2 u_3 \rangle - \langle \omega_3 u_2 \rangle$. That is the 'turbulent force' is due to the coupling between large and small scales (again nonlocality, see chapter 6). The corresponding correlations between velocity and vorticity are small: both $\langle \omega_2 u_3 \rangle$ and $\langle \omega_3 u_2 \rangle$ are of order 10^{-2}. However, this does not mean that the coupling between $\boldsymbol{\omega}$ and $\mathbf{u}$ is insignificant. Indeed, without such a coupling $d \langle uv \rangle / dx_2 = 0$, so that the mean flow would not 'know'

---

[10]*It may also be that such (i.e. very rare and exceptionally strong) events are rather sensitive to details of the physics that do not appreciably affect the character of the majority of events. This does not mean that one should not keep trying, by insight and discernment, to discover useful statistical measures, but rather that statistics will have to be used with that humility and appreciation of the combination of admission of ignorance and decision to ignore detail so successfully used by workers in the past* (Mollo-Christensen, 1973).

anything about turbulent fluctuations at all and therefore would remain as the laminar one.

Moreover, even if a correlation between two quantities is precisely vanishing, this still does not necessarily mean that the interrelation/coupling between these two quantities is not existing or is unimportant. For example, in homogeneous turbulent flows, velocity and vorticity, and vorticity and the rate of strain tensor are precisely uncorrelated, $\langle \boldsymbol{\omega} \times \mathbf{u} \rangle \equiv 0$, $\langle \omega_i s_{ij} \rangle \equiv 0$, but their interaction is in the heart of the physics of any turbulent flow. We will return to this issue in chapter 6.

Single point statistics in may cases may be insufficient and even misleading. For example, single point PDFs of velocity fluctuations are known to be quite close to the Gaussian distribution. In particular, the third moment of velocity fluctuations is close to zero (more precisley its skeweness, $\langle u_1^3 \rangle / \langle u_1^2 \rangle^{3/2} \approx 0$), and the flatness, $\langle u_1^4 \rangle / \langle u_1^2 \rangle^2 \approx 3$, as in a Gaussian field. Similarly other higher order odd moments are small, and even moments assume values close to those of a Gaussian field, e.g. $\langle u_1^6 \rangle / \langle u_1^2 \rangle^3 \approx 15$. However, the conclusion that velocity fluctuations are really almost Gaussian would be a misconception, not to mention the field of velocity derivatives (chapters 6 and 7). This is seen when one looks at *two*-point statistics. For instance, in such a case the odd moments are significantly different from zero (Frenkiel et al., 1979). This is one of the simplest among numerous examples[11] when multipoint (in space and time) statsistics is useful.

3.5. Decompositions/representations

One of the common approaches both in theory and data analysis is a reductionist one, i.e. some decomposition of the flow field.

The first known decomposition was given by Reynolds (1895), in which the flow field is represented as a sum of a mean and a fluctuative, the latter being just the difference between the mean (assumed to exist) and the instantaneous fields. There are attempts to extend this approach to a triple decomposition consisting of a mean, 'coherent' and 'random' contributions. Most of these attempts are of heuristic nature since the 'coherent' contribution is well defined only in rather special cases. It is noteworthy that when means (in some sense) exist the Reynolds decomposition is more physically and mathematically natural than its analogues, such as triple decompositions or those associated with large eddy simulations (LES).

The *formal* decompositions employ some suitable basis of expansion of

---

[11]The widely known two-point correlations for some separation r and/or time t are related to the flow structure(s) larger than $\sim r/t$. An example of application of three-point statistics to structures of the passive scalar in turbulent flow is given by Mydlarski and Warhaft (1998).

the flow field. These are represented by the Fourier decomposition or its 'relatives' such as Fourier-Weiestrass, Gabor, helical, Littlewood-Paley, or any other complete basis depending on the geometry of the flow, wavelets, wavepackets, solitons (Arneodo et al., 1999; Farge et al., 2000; Meneveau, 1991; Zimin and Frick, 1993), and filters (Germano, 1999; Leonard, 1974; Meneveu and Katz, 2000), Karhunen-Loeve or proper orthogonal decomposition (Holmes et al., 1996, 1997; Sirovich, 1997).

In *heuristic* decompositions the flow field is represented as a 'two-fluid' one, e.g. organized and incoherent or deterministic and random (Cantwell, 1991; Farge and Guyon, 1999; McComb and Watt, 1992; She, 1991), or as a collection of some 'simple' objects – vortex filaments, vortons, 'eigensolutions', etc., (Pullin and Saffman, 1998). Another kind of heuristic representations is prompted by the intermittent structure of turbulent flows: breakdown coefficients/multipliers (Novikov, 1971, 1990a) or equivalently (multi)fractals (Frisch, 1995).

There are several difficulties with all decompositions mainly due to the nonlinear and nonlocal nature of turbulence. These difficulties are not trivial and seem to be 'generic'. As mentioned above, it is not even known how to separate random gravity-wave motion (which does not produce vertical transport) and genuine turbulence (which does) in a stably stratified fluid (Stewart, 1959). Under turbulent motion/dynamics the interaction of 'modes', whatever they are, is strong. The resulting structure(s) is(are) not represented by the modes of any known decomposition, for example by a Fourier-decomposition of a flow in a box with periodic boundary conditions. The emergence of structures in such a flow, such as the slender vortex filaments, in a random fashion (at random times with random orientation, and to a large extent random shapes) points to the limitation of utilization of Fourier decomposition (or similar), which does not 'see' these or any other structure(s). Another example is the chaotic regime of a system with few degrees of freedom only, e.g. three as in the Lorenz system (Lorenz, 1963), or four in the forced spherical pendulum (Miles, 1984), but with a *continuous* spectrum. Hence the ambiguity of Fourier decomposition (see Liepmann, 1962; Tennekes, 1976; and Lohse and Müller-Groeling, 1996 for other aspects of Fourier-transform ambiguity). All the attempts to find a 'good' decomposition are related to what Betchov (1993) called the 'dream of linearized physicists', i.e. a *superposition* of some, desirably simple elements (e.g. Germano, 1999). The dream is, of course, to find sets consisting of small number of weakly interacting elements/objects adequately representing the turbulent field. Those known so far are interacting strongly[12]

---

[12]Landau wrote (1960, p. 245) : *It is well known that theoretical physics is at present almost helpless in dealing with the problem of strong interactions.* The situation in turbulence seems to be not better if not worse.

and most of them nonlocally. This is a reflection of one of the central difficulties in 'solving the turbulence problem' as a whole, in general, and the
'closure problem' such as LES and other reduced descriptions of turbulence
(Kraichnan, 1988), in particular, as well as in construction of a kind of
statistical mechanics of turbulence generally (Kraichnan and Chen, 1989).
We return to these matters in chapters 5, 7 and 9.

3.6. Summary

There is no basis for contraposing 'statistical' and 'deterministic', just like
it seems impossible to separate the structure(s) and the 'random structureless' background from the nonrandom processes. With few exceptions there
seem to exist no other way of handling turbulence than by statistical methods. These include not only the 'traditional' things like means/averages
and other simple characteristics, but all kinds of statistics including rather
intricate/exquisite ones such as conditional statistics can be, depending on
the nature of problems in question and the ability/skill of the researcher to
formulate such questions.

The question whether adequate low dimensional description of turbulent
flows is possible depends on the meaning of the term 'adequate'. In the strict
sense, i.e. from the basic point of view, it seems that there does not exist
such a description, though as a (semi) empirical tool it may be definitely
more than satisfactory.

CHAPTER 4

KINEMATICS

Mostly on the behaviour of passive objects

The term *kinematic(s)* is associated with several issues, all of which have in common things which are not directly related to the (Navier-Stokes) dynamics of turbulence. In other words the dynamics of fluid motion, except incompressibility, does not enter into the problems in question. These issues include the following:

* - Kinematic (statistical) properties/structure of real turbulent and artificial (in some sense, e.g. Gaussian) random flows such as (an)isotropy, (in)homogeneity, etc.

* - Passive objects in random flow fields including artificial ones.

* - Kinematic (Lagrangian) chaos.

In other words, by dynamics we mean dynamics of fluid motions *per se*, e.g. the flow properties associated with the dynamics obeying the NSE.

The first issue is of purely technical nature and is covered extensively in several monographs (e.g. Hinze, 1975; Monin and Yaglom, 1971, 1975; Mathieu and Scott, 2000; Pope, 2000). This aspect is also closely related to the so-called eddy structure identification in turbulent flows (Bonnet, 1996). We do not give here any systematic information on these matters. Instead appropriate references and reminding are made in the text in an *ad hoc* manner throughout the book.

The second and the third issues are described mostly to an extent necessary as a background for comparison with the genuine dynamical features of turbulent flows. The material is included in this chapter, since it is essentially of kinematic nature in the sense mentioned above. Its place here underscores the qualitative difference between the behaviour of passive objects in random flow fields (including artificial ones) and kinematic/Lagrangian chaos, and real fluid dynamical turbulence[1]. The relation to and comparison with the dynamics of turbulence is discussed in chapters 6-8.

4.1. Passive objects in random fluid flows

By definition a passive object in a fluid flow has no dynamical effect on the fluid motion itself, and one is interested in the effects of fluid turbulence

---

[1]Especially in view of the claims made in the chaos community that 'the problem of turbulence was solved' with the developments in chaos theory.

(or some artificial random velocity field) on the field of a passive object. Passive objects include passive scalars such as dispersing contaminants, chemical species, temperature, moisture; passive vectors such as material lines, (weak) magnetic field in an electrically conducting fluid; passive surfaces such as material surfaces, and in some cases reacting surfaces and turbulent flames; material volumes.

An essential point is that the evolution of passive objects obeys *linear* equations in which the velocity field does not 'know' anything about the presence of these objects and therefore the velocity field is considered as given *a priori* be it a real fluid flow field or some artificial one. There is no involving phenomenon as pressure[2]. This does not mean that the problems of the evolution of passive objects are simple. The main complication and simultaneously rich variety of phenomena comes from the fact that the velocity field enters as a coefficient in front of the spatial derivatives, i.e. it is due its multiplicative character, so that statistical problems become in a sense nonlinear.

The strongly enhanced mixing properties in a turbulent flow are associated with the fact that the fluid particles wander away from their initial positions even in the absence of any mean flow. Equally, or even more important is the fact that *particles which were originally neighbours move apart as the motion proceeds, so that in a diffusive motion the average value of* l^2/l_0^2 *continuously increases* (Taylor, 1938). Here, $l_0 = l(0)$ is the initial distance between the two fluid particles and $l = l(t)$ is this distance at some subsequent time moment, t. This intuitive idea of the relative diffusion received support in the paper by Cocke (1969). Namely, Cocke proved the following important results[3].

The first result is that the length of an infinitesimal material line element, $l \equiv |\mathbf{l}|$, increases on average in *any* isotropic random velocity field. Similarly, Cocke showed that an infinitesimal material surface element, N, identified by its vector normal, $\mathbf{N}$, increases on average in an *any* isotropic random velocity field as well[4]. We stress again that these are purely kinematic results: the flow does not have to be a real one, i.e. to satisfy the Navier-Stokes equations and/or to be observable in laboratory or elsewhere – the only requirement is that the flow should be random and isotropic. For example, this result is true for a Gaussian velocity field as well, which is

---

[2]Hence 'shocks' in the form of ramp-cliff structures just like in the Burgers equation.

[3]For more details and a review of other related references see Monin and Yaglom (1975), § 24.5; for later references see Bohr et al. (1998), Chaté et al. (1999), Drummond (1993), Girimaji and Pope (1990), Tabor and Klapper (1994), Yeung (1994).

[4]More precisely Cocke showed that $\ln[\langle l(t) \rangle / l(0)] \geq 0$, and $\ln[\langle N(t) \rangle / N(0)] \geq N^2(0)$ for all $t > 0$ with equality holding only if there is no fluid motion at all. Arguments similar to those by Cocke (1969) show that $\langle l^p(t) \rangle \geq l^p(0)$ and $\langle N^p(t) \rangle \geq N^p(0)$ for any $p > 0$ (Monin and Yaglom, 1975, pp. 579 - 580).

important for the purpose of comparison of material line elements, which are passive, and vorticity, which is not (see chapters 6-7). On the qualitative level the results by Cocke were confirmed in a number of DNS experiments both for real and artificial flow fields (Drummond, 1993; Girimaji and Pope, 1990; Huang, 1996; Yeung, 1994) and laboratory experiments (Lüthi et al., 2001).

In view of the fact that $\mathbf{l}$ and $\mathbf{N}$ satisfy the equations (C.1) and (C.2) respectively the results by Cocke mean that $\frac{D\langle l^2(t)\rangle}{Dt} = \langle l_i l_j s_{ij}\rangle > 0$ and $\frac{D\langle N^2(t)\rangle}{Dt} = -\langle N_i N_j s_{ij}\rangle > 0$, and the mean growth rates $\langle \frac{1}{l(t)}\frac{Dl(t)}{Dt}\rangle = \langle l_i l_j s_{ij}/l^2\rangle > 0$ and $\langle \frac{1}{N(t)}\frac{DN(t)}{Dt}\rangle = -\langle N_i N_j s_{ij}/N^2\rangle > 0$. That is, the mean rate of material line (surface) stretching is positive. In other words, there is prevalence of stretching over compressing. It is important to emphasize that $\langle l_i l_k s_{ik}\rangle = 0$ for random s_{ij} and random (and independent of s_{ij}) orientation of l_i. The mean $\langle l_i l_k s_{ik}\rangle > 0$ because though the vector, l_i, and the stretching vector, $W_i^l \equiv l_j s_{ij}$, are completely random in space, they tend to be strictly aligned due to the constraint imposed by the equation (C.1), so that their scalar product $\mathbf{l} \cdot \mathbf{W}^l = l_i l_k s_{ik}$ tends to be positive, and the PDF of the cosine of the angle between these two vectors, $\cos(\mathbf{l}, \mathbf{W}^l)$, is positively skewed (see below). Similarly, the scalar product $-\mathbf{N} \cdot \mathbf{W}^N = -N_i N_j s_{ij}$ tends to be positive, and the PDF of $\cos(\mathbf{N}, \mathbf{W}^N)$, $W_i^N = -N_j s_{ij}$, is positively skewed as well due to the constraint imposed by the equation (C.2).

Let us now look at passive vectors in the presence of molecular diffusivity. There are two kinds of such vectors corresponding in some sense to $\mathbf{l}$ and $\mathbf{N}$. The first one, $\mathbf{B}$, is the vector obeying the equation (C.36), and is frozen into the fluid in the absence of molecular diffusive effects. The well known example is the (weak) magnetic field in an electrically conducting fluid. The second kind of passive vectors, $\mathbf{G}$, is the gradient of some passive scalar, θ, i.e. it is governed by the equation (C.33); in the absence of molecular diffusive effects, the surfaces $\theta = const$ become material surfaces.

Cocke's results allow us to expect that the quantities $\langle B_i B_k s_{ik}\rangle$, $\langle B_i B_k s_{ik}/B^2\rangle$, $-\langle G_i G_k s_{ik}\rangle$ and $-\langle G_i G_k s_{ik}/G^2\rangle$ all should be positive in the presence of molecular diffusive effects as well. It is noteworthy that the similarity in the (statistical) behaviour between the material elements, $\mathbf{l}$, (surfaces, $\mathbf{N}$) and some 'normal' passive vectors, $\mathbf{B}$, (or $\mathbf{G}$) is not at all trivial for two reasons. First, there are many fewer field lines of 'normal' vectors, such as magnetic field lines, than the material ones - at each point there is typically only one such line of $\mathbf{B}$ (or $\mathbf{G}$), but there are infinitely many material lines (surfaces) passing through a point. This may lead to differences in the statistical properties of the two fields. Second, there is a subtle issue of the singular limiting behaviour of the system with the diffusivity tending to zero. It can be very different from the purely diffusionless case when the diffusivity is put to zero at the outset, as in the case of material elements, $\mathbf{l}$, and surfaces, $\mathbf{N}$, though some parameters may be similar (see Childress and Gilbert, 1995; Ott, 1999; and references therein). An additional difference between $\mathbf{l}$ and $\mathbf{B}$ is that the latter is solenoidal ($div\mathbf{B} = \mathbf{0}$), whereas the former, generally, is not. Similarly $\mathbf{G} = \nabla\theta$ is a potential vector, whereas the material surface element $\mathbf{N}$ is not. Nevertheless, there is reasonable evidence that the quantities, such as $\langle B_i B_k s_{ik}\rangle$, $\langle B_i B_k s_{ik}/B^2\rangle$, $-\langle G_i G_k s_{ik}\rangle$ and $-\langle G_i G_k s_{ik}/G^2\rangle$ are positive – a property which can be seen as universal for any random fluid flow, be it real or artificial, such as the Gaussian velocity field. This is indeed the case as observed in numerical simulations (Huang, 1996; Nieuwstadt and Brethouwer, 2000; Ohkitani, 1998; Ruetsch and Maxey, 1991, 1992; Tsinober and Galanti, 2001).

The net positive stretching in all the mentioned cases is associated with two concomitant processes, tilting and folding (and production of curvature), so that any random fluid flow acts in such a way as to create a fine structure in the field of a passive object. For example, in case of a passive scalar, the positiveness of the term $-\langle G_i G_k s_{ik}\rangle$ is associated with two aspects. First, it represents the rate of production of the 'dissipation' $\chi\langle\frac{\partial\theta}{\partial x_i}\frac{\partial\theta}{\partial x_i}\rangle \equiv \chi\langle G^2(t)\rangle$ of a passive scalar (see equation [C.35]), so that the latter is continuously amplified by the stretching process reflected in the term $-\langle G_i G_k s_{ik}\rangle$. Production of the gradients $G^2(t)$ of a scalar field is associated with the fine structure of the passive scalar, θ, itself. Second, the term $-\langle G_i G_k s_{ik}\rangle$ is balanced, at least in part, by the 'dissipation', $-\chi\langle\frac{\partial G_i}{\partial x_k}\frac{\partial G_i}{\partial x_k}\rangle$, of the vector $\mathbf{G}$ itself (see again the corresponding balance equation [C.35]). The consequence is that the gradients $\frac{\partial G_i}{\partial x_k}$, associated with the fine structure of $\mathbf{G}$, are amplified too. An example, of such structure both for the passive scalar itself (concentration of a fluorescent dye) and its dissipation is shown in plate 8.

It is noteworthy that the continuous stretching even of infinitesimal material elements and other passive vectors by random flows is a statistical tendency; it occurs in the mean only, not any individual element is stretched. This is seen, for example, from the PDF of the rates $l_i l_k s_{ik}/l^2$ and $N_i N_k s_{ik}/N^2$ shown in Figure 4.1. Namely, these quantities are *negative* in about 1/3 of the volume occupied by the fluid flow. This latter is associated with the compressing and folding of material elements and the production of their curvature.

Similar behaviour is observed in the case of nonzero diffusivity, i.e. for the quantities $\langle B_i B_k s_{ik}\rangle$, $\langle B_i B_k s_{ik}/B^2\rangle$, $-\langle G_i G_k s_{ik}\rangle$ and $-\langle G_i G_k s_{ik}/G^2\rangle$. As an example, a positively skewed PDF of $B_i B_k s_{ik}$ is shown in Figure 4.2 for the case of statistically stationary velocity field maintained in a cubic domain with periodic boundary conditions by a deterministic forcing in RHS of NSE, and with the vector $\mathbf{B}$ being governed by the equation

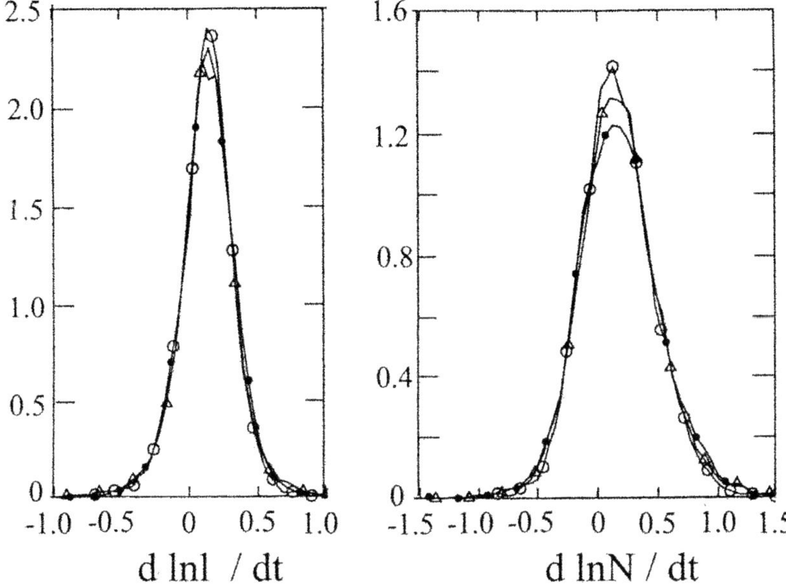

Figure 4.1. PDFs of the mean growth rates of material line and surface elements, adapted from Girimaji and Pope (1990). Note that both PDFs are clearly positively skewed, but contain a considerable negative contribution, corresponding roughly to 1/3 of the volume occupied by the fluid flow. Both distributions correspond to the statistically stationary state of the fluid flow maintained by forcing the flow at large scales in a cubic domain with periodic boundary conditions. Note that even if the rates $\frac{D \ln l(t)}{Dt} = l_i l_j s_{ij}/l^2$ and $\frac{D \ln N(t)}{Dt} = -N_i N_j s_{ij}/N^2$ are statistically stationary (which is approximately the case shown here), e.g. $\langle l_i l_j s_{ij}/l^2 \rangle$ and $\langle -N_i N_j s_{ij}/N^2 \rangle$ are time independent, the 'energy' $\langle l^2 \rangle$ and $\langle N^2 \rangle$ is continuously growing with time as follows from Girimaji and Pope (1990). Their results are valid for moderate Reynolds numbers ($\circ, \bullet$, and $\triangle$ correspond to $Re_\lambda = 38, 63$ and 90) and as Cocke's results (1969) for small material line and surface elements; see also Lüthi et al. (2001).

(C.36) without any forcing. Though the mean kinetic energy of the flow was constant, the mean energy of the field **B** is continuously growing with time during all the time of simulation. This phenomenon is akin to the so called dynamo effect of spontaneous magnetic field amplification by the motion of electrically conducting fluid under certain conditions (see Childress and Gilbert, 1995; Ott, 1999; and references therein)[5]. The growth of **B** is due to the linearity of the equation (C.36) and the absence of the Lorenz force

---

[5]The ABC forcing is strongly helical, $curl\mathbf{F} \parallel \mathbf{F}$, and therefore, along with kinetic energy such a forcing makes an input of helicity into the flow. Presence of nonzero helicity, $\langle \boldsymbol{\omega} \cdot \mathbf{u} \rangle$, is known to aid the process of amplification of magnetic field. However, this is not the only possibility. The amplification of, **B**, was observed with the nonhelical (NH) forcing, $\mathbf{f} \cdot curl\mathbf{f} = 0$, which does not make an input of helicity into the flow.

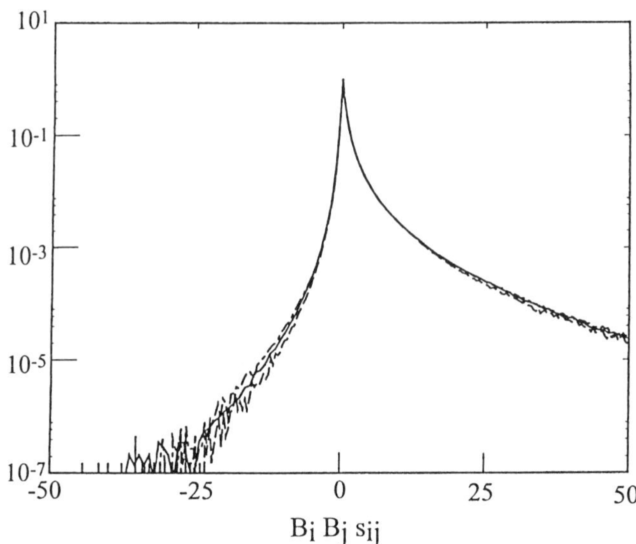

Figure 4.2.   PDFs of $B_iB_ks_{ik}$. Here too, as in figure 4.1, the PDFs are clearly positively skewed, and contain a considerable negative contribution. Shown are three cases: 1) ABC – the forcing has the form $\mathbf{F} = F\{A\sin z + C\cos y,\ B\sin x + A\cos z,\ C\sin y + B\cos x\}, A = B = C$; 2) NH (nonhelical) - $\mathbf{F} = F\{A\cos z\cos y,\ B\cos x\cos z,\ C\cos y\cos x\},\ A = B = C$ and 3) slightly compressible (C), i.e. with $div\mathbf{v} \neq 0$. In all three cases the diffusivity in the equation (C.36) for $\mathbf{B}$ was equal to the viscosity in the NSE and $Re_\lambda \approx 35$. The PDFs correspond to several close time snapshots when the initial conditions were 'forgotten'. The form of the PDFs is the same for all such time moments. Note also that the PDFs for all the three cases are essentially the same. The PDFs of $-G_iG_ks_{ik}$ are similar and also positively skewed (Tsinober and Galanti, 2001).

$\mathbf{j} \times \mathbf{B}$ in the NSE, which when present 'reacts back' and causes nonlinear saturation of the magnetic field growth (Galanti et al. 1992; Brandenburg, 1995; Brandenburg et. al, 1996).

The results of DNS (Tsinober and Galanti, 2001), show that the balance in the equations (C.34, C.37) is dominated by the production terms $-\langle G_iG_ks_{ik}\rangle$ and $\langle B_iB_ks_{ik}\rangle$ respectively, with the forcing term at least an order of magnitude smaller. That is, the main contribution to the formation of small scale structure of passive objects comes mostly from the strain of the velocity field, with smaller (but not negligible) contribution from the external forcing. We discuss this issue and the differences between the fields G_i, B_i, and the vorticity, ω_i in more details in chapters 6 and 7.

The production processes described by the terms $\langle B_iB_ks_{ik}\rangle$, $\langle B_iB_ks_{ik}/B^2\rangle$, $-\langle G_iG_ks_{ik}\rangle$ and $-\langle G_iG_ks_{ik}/G^2\rangle$ are directly associated with the rate of strain, s_{ij}, only. This is because when looking at the energy balance, i.e. of G^2 and B^2, one deals only with the magnitude $\mathbf{G}$ and $\mathbf{B}$. However, the *direction* of $\mathbf{G}$ and $\mathbf{B}$, especially their orientation in respect with the eigenframe, λ_i, of the rate of strain tensor, s_{ik}, does depend on vorticity. This is seen immediately from the vector identity $B_k\frac{\partial u_i}{\partial x_k} = B_k s_{ik} + \frac{1}{2}\epsilon_{ijk}\omega_j B_k$. In other words, many aspects of the behaviour of the vector fields G_i, B_i, such as *alignments* and other geometrical relations, are influenced by vorticity as well[6]. For instance, the alignments reflect important details about how the quantities $-\langle G_i G_k s_{ik}\rangle$ and $\langle B_i B_k s_{ik}\rangle$ and the corresponding rates become positive: the orientation of vectors $\mathbf{G}$ and $\mathbf{B}$ with respect to the eigenframe, λ_i, of the rate of strain tensor, s_{ik} is of utmost importance.

4.1.1. GEOMETRICAL STATISTICS

Gradient of a Passive Scalar
Following Betchov (1956), it is convenient to represent the production term $-G_i G_j s_{ij}$ in the eigenframe, λ_i, of the rate of strain tensor, s_{ij}, as

$$-G_i G_j s_{ij} = -G^2\{\Lambda_i\cos^2(\mathbf{G},\lambda_i), \tag{4.1}$$

Here Λ_i ($\Lambda_1 > \Lambda_2 > \Lambda_3$) are the eigenvalues of the rate of strain tensor and $\cos(\mathbf{G},\lambda_i)$ - is the cosine of the angle between $\mathbf{G}$ and λ_i. In addition, it is useful to represent the relation (4.1) as $\langle -G_i G_j s_{ij}\rangle = \mathbf{G}\cdot\mathbf{W}^G = GW\cos(\mathbf{G},\mathbf{W})$, $W_i^G = G_j s_{ij}$, so that the positiveness of $\langle -G_i G_j s_{ij}\rangle$ should be associated with the strict alignment between $\mathbf{G}$ and the corresponding stretching vector $\mathbf{W}^G$ and positively skewed PDF of $\cos(\mathbf{G},\mathbf{W}) = -\{\Lambda_i\cos^2(\mathbf{G},\lambda_i)\}\{\Lambda_i^2\cos^2(\mathbf{G},\lambda_i)^{1/2}$. The details depend on the mutual orientation of $\mathbf{G}$ and the eigenframe, λ_i, of s_{ij}, and the behaviour of its eigenvalues, Λ_i.

Since $\Lambda_1 + \Lambda_2 + \Lambda_3 = 0$, $\Lambda_1 > 0$, $\Lambda_3 < 0$, it is straighforward to see from (4.1) that the positiveness of $\langle -G_i G_j s_{ij}\rangle$ is associated with the predominant tendency of alignment between the scalar gradient, $\mathbf{G}$, and the eigenvector, $\lambda_\mathbf{3}$, corresponding to the compressive (negative) eigenvalue, Λ_3, of the rate of strain tensor, s_{ij}. Examples demonstrating the above tendency both in DNS and in laboratory experiments are shown in Figure 4.3 and Plate 5 (right).

In other words the amplification of the gradients, G_i, of the passive scalar field (i.e. increase of the equi-surfaces of the passive scalar) is associated with predominant *compression*. The importance of compression in the process of production of the gradients of the passive scalar field is reflected in the structure(s) of this field: it is sheet-like (Chen and Cao, 1997; Flohr, 1999; Frederiksen et al., 1997; Su and Dahm, 1996); see plates 5 and 8. The so-called ramp/cliff structures are also due to the predominant compression (see figure 5 in Warhaft [2000] or figure 3 in Shraiman and Siggia [2000]; also Celani et al. [2001]). This should be contrasted to the predominant

---

[6]Their and similar statistical properties are denoted by the term *geometrical statistics*.

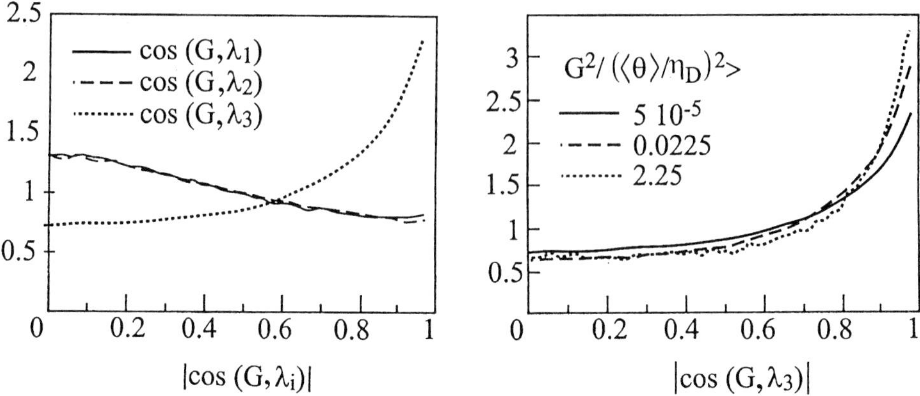

Figure 4.3. PDFs of the cosine of the angles between the scalar gradient, $\mathbf{G} = \nabla\theta$, and the strain eigenframe, λ_i in the far field of a turbulent jet flow at $Re_\lambda \sim 50$, adapted from Su and Dahm (1996). Here $\eta_d = \eta Sc^{1/2}$ is the molecular diffusion length, $\eta = 11.2\delta Re_\delta^{-3/4}$ - a scale proportional to the Komogorov scale and $Sc = \nu/D$, is the Schmidt number with D being the diffusivity of the passive scalar (fluorescent dye). Note the preferential alignment between $\mathbf{G}$ and the λ_3 coresponding to the compressive eigenvalue Λ_3, of the rate of strain tensor, s_{ij}. Qualitatively this alignment in the jet flow is the same as in the DNS of NSE of a flow in cubic box with periodic boundary conditions, see Flohr (1999) and Galant and Tsinober (2000). b) - influence of magnitude of $\mathbf{G}$ on the preferential alignment of the scalar gradient with the compressive eigenvector.

stretching in case of passive vectors of 'frozen' type, B_i, and also in case of vorticity with structure(s) of a different kind, e.g. tube-like. Of course, due incompressibility, stretching in some direction is necessarily accompanied by compressing at least in one another direction. However, the above comparison shows that there is hardly any analogy between the behaviour of passive vectors G_i and B_i. On the difference between passive vectors of 'frozen' type, B_i, and/or material lines and vorticity see chapter 6.

'Frozen' Passive Vectors.
In a similar way it is seen from

$$B_i B_j s_{ij} = B^2 \Lambda_i \cos^2(\mathbf{B}, \boldsymbol{\lambda}_i) = \mathbf{B} \cdot \mathbf{W}^B = BW^B cos(\mathbf{B}, \mathbf{W}^B), \qquad (4.2)$$

that the positiveness of $\langle B_i B_j s_{ij}\rangle$ is associated with the predominant tendency of alignment between the vector, $\mathbf{B}$, and the eigenvector, $\boldsymbol{\lambda_1}$ corresponding to the largest stretching (positive) eigenvalue, Λ_1, of the rate of strain tensor, s_{ij}. Such an alignment was observed in numerical simulations by Drummond and Münch (1990), Girimaji and Pope (1990), and Huang (1996) for material line elements, l_i. On the other hand, Lüthi et al. (2001) observed experimentally preferential alignment of material elements $\mathbf{l}$ both with $\boldsymbol{\lambda_1}$ and $\boldsymbol{\lambda_3}$.

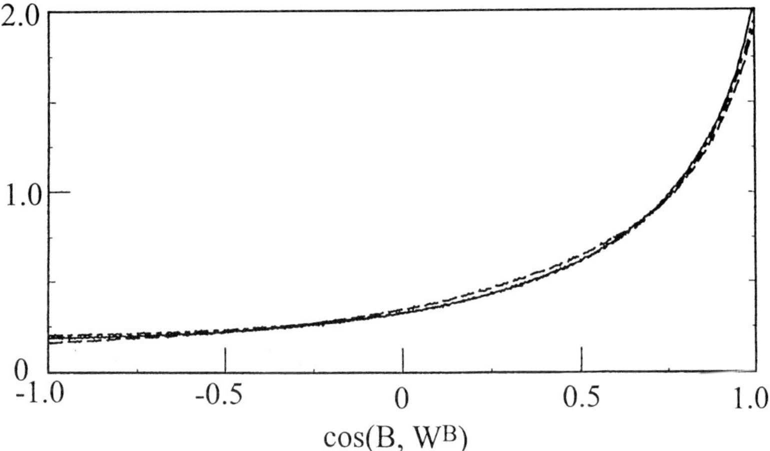

Figure 4.4.   PDFs of $cos(\mathbf{B}, \mathbf{W}^B)$ for the data as in figure 4.2.

However, since in real turbuent flows Λ_2 is positively skewed another possibility would do too: alignment between the passive vector, $\mathbf{B}$, and the eigenvector of the rate of strain, $\boldsymbol{\lambda_2}$ corresponding to the intermediate eigenvalue, Λ_2, of the rate of strain tensor, s_{ij} (as does vorticity, see chapters 6 and 7). The results by Ohkitani (1998) and Tsinober and Galanti (2001) (both with nonzero diffusivity) show such a tendency for alignment between $\mathbf{B}$ and both $\boldsymbol{\lambda_1}$ and $\boldsymbol{\lambda_2}$ with stronger alignment between $\mathbf{B}$ and and $\boldsymbol{\lambda_2}$. Tsinober and Galanti (2001) observed this tendency in all the three cases (ABC, NH and compressible) though with some differences between the three. However, in all the three cases the alignment between $\mathbf{B}$ and the corresponding stetching vector $\mathbf{W}^B$, $W_i^B = B_j s_{ij}$ is essentially the same. This can be seen from the Figure 4.4, which shows the tendency of strict alignment between $\mathbf{B}$ and $\mathbf{W}^B$ corresponding to the positiveness of $\langle B_i B_j s_{ij} \rangle = \langle \mathbf{B} \cdot \mathbf{W}^B \rangle$ and the positively skewed PDFs of $B_i B_j s_{ij} = \mathbf{B} \cdot \mathbf{W}^B$, see (4.2) and Figure 4.2. Note that, if Λ_2 is positively skewed, the PDF of $\cos(\mathbf{B}, \mathbf{W}^B) = \{\Lambda_i \cos^2(\mathbf{B}, \boldsymbol{\lambda}_i)\}/\{\Lambda_i^2 \cos^2(\mathbf{B}, \boldsymbol{\lambda}_i)\}^{1/2}$ ~ 1 in both cases, when either $\cos^2(\mathbf{B}, \boldsymbol{\lambda}_1) \sim 1$, or $\cos^2(\mathbf{B}, \boldsymbol{\lambda}_2) \sim 1$.

The above differences may be due to the differences discussed above, which include differences in forcing, lack of molecular diffusivity in case of l and that the vector $\mathbf{B}$ is solenoidal, whereas the vector l_i is not. More work is necessary to clarify the issue.

An important point is, that though the *rate* of production of B^2 (and similarly of G^2), i.e. $B_i B_j s_{ij}/B^2 = \Lambda_i \cos^2(\mathbf{B}, \boldsymbol{\lambda}_i)$, does not *directly* depend on the magnitude of B, it depends on B via the dependence of alignments, i.e. $\cos^2(\mathbf{B}, \boldsymbol{\lambda}_i)$, on the magnitude of $\mathbf{B}$. This dependence is such that the preferential alignments become stronger at larger magnitudes of $\mathbf{B}$ and corresponding ly of $\mathbf{G}$, see Figure 4.3.

Two-dimensional flows

It is well known that the behaviour of dynamical variables (velocity, vorticity, etc.) is qualitatively different in three and two dimensions (see chapter 8 and appendix C). When dealing with, e.g. passive scalars, it is believed that in many respects their behaviour is similar in three and two dimensions, *since the conservation properties of the scalar do not depend on dimensionality* (see Holzer and Siggia, 1994; Warhaft, 2000 and references therein). However, there are also differences some of which can be seen via geometrical statistics, e.g. alignments. In two dimensions the rate of strain tensor has only two eigenvectors: one positive, and one negative - there is no intermediate one. Therefore, in the two-dimensional case the positiveness of quantities like $\langle l_i l_j s_{ij} \rangle$, $(\langle B_i B_j s_{ij} \rangle)$, $\langle N_i N_j s_{ij} \rangle$ $(\langle G_i G_j s_{ij} \rangle)$ and the corresponding rates is always due to the tendency of alignment of the corresonding vector with the eigenvector corresponding to the positive (negative) eigenvalue.

More details on the behaviour of the alignments can be seen from the equations for the evolution of vectors $a_i^G = \cos(\mathbf{G}, \boldsymbol{\lambda}_i)$ and $a_i^B = \cos(\mathbf{B}, \boldsymbol{\lambda}_i)$. This approach allows us to distinguish between pure strain effects and rotation. The latter consists of *two* contributions: vorticity and the rotation of the eigenframe, $\boldsymbol{\lambda}_i$, of the rate of strain tensor, s_{ij} (see Lapeyre et al., 2000; Nieuwstadt and Brethouwer, 2000; Nomura and Post, 1998; Tabor and Klapper, 1994; and references therein).

A final remark is that the viscous effects also influence the alignments. However, practically nothing is known about this influence so far.

4.2. Kinematic/Lagrangian chaos/advection

In the previous section the velocity field was assumed random, so that the chaotic behaviour of passive objects was mainly due to the random nature and the multiplicative character of the velocity field and, as in some problems, due to the random forcing.

Since the equations describing the evolution of passive objects are *linear*, it may seem that there is no place for chaotic behaviour of passive objects if the velocity field is not random and is regular and fully laminar, because the chaotic behaviour appears/shows up in *nonlinear* systems. There is, however, no real contradiction or paradox. This apparent contradiction is resolved via the following observations.

Until now we used the so called Eulerian description, in which the observation of the system is made in a *fixed* frame as the fluid goes by. In this case the motion is characterized by the velocity field $\mathbf{u}(\mathbf{x},t)$ as a function of position vector, $\mathbf{x}$, and time, t. Another way to characterize the fluid flow is the Lagrangian description in which the observation is made *following* the fluid particles wherever they move. Here the dependent variable is the position of a fluid particle, $\mathbf{X}(\mathbf{a},t)$, as a function of the particle label, $\mathbf{a}$, (usually it's initial position, i.e. $\mathbf{a} \equiv \mathbf{X}(0)$) and time, t. The relation between the two ways of description is given by the following equation[7]

$$\frac{\partial \mathbf{X}(\mathbf{a},t)}{\partial t} = \mathbf{u}[\mathbf{X}(\mathbf{a},t); t], \qquad (4.3)$$

i.e. the Lagrangian velocity field, $\mathbf{v}(\mathbf{a},t) = \frac{\partial \mathbf{X}(\mathbf{a},t)}{\partial t}$, is related to the Eulerian velocity field, $\mathbf{u}(\mathbf{x},t)$, as $\mathbf{V}(\mathbf{a},t) \equiv \mathbf{u}[\mathbf{X}(\mathbf{a},t); t]$.

If the Eulerian velocity field is known/given – as in all problems of kinematic nature – then the equation (4.3) serves for determination of the trajectory of a fluid particle with the initial position $\mathbf{X}(0) \equiv \mathbf{a}$. This equation is *nonlinear* (for almost all) even for very simple fluid flows and is generically non-integrable for all such flows.

One of the important developments of the so called deterministic chaos is that (even) simple systems governed by purely deterministic nonlinear set of equations as a rule exhibit irregular apparently random/stochastic behaviour (see, e.g. Mullin, 1993 and references therein). In particular it is well established that the trajectories of fluid particles – their motion is governed by the equation (4.3) – exhibit chaotic behaviour even when the Eulerian velocity field is not random, is regular and fully laminar. This is generally true of two-dimensional time dependent flows, three-dimensional time-independent flows, and, of course, time-dependent ones. In other words, many fluid flows which are laminar in the Eulerian sense (E-laminar) exhibit the so called Lagrangian (kinematic) chaos or Lagrangian turbulence (chaotic advection), i.e. they are L-turbulent. This chaotic Lagrangian property of simple Eulerian fluid flows leads to enhanced mixing properties of such flows at any, even very small, Reynolds numbers, i.e. enhanced transport of passive objects occurs not only in genuinely turbulent (E-turbulent) flows, but also in simple E-laminar flows possessing the property of La-

---

[7]The relation between the two ways of description can be seen also by looking at any conservative property of fluid particles (i.e. a nondiffusive passive scalar) such as nondiffusive 'dye' or any other (e.g. radioactive) label. Due its conservative character it is time independent in the Lagrangian description, i.e. has the form $\vartheta(\mathbf{a})$, but is time dependent in some fixed point of space, $\mathbf{x}$, i.e. in the Eulerian description, and has the form $\theta(\mathbf{x},t)$. Hence, both are related via $\vartheta(\mathbf{a}) = \theta[\mathbf{X}(\mathbf{x},t), t]$. Since $\frac{\partial \vartheta(\mathbf{a})}{\partial t} = \frac{D\theta}{Dt} = 0$ it follows that $\frac{\partial \theta}{\partial t} + u_k \frac{\partial \theta}{\partial x_k} = 0$, which is just an expression of the fact that the material derivative of any Lagrangian conservative property should vanish (see Monin and Yaglom, 1971).

grangian chaos (L-turbulent)[8]. An example of such enhanced mixing at $Re \sim 1$ only is shown in plate 9 (see also Fountain et al., 2000).

The enhanced transport of passive objects by L-laminar but L-turbulent flows is closely related to the property that (almost all) nearby fluid elements separate exponentially in time. That is, the material elements are stretched on average in such flows too, $\frac{D\langle l^2(t)\rangle}{Dt} = \langle l_i l_k s_{ik}\rangle > 0$ and $\left\langle \frac{1}{l(t)} \frac{Dl(t)}{Dt} \right\rangle = \langle l_i l_k s_{ik}/l^2 \rangle > 0$. This stretching property appears to be necessary for the same behaviour of passive vectors with nonzero diffusivity, in spite of the qualitative difference between the two cases (Childress and Gilbert, 1995; Ott, 1999; Zeldovich et al., 1983, 1990; and references therein. As mentioned, this is a subtle issue, as are other issues associated with the behaviour of systems described by differential equations in which the diffusivity (i.e. the coefficient in front of the highest derivative) is small and tends to zero: the singular limiting behaviour of the system is usually qualitatively different from the case when the diffusivity (viscosity) is put to zero at the outset. First, one of the reasons is easily seen in the case of incompressible flows. Stretching in one (or two) direction(s) results in compressing in, at least, one other direction bringing fluid elements very close one to another, so that the diffusive (viscous) effects become important for whatever small diffusivity (viscosity). This does not happen if the diffusivity (viscosity) is precisely vanishing at the outset. Second, it is typical for chaotic flows (i.e. L-turbulent, but both E-laminar and/or, of course, E-turbulent) that as the diffusivity becomes smaller passive objects develop more fine scale structure, so that the advective and diffusive terms in the corresponding equations remain of the same order in the spot-like (intermittent) regions throughout the whole flow domain. The formation of the small scale structure is both due to the inviscid production of the type $\langle B_i B_k s_{ik}\rangle$, i.e. stretching, and diffusive effects as well as their interaction. The small scale structure developing in diffusionless systems, e.g. as a consequence of Lagrangian chaos with zero diffusivity, is, generally, qualitatively different from that for whatever small but nonzero diffusivity.

It seems well established that the positiveness of quantities such as $\langle l_i l_k s_{ik}\rangle$ and $\langle l_i l_k s_{ik}/l^2 \rangle$, and $\langle B_i B_k s_{ik}\rangle$ and $\langle B_i B_k s_{ik}/B^2 \rangle$ can be seen as a universal qualitative property not only of any Eulerian random fluid flow (be it real fluid turbulence or artificial), but of any fluid flows that are Lagrangian chaotic, many of which are simple laminar in the Eulerian sense. These and similar quantities are closely related to the positiveness of the so-called Liapunov exponents associated with exponential stretching at least

---

[8]This chaotic property of the trajectories of the fluid particles makes it more difficult to follow them, i.e. much more difficult to utilize the Lagrangian description of even the simplest fluid flows which exhibit Lagrangian chaos.

in some part of the flow region[9]. However, the structure or even simpler properties of flow regions where the Liapunov exponents are or should be positive is not known even for simple flows, e.g. three-dimensional time-independent flows. Note that the linearization of the equation (4.3) for a small disturbance l (i.e. difference of the equation (4.3) for $\mathbf{X} + \mathbf{l}$ and for $\mathbf{X}$) is precisely the equation (C.1) $\frac{Dl_i}{Dt} = l_k \frac{\partial u_i}{\partial x_k}$, and the RHS of the equation for l^2 is precisely $l_i l_k \frac{\partial U_i}{\partial x_k} \equiv l_i l_k s_{ik}$.

4.3. On the relation between Eulerian and Lagrangian fields

Given the marker dispersion the problem is to determine the source(s) of agitation. In general, owing to chaotic advection, this inverse problem is impossible to solve (Aref , 1984).

... the possession of such relationship would imply that one had (in some sense) solved the general turbulence problem. Thus it seems arguable that such an aim, although natural, may be somewhat illusory (McComb, 1990).

What one sees is real. The problem is interpretation.

The relation between Eulerian and Lagrangian fields is a long-standing and most difficult problem. The general reason is because the Lagrangian field is an extremely complicated non-linear functional of the Eulerian field. The complexity of the relation between the Lagrangian and Eulerian fields is seen in the example mentioned in the previous section of Lagrangian (kinematic) chaos or Lagrangian turbulence (chaotic advection) with *a priori* prescribed and *not* random Eulerian velocity field (E-laminar)[10]. In such E-laminar but L-turbulent flows the Lagrangian statistics has no Eulerian counterpart, as in the flow shown in the plate 7. Indeed, though the Eulerian velocity field, $\mathbf{u}(\mathbf{x}; t)$ is not chaotic and is regular and laminar, the Lagrangian velocity field $\mathbf{v}(\mathbf{a},t) \equiv \mathbf{u}[\mathbf{X}(\mathbf{a},t); t]$ is chaotic because $\mathbf{X}(\mathbf{a},t)$ is chaotic. This shows that, in general, there does not exist a unique relation between Lagrangian and Eulerian statistical properties in genuine turbulent flows as was foreseen by Corrsin in 1959 : *in general, there is no reason to expect that* L_{ik} (the Lagrangian two point velocity correlation tensor) *and* E_{ik} (the Eulerian two point velocity correlation tensor) *will be uniquely related.* In other words it may be meaningless to look for such a relation.

---

[9] See, Acrivos (1991), Bohr et al. (1998), Childress and Gilbert (1995), Ott (1999), Zeldovich et al. (1990) and references therein.

[10] This is why Lagrangian description - being physiclly more transparent - is much more difficult than the Eulerian description, see the citation from Lagrange at page 22.

4.4. On analogies and relations between passive and active fields

> *Passive contaminants are transported by turbulent motions in much the same way as momentum...Momentum is not a passive contaminant; "mixing" of mean momentum relates to the dynamics of turbulence, not merely its kinematics* (Tennekes and Lumley, 1972).
>
> *The advection-diffusion equation, in conjunction with a velocity field model with turbulent characteristics (prescribed a priori), ...serves as a simplified prototype problem for developing theories for turbulence itself* (Majda and Kramer, 1999).

The intricacy of the relation between the Eulerian and Lagrangian fields of the *same* fluid flow has a number of other important consequences. This is a part of a broader question. Namely, what can be learnt about the properties and especially dynamics of real turbulence from studies of passive objects (scalars, vectors)? In particular, what can be learnt about the velocity field and other dynamical variables in real turbulence from comparison of the behaviour of passive objects in real and some 'synthetic' turbulence?

The first issue includes the flow visualisation in which the relation between the field of velocity and that of a passive scalar (dye, very small particles) plays a crucial role, as does the interpretation of this relation. It appears that the meaning of 'seeing' turbulent flow is not so simple.

We have seen that the structure of a passive tracer can be (and usually is) very complicated (plate 9), whereas the corresponding velocity field is rather simple. Another example is shown in figure 4.5. It is seen that the passive tracer has structure at locations where the velocity field has none.

One more example is shown in the plate 10. After the two vortices merged, the structure of the passive markers is pretty complicated as compared to the initial state, but the structure of the flow field is essentially the same as in each individual vortex before the merging.

These examples show that flow visualizations used for studying the structure of dynamical fields (velocity, vorticity, etc.) of turbulent flows may be quite misleading. The general reason is that the passive objects do not 'want' to follow the dynamical fields (velocity, vorticity, etc.)[11] – along with some common features the mechanisms of formation of structure(s) are essentially different for the passive objects and the dynamical variables. As mentioned one of the reasons is the presence of Lagrangian chaos, which is manifested as rather complicated structure of passive objects even in very

---

[11] Formally, it is obvious since the passive objects and active fields obey *different* equations. In particular the problems associated with the former are essentially linear, whereas the problems involving the latter are genuinely nonlinear.

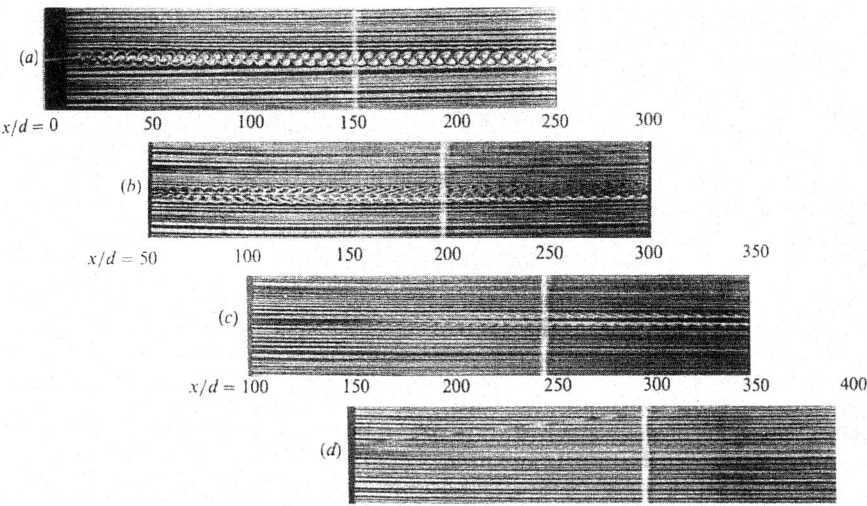

Figure 4.5. Smoke visualization of a wake past a circular cylinder at $Re = 90$ (Cimbala et al., 1988). The velocity field in figures a) – d) is the same. The difference is in the location of the smoke release. The figure d) clearly indicates that beyond the distance $x/d = 150$ there is no Karman vortex street in the velocity field at all, whereas looking at the figure a) only would imply quite the opposite.

simple regular velocity fields. On the other hand the ramp-cliff structures of a passive scalar are observed in pure Gaussian 'structureless' random velocity field (Holzer and Siggia, 1994), just like those in a variety of real turbulent flows practically independently of the value of the Reynolds number (Warhaft, 2000 and references therein. Therefore one can expect that the structure of passive objects in turbulent flows arises from two (essentially inseparable) contributions: one due to the Lagrangian chaos and the other due to the random nature of the velocity field itself. Among other reasons are the difference in sensitivity to initial (upstream) conditions, as in the case shown in figure 4.4, the difference in 'symmetries', e.g. the velocity field may be locally isotropic, whereas the pasive scalar may not be (see references in Celani et al., 2001; Villermaux, et al. 2001; and Warhaft, 2000), and some others. This does not mean that qualitative study of fluid motion by means of colour bands (Reynolds, 1884) is impossible or necessarily erroneous. However, watching the dynamics of material 'coloured bands' in a flow may not reveal the nature of the underlying motion, and even in the case of right qualitative observations the right result may come not necessarily for the right reasons. The famous verse by Richardson belongs to this kind of observation.

The second issue is how sensitive is, e.g. the field of a passive scalar to the properties of dynamical fields (velocity, vorticity, etc.). It appears that many, especially qualitative but also quantitative, characteristics of passive scalars are insensitive to the details of the velocity field, (Kraichnan, 1968; Warhaft, 2000), as long as the velocity field is random. For example, the 4/3 Richardson law is observed in a purely Gaussian and two-dimensional field (Elliott and Majda, 1996). This is the reason that many properties of passive scalars are essentially the same for a Gaussian velocity field and for real turbulent flows (see Holzer and Siggia, 1994; Majda and Kramer, 1999; Warhaft, 2000; Shraiman and Siggia, 2000; and references therein). This is true not only of various statistical properties, but also of structural details such as formation of the mentioned above ramp-cliff structures – a name given to the sheetlike fronts (shocks) with sharp gradients of a passive scalar across them. They were observed both in numerical simulations of (two-dimensional) passive scalar in Gaussian velocity field (Holzer and Siggia, 1994), and in various experiments and field observations as well as in numerical simulations of the Navier-Stokes equations together with the advection-diffusion equation in a periodic box[12]. In other words nontrivial statistical behaviour of passive objects is expected for *any* random velocity field independently of the nature of this randomness. This is not very surprising, since – as mentioned – the Lagrangian field is an extremely complicated non linear functional of the Eulerian field.

On the other hand there are properties of passive objects which do depend on the details of the velocity field. For example, the PDF of a passive scalar depends on a variety of factors, such as Reynolds number, the presence of mean shear flow and /or mean scalar gradient, and many others (see Majda and Kramer, 1999; also Falkovich, 1999). Just these very properties can be effectively used to study the differences between the real turbulent flows and the artificial random fields. More precisely the essential differences in the behaviour of passive objects in a real and synthetic turbulence may be exploited in order to gain more insight into the dynamics of real turbulence. At present, however, the knowledge necessary for such a use is very far from being sufficient. With few exceptions it is not even clear what can be learnt about the dynamics of turbulence from studies of passive objects (scalars and vectors) in real and 'synthetic' turbulence. This requires systematic comparative studies of both.

4.5. Summary

The predominance of stretching over compressing of passive objects and formation of intricate structure can be seen as a universal qualitative property of any Eulerian random fluid flow, be it real fluid turbulence or some

---

[12]For references and further details see Celani et al., (2001), Majda and Kramer (1999), Overholt and Pope (1996), Gawedzki and Vergassola (2000), Shraiman and Siggia (2000) and Warhaft (2000).

artificial random field. In the latter case, the non-Gaussianity of a passive field possessing structure arises from a simple Gaussian 'structureless' velocity field. This is a kind of irreversible effect of the randomness of velocity field on passive objects independent of the nature of this randomness.

The predominance of stretching over compressing of passive objects also occurs in all fluid flows that are Lagrangian chaotic (L-turbulent) a set that includes most of the simple laminar flows in the Eulerian sense (E-laminar). Hence, generally, there is no one-to-one relation between the Lagrangian and Eulerian statistical properties in turbulent flows, just like there may be no correspondence between the structure(s) of a passive object (dye) and the field of a dynamically active variable (velocity, vorticity) in the same fluid flow.

Thus the essential differences in the behaviour of passive and active fields and the intricacy of the relation between them require caution in promoting analogies between the two (e.g. see section 6.8, p. 142). We return to this issue in chapters 5-8.

CHAPTER 5

PHENOMENOLOGY

and the Kolmogorov 4/5 law

Correlations after experiments done is bloody bad. Only prediction is science. (Fred Hoyle, 1957, *The Black Cloud*, Harper, N-Y.)

Phenomenology - The branch of a science that classifies and describes its phenomena without any attempt at explanation. (Webster's New World Dictionary, College edition, 1962)

In our present state of understanding, these simple models will be based, in part on good physics, in part on bad physics, and in part on shameless phenomenology. (Lumley, 1992).

Our present understanding of anything turbulent is at best phenomenological...(Siggia, 1994).

5.1. Introductory notes

There is no definition of what is phenomenology of turbulent flows. In a broad sense, it can be defined by a statement of impotence: it is almost everything 1 except the direct experimental results (numerical, laboratory and field) and/or results (a very small set indeed), which can be obtained from the first principles, e.g. NSE. Phenomenology involves use of dimensional analysis, variety of scaling arguments, symmetries, invariant properties and various assumptions, some of which are of unknown validity and obscured physical and mathematical justification. Thus in the broad sense phenomenology includes also most of the semi-empirical approaches and turbulence modelling[1]. Doing all this requires insight into the basic physics of turbulence, hard experimentation and painful efforts of interpretation. The latter may be quite problematic, especially in models having enough free parameters to guarantee the right results not necessarily for the right reasons.

[1]Most of this enormous material is beyond the scope of this small book. Relevant references can be found in Frisch (1995), Lesieur (1997), McComb (1990), Meneveau and Katz (2000), Mathieu and Scott (2000), Monin and Yaglom (1971, 1975), Pope (2000), Tennekes and Lumley (1972).

It is often claimed that in turbulence research, phenomenology helps to explain some features of turbulent flows, so that there is such a thing as phenomenological understanding of turbulence. This seems too ambitious; phenomenology is mostly a kind of description of some statistical aspects of turbulent flows, which is based on or motivated by some experimental data. The best what is achieved by phenomenology is formulation of some plausible *a priori* hypotheses, i.e. those *before* experiments done. The famous Kolmogorov hypotheses belong to this category. We start with what is called Kolmogorov phenomenology, though it has been discussed in many books, reviews and numerous papers. There are several reasons to do so. First, we will need this material as background to several discussons below and the following chapters. Second, Kolmogorov is in many cases ascribed things he did not write, so our exposition is based exclusively on his original papers in Russian, (Kolmogorov, 1941a,b). This is followed by a discussion of cascade – one of the central phenomenological ideas – and related matters. Other applications of phenomenological reasoning are mentioned in the subsequent chapters.

5.2. Kolmogorov phenomenology and related subjects

The well known are the two similarity hypotheses, which are frequently called universality hypotheses (e.g. Frisch, 1995). However, it is to be stressed that before putting forward his two similarity hypotheses, Kolmogorov formulated the *hypothesis of local isotropy* based on the definitions of local homogeneity and isotropy. Together with his definition of local isotropy[2] this hypothesis postulates that at large Reynolds numbers all the symmetries of the Navier-Stokes equations are restored in the statistical sense[3]— except for one involving scaling: ...*we think it rather likely that in an arbitrary turbulent flow with sufficiently large Reynolds number* $Re = \frac{LU}{\nu}$ *the hypothesis of local isotropy is realized with good approximation in sufficiently small regions* G *of the four-dimensional space* (x_1, x_2, x_3, t) *not lying close to the boundaries of the flow or its other special regions. The*

---

[2]The Kolmogorov (1941a) definition of local isotropy states that :

The turbulence is called locally isotropic in the domain G, *if it is homogeneous and if, besides, the distribution laws mentioned in Definition 1 are invariant with respect to rotations and reflections of the original system of coordinate axes* $(x_1, x_2, x_3,)$.

Local homogeneity as defined in definition 1 states the invariance of the distributions F_n on space- and time translations and Galilean transformations. Here F_n is a $3n$-*dimensional distribution law of probabilities* for the n velocity increments $\mathbf{w}(P^{(n)}) = \mathbf{u}(P^{(n)}) - \mathbf{u}(P^{(0)})$ between n points $P^{(n)}(x_1^{(n)}, x_2^{(n)}, x_3^{(n)})$ and a certain fixed point $P^{(0)}(x_1^{(0)}, x_2^{(0)}, x_3^{(0)})$.

[3]Frisch (1995) presents this in the form of his hypothesis **H1** (p.74), but omits to mention that it is due to Kolmogorov: there is no presentation of the hypothesis of local isotropy in his book.

original text in the English translation of the Kolmogorov papers published in Friedlander and Topper (1962) is reprinted in a slightly less satisfying version in the special issue of the Proc. Roy. Soc. London (1991), **A434**: *in turbulent flow with sufficiently large Reynolds number in sufficiently small regions G of the four- dimensional space* (x_1, x_2, x_3, t) *not lying close to the boundaries of the flow or other singularities of it.* This translation is not very accurate. Specifically, the term *singularities* used for the Russian *osobennosti* is inadequate in this particular case. Therefore the last four words in the above citation were replaced by *its other special regions.*

Naturally, this hypothesis, which we have put forward in such a general and somewhat indefinite form, cannot be proved rigorously. This is followed by an exposition of the turbulence cascade process on a 2/3 page footnote as a *qualitative* justification of the suggested hypothesis. In order *to make its experimental justification possible for individual special cases,* Kolmogorov presents a *number of consequences of the hypothesis of local isotropy* and only then turns to his similarity hypotheses. The first similarity hypothesis states that for *the locally isotropic turbulence the distributions* F_n *are uniquely determined by the quantities* ν *and* ϵ. The second similarity hypothesis states that, if the separations between the points *are large in comparison with* $\eta = \nu^{3/4}/\langle\epsilon\rangle^{1/4}$, *thEn the distributions* F_n *are uniquely determined by the quantity* $\langle\epsilon\rangle$ *and do not depend on* ν. It is Kolmogorov's second similarity hypothesis which is introduced in order to cope with the *possible* scale invariance symmetry of NSE[4] at $Re \gg 1$ and which allowed to find the 2/3 famous exponent; Kolmogorov used the term hypotheses of *similarity,* i.e. scale invariance, intentionally, and avoided using the term *universality.* Namely, straightforward dimensional analysis applied to the second order structure function in the so-called *inertial range* of scales r, $L \gg r \gg \eta$, resulted in

$$S_2^{\|}(r) \propto C_2 \langle\epsilon\rangle^{2/3} r^{2/3}, \tag{5.1}$$

Here $S_p^{\|}(r) = \langle(\Delta u_{\|})^p\rangle$, $p = 2$, - is the second order structure function of the longitudinal velocity increment $\Delta u_{\|} \equiv [\mathbf{u}(\mathbf{x} + \mathbf{r}) - \mathbf{u}(\mathbf{x})] \cdot \mathbf{r}/r$, and C_2 is an 'absolute' constant; $\langle\epsilon\rangle$ is the mean rate of energy dissipation, ν - kinematic viscosity, and $\eta = \nu^{3/4}\langle\epsilon\rangle^{1/4}$ is the Kolmogorov dissipation scale.

It is noteworthy that Kolmogorov never worked in Fourier space. This was done by his Ph. D. student A. Obukhov, who formulated the $-5/3$ law for the energy spectrum

$$E(k) = C_K \langle\epsilon\rangle^{2/3} k^{-5/3}, \tag{5.2}$$

[4]Kolmogorov was aware of the self-similarity before the completion of the first K41 paper, but '*did not know how to determine the exponent m*' (Yaglom A.M., 1994, *Annu. Rev. Fluid Mech.*, **24**, p. 8).

which in some senSe (only) is equivalent to (5.1).

The 2/3 and 5/3 laws received considerable experimental support[5].

As mentioned in his 1941a paper Kolmogorov only explicitly treated the second order function , though all his hypotheses were formulated for any order. Instead, the 1941a paper was followed by the most remarkable quantitative prediction, perhaps the only prediction in the theory of turbulence made so far, of his -4/5 law obtained as a direct consequence from the Navier Stokes equations (Kolmogorov 1941b) for the inertial range $L \gg r \gg \eta$

$$S_3^{\|}(r) = -4/5\langle \epsilon \rangle r, \tag{5.3}$$

in which the constant $C_3 = -4/5$. However, this relation was obtained for globally and not for locally isotropic turbulence. A number of justifications for various versions of locally isotropic and/or locally homogeneous turbulent flow was made by Danaila et al. (1999), Hill (1997), Lindborg (1996, 1999), and Mann et al. (1999); see also references in these papers. A rigorous proof that $\lim_{r/L,\,\eta/r \to 0} S_3^{\|}(r)/\langle \epsilon \rangle r = -4/5$ was recently obtained by Nie and Tanveer (1999) without any assumptions on local homogeneity, isotropy and stationarity with $S_3^{\|}(r)$ defined by intergrating in space/time and over all possible orientations of $\mathbf{r}$, though they were unable to check their results numerically for a number of reasons: low Reynolds number ($Re_\lambda = 155$) and prohibitive computational expense to compute over all orientations of $\mathbf{r}$ (solid angle integrations), which is necessary due to possible anisotropy of the flow.

The Kolmogorov papers (1941a,b) raised a number of basic issues which have kept the turbulence community quite busy until now. We address briefly three such issues.

The first issue is about the validity of the 4/5 law. Since the four fifth law is a consequence of the Navier-Stokes equations, it should be possible, at least in principle, to obtain it from experiments (laboratory, DNS or even field) in a rather clean way. However, even at rather high Reynolds numbers, the results exhibit large variability for the range of r in which $K = S_p^{\|}(r)/(-4/5\varepsilon r) \approx 1$. For example, in the experiments by Gagne (1987) in the S1 wind tunnel in Modane at $Re_\lambda \sim 3000$, this range was less than one decade, whereas in similar experiments by Malecot (1998) at $Re_\lambda \sim 2500$ and in experiments by Mydlarski and Warhaft (1996) with the turbulent grid flow even at $Re_\lambda \sim 450$ there was almost no such range at all. In experiments by Praskovsky (1998) at similar Re_λ, the range of r in which $K \approx 1$ was about one and half decades for two flows (mixing layer

---

[5]See references in Monin and Yaglom (1975), Frisch (1995), Saddoughi (1997), Sreenivasan and Antonia (1997). In a recent experiment by Kholmyansky et al. (2001) the -5/3 law spans over three decades for all the three velocity components.

and return channel) in the large wind tunnel of TSAGI (near Moscow). Even at $Re_\lambda \approx 10^4$ in experiments by Sreenivasan and Dhruva (1998) and Kholmyansky et al. (2001) in the atmospheric surface layer, this range was only about two decades. However, attempting to define the scaling range in a more precise manner via local slopes of K causes this range to become much shorter and in some experiments even to disappear. In some cases when the large scales are approximately isotropic the four-fifths law is observed even at $Re_\lambda \sim 220$, though in a rather limited range slightly more than half a decade, as in the DNS by Chen and Cao (1997).

Among the possible causes for deviations from the four-fifths law, Frisch (1995, p. 129) lists the following: *lack of asymptoticity (e.g. contamination by the dissipation range), lack of homogeneity and/or isotropy, violations of Taylor hypothesis, violation of the hypothesis ... of the finiteness of energy dissipation, inaccurate determination of the dissipation rate, and poor quality of the data.*

It is noteworthy that these and/or possibly other causes (see e.g. Moisy et al., 1998; Danaila et al., 1999; and references therein), result in considerable deviations from the 4/5 law, whereas these same causes, whatever they are, have little effect on the 2/3 (5/3) law. This seems surprising, since the four fifths law is a consequence of Navier-Stokes equations, while the 2/3 law is only a consequence of dimensional arguments (dimensional necessity) supplemented by the above mentioned hypotheses. So far there is no clear answer to this issue. One of the possible reasons is the lack of isotropy in two meanings. First, the 4/5 law applies strictly to globally isotropic flows. Second, flows even with very large Reynolds number may lack local isotropy[6].

There were many attempts to test the hypothesis of local isotropy in a variety of ways, all of which exploit some consequences of this hypothesis. These include testing the kinematic relations between various quantities both in physical and in Fourier space, and many other such approaches. The issue has a long history and cannot be reviewed here in full; for a partial list of references, see Ferchichi and Tavoularis (2000), Saddoughi

---

[6]The difficulty in a clean experimental confirmation of the four fifths law at large Reynolds numbers is two-fold. First, it is extremely difficult, if not impossible, to set up experimentally a large Reynolds number flow which is isotropic in large scales. Second, if one gives up the isotropy in the large scales, then one has to use the result obtained by Nie and Tanveer (1999). That is in order to determine the third order structure function it is necessary to perform both space/time and solid angle averaging, i.e. over all possible orientations of the separation vector **r**. With the existing techniques this is impossible in high Reynolds number physical experiments where the averages are performed over time only. As for direct numerical simulations, that may be possible for moderate Reynolds numbers with rather large amount of computations. Until then *we can hold strong opinions either way* (Feynmann, 1963), though there is little doubt about the validity of the four fifths law in a globally isotropic turbulent flow.

(1997), Shen and Warhaft (2000), Tsinober (1993, 1998a), Yeung et al. (1995) and Zhou and Antonia (2000). We mention here only the turbulent shear flows as those, for which considerable evidence has accumulated since the early fifties *against* the hypothesis of local isotropy at rather high Reynolds numbers. The recent ones are the experimental results obtained in an approximately homogeneous shear flow by Garg and Warhaft (1998) and Shen and Warhaft (2000) for $Re_\lambda \sim 10^3$ (see also Ferchichi and Tavoularis, 2000)[7]. On a qualitative level their results show that local isotropy holds for the second order statistics, but is violated for higher order statistics[8] in agreement with the results mentioned above regarding the deviations from the four fifths law. One of their quantitative results is that the skewness, $\langle (\Delta u_{12})^3 \rangle / \langle (\Delta u_{12})^2 \rangle^{3/2}$, of the longitudinal velocity increments, $\Delta u_{12} = u_i(x_1, x_2 + r, x_3) - u_1(x_1, x_2, x_3)$, along the mean velocity gradient, is ~ 0.5 in the inertial range, whereas it should be close to zero if the flow is locally isotropic in this range. This effect is much stronger for the superskewness $\langle (\Delta u_{12})^5 \rangle / \langle (\Delta u_{12})^2 \rangle^{5/2}$. These results, compared to those for flows without mean shear, imply that there is a *direct* influence of mean shear on the small scales – an effect inferred already by Townsend (1954) among others. Such an effect is possible due to the permanent bias of the mean shear to which the field of fluctuations is exposed due to its very large *residence* time in the mean shear. More generally, the anisotropy in large scales seems to be felt down to the smallest scales in the inertial range and may be (but may also not be, see chapter 6) 'forgotten' only in the range of scales comparable with the Komogorov scale, η. In this sense the 'cascade' (see next section) is not an information loosing process. In other words, there is quite reliable experimental evidence, that, at least in turbulent shear flows, the hypothesis of local isotropy is violated[9], and this seems to be the case with many other (but not all) flows, which are anisotropic in the large scales[10]. One of the main possible reasons for the violation of local isotropy (and the so-called anomalous scaling) even at the largest accessible Reynolds numbers is the direct and bidirectional coupling between large and small scales when the large scales are anisotropic – one of the manifestations of nonlocality of turbulent flows. This matter is taken up in chapter 6 in the section on nonlocality. We add here that violation of

---

[7]Ferchichi and Tavoularis (2000) drawe contrasting conclusion to that of Shen and Warhaft (2000). This seems to be a matter of interpretation: closer inspection of the results shows that both support the conclusions of Shen and Warhaft (2000).

[8]Anisotropy of velocity derivatives at the level of second order statistics was observed in jet flows in the form of locally axisymmetric turbulence; see Hussein (1994) and references therein.

[9]Similar evidence exists since Stewart (1969) for passive scalars in presence of mean gradient of passive scalar; see references in Warhaft (2000) and Villermaux et al. (2001).

[10]This is one of the reasons we chose to discuss the 4/5 law in this section.

local isotropy means violation of the basic hypothesis of the so-called fully developed turbulence, that all the symmetries of the Navier-Stokes equations are restored in the statistical sense locally in time and space (see, e.g. Frisch, 1995).

The second issue is about the behaviour of dissipation of turbulent flows at very large Reynolds numbers. It is widely thought that the Kolmogorov similarity hypotheses imply that the mean dissipation, $\langle \epsilon \rangle$, remains *finite* as $Re \to \infty$. More precisely the issue is whether the normalized mean dissipation $\varepsilon = U^3 L^{-1} \langle \epsilon \rangle$ tends really to a finite limit as $Re \to \infty$ ($\nu \to 0$), or is it Re dependent even at very large Reynolds numbers[11]. There is much speculations about this subject, while the experimental evidence favouring the former is extremely limited, though in the engineering practice this fact has been recognized long ago in a great variety of flow configurations (see e.g. Idelchik, 1996; and also figure 1.8). Recent results obtained using glycerol, water and low temperature helium gas (Cadot et al., 1997; see also references therein), show that $\varepsilon = const$ within the range of Reynolds numbers varying over more than three decades, $3 \cdot 10^3 < Re < 7 \cdot 10^6$ (see figure 5.1) when the flow is forced by very 'rough' moving boundaries. However, when the moving boundaries were smooth $\varepsilon(Re) \neq const$, and was a decreasing function of Reynolds number similar to such Re-dependence in other configurations, e.g. in pipes with smooth walls. Nevertheless, the *bulk* of the flow exhibited clear Re-independent behaviour, indicating that the main difference between the two cases is due to the dissimilarity in the coupling between the boundaries and the bulk of the flow, or more generally due to the difference in the mechanisms of turbulence production[12].

It is noteworthy that rigorous upper bounds of ε are independent of Reynolds number at large Re (see Doering, 1999; Kerswell, 1999; and references therein) and thereby are consistent with the experimental results[13]. The Reynolds number independent behaviour of some global characteristics of turbulent flows at large Re, such as the total dissipation in the above example, the drag of bluff bodies (fig 1.8) and resistance in many other configurations, comprises one of the *quantitative* universal properties of turbulence at large Re. This is distinct from *some possibly* universal (mostly scaling) properties of small scale turbulence.

Since the finite limit of the mean dissipation $\varepsilon = U^3 L^{-1} \langle \epsilon \rangle$ at $Re \to \infty$ defines a unique scaling exponent in the two thirds law (5.1) and since the four fifths law (5.3) is a consequence of the Navier-Stokes equations, the

---

[11]The scaling $U^3 L^{-1}$ for the mean dissipation was obtained by Taylor (1935).

[12]It should be emphasized that the independence of some parameter of viscosity at large Reynolds numbers does not mean that viscosity is unimportant. It means *only* that the (cumulative) effect of viscosity is Reynolds number independent.

[13]Unfortunately the *lower* bounds are not as good, since they correspond to the values for laminar flows.

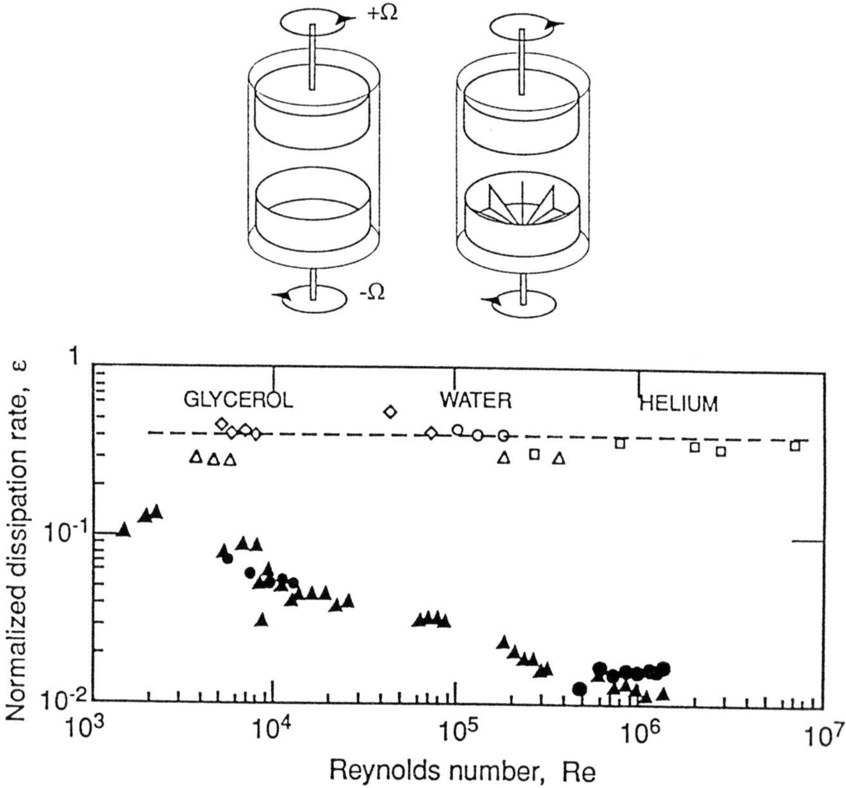

Figure 5.1. Reynolds number dependence of normalized dissipation rate of energy ε in a turbulent flow in a circular tank forced by counter rotating top and bottom. The schematic of the water/glycerol and low temperature helium gas facilities are shown in the upper part of the figure. Adapted from the Ph. D. Thesis by O. Cadot (1995), Laboratoire de Physique Statistique de l'Ecole Normale Supérieure, Université Paris VII. Similar results were obtained in the Couette-Taylor flow, on both see Cadot et al. (1997).

two scaling exponents, $\zeta_2 = 2/3$ and $\zeta_3 = 1$, possess a special status.

The third issue is about the scaling exponents, ζ_p, of order higher than 3, $p > 3$. The attempts of experimental verification of the scaling exponents for the higher order structure functions showed that instead of

$$S_p^{\|}(r) \propto r^{p/3}, \tag{5.4}$$

one has (see figure 7.2)

$$S_p^{\|}(r) \propto r^{\zeta_p}, \tag{5.5}$$

with $\zeta_p = p/3 - \mu_p < p/3$, which means that in order to make the latter relation dimensionally correct it seems to be necessary to extend the list of governing parameters beyond the mean dissipation rate, $\langle \epsilon \rangle$ (e.g. Kuznetsov et al., 1992). For example, such a parameter may arise due to the lack of local isotropy discussed above, an effect which is not 'felt' at low order scaling exponents. In such a case, an additional parameter may be the degree of anisotropy and/or some characteristic large scale, so that 'simple' dimensional analysis is not simple anymore. Another commonly believed alternative is that the observed deviation of the scaling exponents in relations of the type (5.5) for structure functions $S_p^{\parallel}(r)$ for $p > 3$ from the values implied by the Kolmogorov theory (i.e. anomalous scaling) is a manifestation of the so-called small scale intermittency. This matter is addressed in chapter 7, which deals with some aspects of intermittency and the turbulence structure.

5.3. Cascade

One gets an impression of little, randomly structured and distributed whirls in the fluid, with the cascade process consisting of the fission of the whirls into smaller ones, after the fashion of the Richardson poem. This picture seems to be drastically in conflict with what can be inferred about the qualitative structure of high-Reynolds-number turbulence from laboratory visualization techniques and from plausible application of Kelvin circulation theorem (Kraichnan, 1974).

The notion that turbulent flows are hierarchical and involve entities... of varying sizes is a common idea... This common notion underlies the concept of cascade, the third key element of turbulence theory (Frisch and Orszag, 1990).

All this cascade in Fourier space is a dream of linearized physicists Betchov, 1993).

The tree examples [jet, boundary layer, and wake]... show that there is something wrong with this idea (the Richardson poem). In each case turbulence begins at small scales and grows larger: not the other way around (Gibson, 1996).

The conceptual picture is that of a cascade organized by wall distance and by eddy size, where energy is transferred to smaller scales at any given location, and to larger ones away from the wall (Jimenez, 1999).

This suggests that the Komogorov cascade process is basically incorrect, albeit an excellent approximation (Shen and Warhaft, 2000).

5.3.1. INTRODUCTION

The cascade[14] picture of turbulent flows takes its origin from Richardson (1922, p. 66):*... we find that convectional motions are hindered by the formation of small eddies resembling those due to dynamical instability. Thus C.K.M. Douglas writing of observations from aeroplanes remarks: "The upward currents of large cumuli give rise to much turbulence within, below, and around the clouds, and the structure of the clouds is often very complex". One gets a similar impression when making a drawing of a rising cumulus from a fixed point; the details change before the sketch is completed. We realize thus that: big whirls have little whirls that feed on their velocity, and little whirls have lesser whirls and so on to viscosity — in the molecular sense*[15].

The last sentence became very popular, though Richardson makes no further use of this 'cascade' picture at any time. Instead Richardson proposes to use the eddy viscosity approach, since he realized the inherent difficulty of the most problematic issue of decomposition of a turbulent field (see chapter 3): *Thus, because it is not possible to separate eddies into clearly defined classes according to the source of their energy..., therefore a single coefficient is used to represent the effect produced by eddies of all sizes and descriptions. We have then to study the variations of this coefficient.*

As mentioned above, the cascade picture supplemented by the assumption about the chaotic nature of cascade was used by Kolmogorov (1941a) in a $2/3$ page footnote as a *qualitative* justification of his hypothesis on the local isotropy of turbulent flows for very large Reynolds numbers.

The cascade picture is based on the intuitive notion that turbulent flows possess a hierarchical structure consisting of 'eddies' (Richardson's 'whirls', Kolomogov's 'pulsations', etc.) as a result of succesive instabilities. The essence of this picture is in its successive hierarchical process, and in this sence it is the same as the Landau-Hopf picture of transition to turbulence as a 'cascade' of successive instabilities. The difference is that the Richarson-Kolomogov cascade refers to a process at some fixed Reynolds number, whereas the Landau-Hopf picture describes the process of changes occurring as the Reynolds number increases (see chapter 2).

The cascade picture of turbulence seems to be more a reflection of the hierarchic structure of various *models* of turbulent flows rather than reality. Most of these models have no connection with Navier-Stokes equations (e.g.,

---

[14]The term 'cascade' comes from Onsager (1945, 1949).

[15]Note that this observation was made by looking at the structure of clouds, i.e. condensed water vapour, at the interface between laminar and turbulent flows in their bulk, which do not necessarily reflect the structure of the underlying velocity field; see chapter 4.

see Jimenez, 2000 and references therein). Hence the term 'phenomenolog-ical' models.

5.3.2. IS THERE CASCADE IN PHYSICAL SPACE?

The Richardson-Kolmogorov cascade picture was formulated in *physical* space and is used frequently without much distinction both in physical and Fourier space, as well as some others. However, it was Neumann (1949) (see also Onsager, 1949) who recognized that this process occurs not in physical space, but in Fourier space: *... the system is "open" at both ends, energy is being supplied as well dissipated. The two "ends" do not, however, lie in ordinary space, but in its Fourier-transform. More specifically: The supply of energy occurs at the macroscopic end — it originates in the forced motions of macroscopic (bounding) bodies, or in the forced maintenance of (again macroscopic) pressure gradients. The dissipation, on the other hand, occurs mainly at the microscopic end, since it is ultimately due to molecular friction, and this is most effective in flow-patterns with high velocity gradients, that is, in small eddies... Thus the statistical aspect of turbulence is essentially that of transport phenomenon (of energy) — transport in the Fourier-transform space.* That is, the nonlinear term in the Navier-Stokes equation redistributes energy among the *Fourier modes*[16] not scales as is frequently claimed, unless the 'scale' is defined just as an inverse of the magnitude of the wave-number of a Fourier mode, which is not easy for everybody to swallow. A natural question is then what does the nonlinear term in physical space do? Is energy transferred from large to small scales in physical space? The answer to the last question depends on the definition of what is a 'scale' in physical space. First, we recall that that there is no contribution from the nonlinear term in the total energy balance equation (and in a homogeneous/periodic flow it's contribution is null in both the total and the mean), since the nonlinear term in the energy equation (C.8) has the form of a spatial flux, $\partial\{...\}/\partial x_j$. In other words the nonlinear term redistributes the energy in physical space, but does it do more than that? It is straightforward to see that in a statistically homogeneous turbulent flow the mean energy of volume of *any scale* (Lagrangian and/or Eulerian)[17] is changing due to external forcing and dissipation only – there is no contribution in the mean of the nonlinear term, which includes the term with the pressure. That is, if one chooses to define a 'scale', l, in physical space as a fluid (or a fixed) volume, say, of order $\sim l^3$, then in a statistically homogeneous flow there is no cascade in physical space in the sense that,

---

[16]See section 2.4 and 6.2.4 in Frisch (1995) for a demonstration of this process.

[17]Following a Lagrangian volume for a reasonable time in turbulent flows is not a simple matter, since Lagrangian volumes 'lose their identity' very fast in turbulent flows, (Pumir, 1999, private communication).

in the mean, there is no energy exchange between different scales. This happens because the nonlinear term in the energy equation (C.8) has the form of a spatial flux, $\partial\{...\}/\partial x_j$, i.e. there is conservation of energy by non linear terms. In other words, the nonlinear term redistributes the energy in physical space if the flow is statistically nonhomogeneous. So, generally, it is a misconception to interpret this or any other process involving spatial fluxes, $\partial\{...\}/\partial x_j$ (e.g. momentum flux), as a 'cascade' in physical space[18].

On the other hand, in a statistically stationary state $\int \mathbf{F} \cdot \mathbf{u} d\tau = \int \epsilon d\tau$, i.e. energy input, which is associated with large scales, equals dissipation, which occurs mostly in the small scales. Is there a contradiction? Does the equality $\int \mathbf{F} \cdot \mathbf{u} d\tau = \int \epsilon d\tau$ (or similar, see appendix C, the text following the equation (C.49)) mean that the energy should be somehow 'transferred' from large to small scales via some multistep process? Not necessarily – for example, two big neighbouring eddies can dissipate energy directly through encounters with each other at small scale - much smaller than their own scales. Such a process still will look in Fourier space as continuous energy transfer from modes with small to modes with large wave numbers. The resolution of the apparent contradiction lies in clarifying the meaning of the term 'scale', which mostly is understood as an inverse of the wavenumber magnitude in a Fourier representation, and what is the meaning of 'transfer of energy (or whatever) from large to small scales' in physical space. This issue is directly related to the decomposition/representation of the turbulent flow field. Indeed, the reason for the above result on the absence of energy exchange between different scales in physical space is because no decomposition is involved in the above 'definition of scale'. *Any* decomposition brings the 'cascade' back to life. For example, there are various ways of filtering the flow field widely used in large eddy simulations. However, one of the problems with decompositions is that the nonlinear term redistributes the energy among the components of a particular decomposition in a *different* way for *different* decompositions, i.e. the energy exchange/transfer is decomposition dependent[19]. Therefore quantities like energy flux are not well defined. In other words the term 'cascade' corresponds to a process of interaction/exchange of (not necessarily only) energy between components of some *particular* decomposition/representation of a turbulent field associated with the nonlinearity and the nonlocality of the turbulence phenomenon, two of the three N's: nonlinearity, nonlocality and nonintegrability, which make the problem so impossibly difficult (chapter 1). On the other hand, energy transfer, just like any physical process,

---

[18]See Appendix D for a collection of some other major misconceptions.

[19]For instance, Fourier, wavelets, POD, filtering and so on: Frick and Zimin (1993), Germano (1999), Holmes et al (1996), Mahrt & Howell (1994), Meneveau (1991), Sirovich (1997), Borue & Orszag (1998).

should be invariant of particular decompositions/representations of a turbulent field. In this sense Kolmogorov's choise of dissipation (and energy input) are well defined and decomposition independent quantities, whereas the energy flux is (generally) not, since it is decomposition dependent. After all Nature may and likely does not know about *our* decompositions.

It is noteworthy that the 'cascade' arising from a decomposition of the flow field viewed as a process of exchange of energy, momentum, etc. between the components of this decomposition is a dynamical process. This should be distinguished from 'cascading processes' resulting from a decomposition of some quantity, e.g. dissipation, usually of its surrogate $(\partial u_1/\partial x_1)^2$, obtained from experimental signals (for recent examples see Frederiksen et al., 1998; Arneodo et al., 1999; Renner et al., 2001; and references therein). The former is a *dynamical process*, whereas the latter is a *representation* characterizing some aspects of the *spatial and/or temporal structure of some flow characteristics*. In other words, 'structure' is not synonymous of 'process': it is the result of a process. Therefore, generally it is impossible to draw conclusions about the former from the information about the latter, though this is done quite frequently. For example, simple chaotic systems with few degrees of freedom only (e.g. three as in the Lorenz [1963] system or four in the forced spherical pendulum, Miles [1984], also Mullin [1993]), produce also 'fine structure', e.g. continuous spectrum, but there is no 'cascade' whatsoever, though, of course, the signal with the continuous spectrum can be cast in a multiplicative representation. Another example, is the complicated structure of a passive object[20] arising in a simple fluid flow via a single instability only(!) (Ott, 1999). This is true also of the vorticity field resulting from a linear instability of such a flow (see also chapter 4).

5.4. What are the 'small scales' in turbulent flows?

We have seen that there is an ambiguity in defining the meaning of the term 'small scales' (or more generally 'scales' or 'eddies', see appendix C) and consequently the meaning of the term 'cascade'. The specific meaning of this term and associated interscale energy exchange/'cascade' (e.g. spectral energy transfer) is essentially decomposition/reperesentation dependent[21].

---

[20]With power law spectrum, (multi)fractality and significant variations down to very small scale can be produced by a single instability at much larger scale without any 'cascade' of successive instabilities.

[21]Indeed, the meaning of *(small) scales* is different for different representations: it is not the same for *Fourier* ('regular' and helical) and similar (Littlewood-Paley), *Wavelets* (wavepackets, solitons), *POD, LES*. It is also different in various heuristic representations, e.g. 'two-fluid' (Organized/ Incoherent, Deterministic/ Random and some other two-fluid models); intermittency-prompted (breakdown coefficients/ multipliers, (multi)fractals); Moffat's 'smart decomposition', Moffatt (1990) and the 'punctuated'

Perhaps, the only common thing in all decompositions/representations (D/R) is that the small scales are *always* associated with the field of velocity derivatives[22]. Therefore, it is naturally to look at this field as the one objectively (i.e. D/R independent) representing the small scales. Indeed, the dissipation is associated precisely with the symmetric part of the velocity derivative tensor $\partial u_i/\partial x_j$ – the rate of strain field s_{ij} both in Newtonian and non-Newtonian fluids, whereas vorticity $\omega_i = \epsilon_{ijk}\partial u_j/\partial x_k$ is, in fact, its anti-symmetric part. Before proceeding with the small scales let us mention that the large scales are naturally characterized by the velocity field itself, **u**. This is justified also by the fact that sustaining turbulent flows requires energy input into flow, e.g. in case of a prescribed force, **F,** the power input is associated with this force is $\int \mathbf{F} \cdot \mathbf{u}dV$, i.e. with the velocity field, **u**.

The advantage of the above 'definition' of small scales can be seen from the following.

While the mean contribution of the nonlinear term in the energy balance is vanishing, the nonlinearity definitely is producing vorticity and strain in physical space, since the mean enstrophy and strain production are strictly positive[23]. As mentioned above it is naturally and justified from the physical point of view to associate the field of velocity derivatives with small scales. It is immediately seen that 3-D turbulent flows have a natural tendency to create small scales[24]. Namely, the velocity field (and its energy) arising in the process of (self) production of the field of velocity derivatives is the one which is associated with small scales. This process is what can be called as energy (and not only energy) transfer from large to small scales in physical space. The latter are not necessarily created via a stepwise turbulent 'cascade': it can be bypassed, and most probably is so in turbulent flows, for example via broad-band instabilities with highest growth rate at short wavelengths (Pierrehumbert and Widnall, 1982; Smith and Wei, 1994) or some other approximately single step process (Betchov, 1976; Douady et al., 1991; Ott, 1999; Shen and Warhaft, 2000; Vincent and Meneguzzi, 1994). The problem goes back to Townsend (1951): *...the postulated process differs from the ordinary type of turbulent energy transfer being fundamentally a single process* (see also Corrsin, 1962; and Tennekes, 1968). Indeed, as mentioned two large neighbouring eddies can dissipate energy directly by

conservative dynamics.

[22]Differentiation is a kind of high-pass filtering, so one can use also higher order derivatives, especially those appearing in the NSE and their consequences, such as velocity Laplacian, etc.

[23]This question involving also the relation between 'cascade' and the process of the self-production of the field of velocity derivatives is addressed in more detail in chapters 6 and 7, see also appendix C.

[24]This is why the cascade is usually associated with vortex stretching and enstrophy production. We shall see in chapter 6 that this is far from being a precise statement.

encountering each other on a very small scale. Such a view goes back to the observation made by Batchelor and Townsend in 1949: *the mean separation of the visible activated regions is comparable with the integral scale of the turbulence, i.e. with the size of the energy-containing eddies* (p. 253)[25].

The process of vorticity production is not just creation of the field of velocity derivatives. It literally involves creation of small scale *structure* in the following sense. Namely, an inevitable concomitant process to vortex stretching is tilting and folding of vorticity due to the energy constraint (Orszag, 1977; Chorin, 1982; Tsinober, 1998a). Simultaneously, the strain field is built up at the same rate too (see chapter 6). This together with limitations on the volume scale leads to formation of fine small scale structure.

The above 'definition' of small scales has a variety of consequences. For example, since the whole flow field (including velocity, which is mostly a large scale object) is determined entirely by the field of vorticity, i.e. the velocity field is a functional of vorticity $\mathbf{v} = F\{\boldsymbol{\omega}(\mathbf{x}, t)\}$[26], the production of vorticity 'reacts back' in creating the corresponding velocity field[27], i.e. the small scales are not just 'swept' by the large ones. Similarly, the velocity field is a functional of the strain tensor, $\mathbf{v} = G\{s_{ij}(\mathbf{x}, t)\}$[28], so that production of strain 'reacts back' on the velocity field as well. Therefore from the physical point of view it seems incorrect to treate the small scales as a kind of passive objects swept by the large scales or just 'slaved' to them (Novikov, 2000). Similarly it seems impossible to 'eliminate' the small scales (as is done in many theories) reducing their reaction back to some eddy viscosity or similar things only. It is noteworthy that due to nonlocality of the relations $\mathbf{v} = F\{\boldsymbol{\omega}(\mathbf{x}, t)\}$, $\mathbf{v} = G\{s_{ij}(\mathbf{x}, t)\}$ mostly small scale vorticity and strain are, generally, creating also some large scale velocity. This and other aspects of nonlocality (see chapters 1 and 6) contradict the idea of cascade in physical space, which is local by definition (e.g. see Frisch, 1995, p.104).

In view of the above arguments it seems that in physical space the energy

---

[25]Note also the quite convincing 'anti-cascade' arguments by Chorin (1994, p. 56) in Fourier space. He makes an attempt to 'rescue' the cascade picture, but the reader is invited to see whether the arguments given really help.

[26]This is seen from the equation $\nabla^2 \mathbf{u} = -curl\boldsymbol{\omega}$, which together with boundary conditions defines uniquely the velocity field. In case of an infinite flow domain this results in the Bio-Savart relation $\mathbf{u} = 1/(4\pi) \int (\boldsymbol{\omega} \times \mathbf{r})/r^3 dr$ (see appendix C).

[27]This essentail process is fully ignored in (quasi) linear approaches like rapid distortion theory (RDT) (Savill, 1987; Hunt and Carruthers, 1990).

[28]This is seen from the equation $\nabla^2 u = 2\partial s_{ik}/\partial x_k$, which together with boundary conditions defines uniquely the velocity field, see equation (C.14') in the appendix C. Of course, the vorticity and strain are not independent and are functionals of each other. See chapter 6, section 6.6 on nonlocality and appendix C, equations (C.15) and the following text.

is dissipated not necessarily via a multistep cascade-like process[29]. Instead, there is an exchange of energy (and everything else) in both directions, whereas the dissipation occurs in 'small scales'.

Therefore, it is quite possible and quite plausible that, insofar as physical space is concerned, the mentioned above Richardson's famous verse (1922, p.66) on the hierarchy of 'whorls'

> *Big whirls have little whorls,*
> *Which feed on their velocity.*
> *And little whorls have lesser whorls*
> *And so on to viscosity –*
> *In the molecular sense*

should be replaced by Betchov's (1976, p. 845)

> *Big whirls lack smaller whirls,*
> *To feed on their velocity.*
> *They crash and form the finest curls*
> *Permitted by viscosity.*

5.5.   Cascade of passive objects?

It is rather common, since Obukhov (1949) and Corrsin (1951), to speak about cascade in case of a passive scalar and more recently passive vector[30]. The main argument is from *some* analogy. Indeed, as mentioned in Chapter 4, for instance in any random isotropic flow the rate of production of 'dissipation' (i.e. corresponding field of derivatives) of both passive scalars and passive vectors is essentially positive (see equations C.35, C.38 in appendix C), which can be interpreted as a sort of 'cascade'. However, the equations describing the behaviour of passive objects are *linear*. Hence, there is no interaction between modes of whatever decomposition of the field of a passive object: the princilple of superposition is valid in case of passive objects[31]. Therefore, it seems more appropriate to describe the pro-

---

[29]Linear stability analysis of a vorticity field in a smooth velocity field has shown that a power law and fractality are produced by a single instability. In other words significant variations down to very small scale can be produced by a single instability at much larger scale without any 'cascade' of successive instabilities (Ott, 1999). This shows also that nonlinearity in the Lagrangian representation cannot be interpreted in terms of some cascade.

[30]See references in Monin and Yaglom (1971, 1975) and Warhaft (2000).

[31]Here 'mode' is meant as a solution of the appropriate equation, e.g. of the advection-diffusion equation (C.30). Of course, there are many ways to use 'modes' that are *not* solutions of this equation, such as Fourier modes. In this case the Fourier modes do interact, since one of the coefficients of the advection-diffusion equation, the velocity field, is not constant. This interaction is interpreted frequently as a 'cascade' of passive objects. But, as mentioned, this interaction is decomposition dependent, and therefore is not ap-

cess in terms of production of the field of derivatives of the passive object, which is performed by the *velocity* straining field, just like it is proposed above for the velocity field. Hence the extension of Kolmogorov arguments and phenomenology to passive objects seems to be much less justified. No wonder that the phenomenological paradigms for the velocity field failed in most cases when applied to passive objects[32]. We are reminded that the 'analogy' between the passive objects and the active variables is, at best, very limited for several reasons, the main of which is the Lagrangian chaos, the 'irreversible' effect of the randomness of the velocity field on passive objects independent of the nature of this randomness, e.g. even a Gaussian one, and the one-way interaction between the velocity field and the field of a passive object (see chapters 4, 6 [e.g., section 6.8 p.148] and 7).

5.6. Summary

Kolmogorov *a priori* phenomenological hypotheses include the hypothesis of local isotropy and two similarity hypotheses. The hypothesis of local isotropy states that at large Reynolds numbers all the symmetries of the Navier-Stokes are restored in the statistical sense (more precisely all the statistical properties of velocity increments) locally in time and space in regions far enough from the boundaries of any turbulent flow or its other special regions. That is, high-Reynolds-number turbulent flows are locally homogeneous, isotropic and stationary. It appears that turbulent shear flows do not conform with this hypothesis, so that regions with mean shear should be considered as 'special' in the sense of Kolmogorov. However, the problem seems to be more serious, since there is also evidence that other kinds of anisotropy in the large scales can result in anisotropy in the small scales. The most likely reason is the nonlocality of turbulence, which, among other things is manifested in direct and bidirectional coupling between large and small scales. Consequently, problems arise also with the similarity hypotheses, which state that all statistical properties of velocity increments for small separations in space/time and large Reynolds numbers are independent of the large scale properties of turbulent flows, except for the mean energy dissipation or energy input. Thus, there are serious reasons to doubt the restoring of all the symmetries (including the scale invariance; see chapter 7) of the Navier-Stokes equations locally in time in space in the statistical sense.

The hypothesis of the finite dissipation limit with increasing Reynolds

propriate for description of physical processes, which are invariant of our decompostions.

[32]Recent experiments by Villermaux et al. (2001) clearly show that this is the case. The behaviour of passive scalar in their experiments is distinctly nonlocal in the sense that the main mechanism responsible for mixiing involves direct interaction between large and small scales 'bypassing' the (nonexistent) cascade.

number seems to have reasonable, but still limited, experimental support.

The notion that turbulent flows are hierarchical, which underlies the concept of the cascade, though convenient, is more a reflexion of the unavoidable (due to the nonlinear nature of the problem) hierarchical structure of models of turbulence and/or decompositions rather than reality. This is emphasized in the case of passive objects, whose evolution is governed by linear equations, with the velocity field entering multiplicatively in these equations, thus making them 'statistically nonlinear'.

CHAPTER 6

DYNAMICS

With the emphasis on velocity derivatives

6.1. Introduction

The true problem of turbulence dynamics is the problem of its origin(s) and successive development from some initial conditions and at some boundary conditions to an ultimate (statistical) state[1]. However, since this route is extremely complicated and involved, a second approach is used quite frequently. Namely, turbulent flows are studied 'as they are' disregarding their origin, and without looking into the details of the mechanisms of turbulence production and sustaining. In both approaches, at least some aspects of the time evolution are of central importance, since turbulence dynamics is a process.

One of the simplest examples of the second approach is when turbulence is produced numerically in a box with periodic boundary conditions by some forcing in the right hand side of the Navier-Stokes equations. If the Reynolds number is not too small almost any forcing will do – random and deterministic – the flow will be turbulent. In both cases it is possible to produce a turbulent flow which is approximately (only) homogeneous and isotropic, and if the forcing is statistically stationary or just time independent the resulting turbulent flow is statistically stationary. In such a case the overall energy balance is described by the equation similar to (C.49), and in this particular case the total turbulent energy balance for the whole flow domain is

$$\frac{d\mathcal{E}_T}{dt} = \mathcal{W}_f - \mathcal{D} \qquad (C.49a)$$

where $\mathcal{E}_T = \int e_T dV$ is the total kinetic energy of turbulent fluctuations, $\mathcal{W}_f = \int u_i f_i dV$ is the total rate of production of energy of turbulent fluctuations by the external forces, and $\mathcal{D} = 2\nu \int s_{ij} s_{ij} dV$ - is the total rate of dissipation (simply dissipation) of energy of turbulent fluctuations by viscosity. If the flow is statistically stationary the production equals dissi-

---

[1]With the hope (based on rich experimental data) that, at least with statistically stationary boundary conditions and forcing, such a state does exist. This does not mean that it is impossible to define a kind of 'ultimate state' for statistically nonstationary turbulent flows.

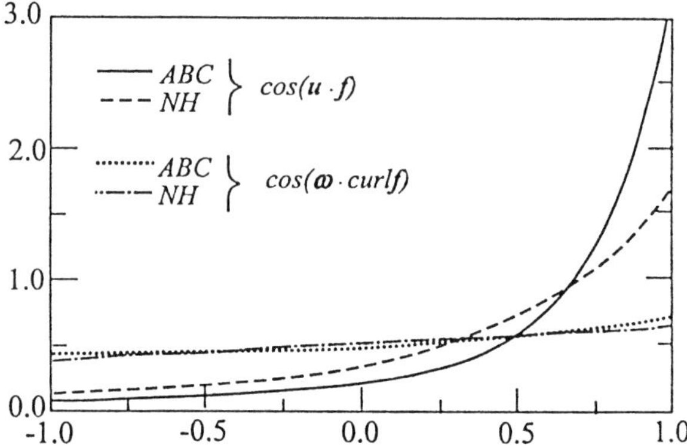

Figure 6.1.   PDFs of the cosine of the angle between velocity and force, $cos(\mathbf{u}, \mathbf{f})$ and between vorticity the *curl* of the force, $cos(\omega, curl\mathbf{f})$.

pation $\mathcal{W}_f = \mathcal{D}$, i.e. $\mathcal{W}_f > 0$. This implies that there should be a tendency for alignment between the velocity vector, $\mathbf{u}$, and the force, $\mathbf{f}$. Indeed, this is what is observed (see figure 6.1).

These results (Galanti and Tsinober, 2000) were obtained with a deterministic forcing corresponding to the so called ABC flow, $\mathbf{f} = f\{A\sin z + C\cos y,\ B\sin x + A\cos z,\ C\sin y + B\cos x\}$, $A = B = C$. This choice was mainly due to the strong instability of ABC flows guaranteeing fast transition to turbulence at a rather low Reynolds number (Galanti et al., 1992; Podvigina and Pouquet, 1994; and references therein). It should be emphasized that the results are practically the same in cases where the coefficients A, B, C are random functions of time. The ABC forcing is strongly helical, $curl\mathbf{f} \parallel \mathbf{f}$, and therefore along with kinetic energy such a forcing makes an input of helicity into the flow. Very similar results were obtained also for other kinds of forcing, e.g. with a force in the form $\mathbf{f} = f\{A\cos z\cos y,\ B\cos x\cos z,\ C\cos y\cos x\}$, $A = B = C$. This forcing, denoted in the sequel as NH, is nonhelical, $\mathbf{f} \cdot curl\mathbf{f} = 0$, and a mixed random forcing was also used consisting of weighted sums of the ABC and NH forcings. In all the above-mentioned cases the Taylor microscale Reynolds number, $Re_\lambda \approx 110$. In addition some runs were made at different Reynolds numbers in order to have a qualitative impression of the Reynolds number effects.

The results shown in figure 6.1 are not unexpected: there should be some alignment between velocity and force in order to have the work performed by the force, $\mathbf{f}$, positive. A more interesting result is that both the energy input, $\mathcal{W}_f = \int u_i f_i dV$, and dissipation, $\mathcal{D} = 2\nu \int s_{ij}s_{ij}dV$, are far from being instantaneously equal, and exhibit quite large fluctuations in time,

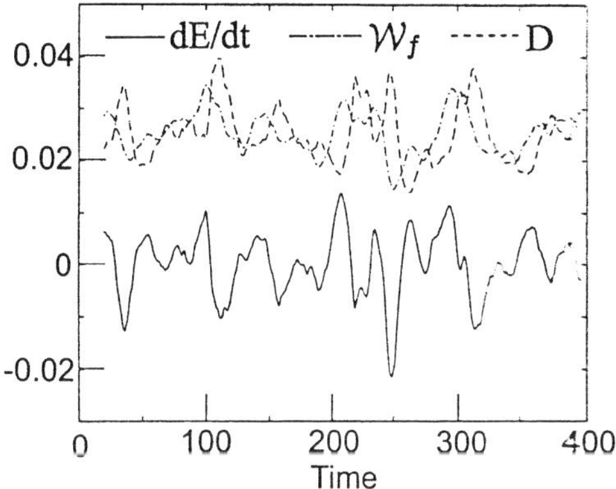

Figure 6.2. Time behaviour of the energy input, $\mathcal{W}_F = V^{-1} \int u_i F_i dV$, and dissipation, $\mathcal{D} = 2\nu V^{-1} \int s_{ij} s_{ij} dV$ for the case of ABC forcing at the resolution 64^3, corresponding to $Re_\lambda \approx 60$. Time is measured in turnover times.

up to 40% of their long time averages (see figure 6.2)[2]. The latter are equal within less then 0.1%. The equality $\int_0^T \mathcal{W}_f dV = \int_0^T \mathcal{D} dV$ occurs at rather large turnover times, $T \approx 500$, even at rather low Reynolds number.

The result shown in figure 6.2 is similar to the one obtained by Pinton et al. (1999) in an experiment in a 'French washing machine' similar to the one mentioned in chapter 5, figure 5.1.

6.2. Why velocity derivatives?

Velocity derivatives, $A_{ij} = \partial u_i / \partial x_j$, play an outstanding role in the dynamics of turbulence for a number of reasons. Their importance has become especially clear since the papers by Taylor (1937, 1938a)[3] and Kolmogorov (1941a,b). Taylor emphasized the role of vorticity, i.e. the antisymmetric part of the velocity gradient tensor $A_{ij} = \partial u_i / \partial x_j$, whereas Kolmogorov stressed the importance of dissipation, and thereby of strain, i.e. the symmetric part of the velocity gradient tensor.

It is noteworthy that the whole (incompressible) flow field is fully de-

---

[2]With prescribed force the energy input and consequently the dissipation depend on the velocity field. Therefore, generally, they cannot be prescribed independently.

[3]Taylor (1937, 1938a) was motivated by the assumption of von Karman (1937) *that the expression* $\sum_i \sum_k \omega_i \omega_k \frac{\partial u_i}{\partial u_k}$ (i.e. enstrophy production, see equation (C.16)) *is zero in the mean and that he (vK) cannot see any physical reason for such a correlation.* Taylor (1937) conjectured that there is a *strong correlation between* ω_3^2 *and* $\frac{\partial u_3}{\partial u_3}$ *so that* (the mean of) $\omega_3^2 \frac{\partial u_3}{\partial u_3}$ *is not equal to zero* (x_3 is directed along vorticity) and showed that this is really the case, Taylor (1938a). He also expressed the view that *stretching of vortex filaments must be regarded as the principal mechanical cause of the higher rate of dissipation which is associated with turbulent motion.*

termined by the fields of vorticity *or* strain with appropriate boundary conditions, see appendix C.

Apart from vorticity and strain/dissipation, there are many other reasons for special interest in the characteristics of the field of velocity derivatives, $A_{ij} = \partial u_i/\partial x_j$, in turbulent flows. For example,

• – The field of velocity derivatives is much more sensitive to the non-Gaussian nature of turbulence or more generally to its structure, and hence reflects more of its physics (see Tsinober, 2000a, and references therein).

• – The possibility of singularities being generated by the Euler and the Navier-Stokes equations (NSE) and possible breakdown of NSE are intimately related to the field of velocity derivatives (Constantin, 1996; Doering, 1999; Doering and Gibbon, 1995).

• – In the Lagrangian description of fluid flow in a frame following a fluid particle, each point is a critical one, i.e. the direction of velocity is not determined. So everything happening in its proximity is characterized by the velocity gradient tensor $A_{ij} = \partial u_i/\partial x_j$. For instance, local geometry/topology is naturally described in terms of critical points terminology[4], (see Chacin and Cantwell, 2000; Chertkov et al., 1999; Martin et al., 1998; Ooi et al. ,1998; and references therein).

• – There is a generic ambiguity in defining the meaning of the term *small scales* (or more generally scales) and consequently the meaning of the term *cascade* in turbulence research. As mentioned in chapter 5, the specific meaning of this term and associated interscale energy exchange/'cascade' (e.g. spectral energy transfer) is essentially decomposition/representation dependent. Perhaps, the only common element in all decompositions/representations (D/R) is that the small scales are associated with the field of velocity derivatives. Therefore, it is natural to look at this field as the one *objectively* (i.e. D/R independent) representing the small scales. Indeed, the dissipation is associated precisely with the strain field, s_{ij}, both in Newtonian and non-Newtonian fluids.

There is a number of more specific reasons why studying the field of velocity derivatives is so important in the dynamics of turbulence. This is one of the main themes of this chapter. Additional emphasis is given to several relatively new aspects, such as geometrical statistics.

6.2.1. VORTEX STRETCHING AND ENSTROPHY PRODUCTION

One of the most basic phenomena and distinctive features of three dimensional turbulence is the predominant vortex stretching, which is manifested in positive net enstrophy production, $\langle \omega_i \omega_j s_{ij} \rangle > 0$. As mentioned, this was discovered by Taylor (1938), and was confirmed subsequently both exper-

[4]This approach was initiated by Perry and Fairlie (1974).

imentally and numerically in a number of investigations (see references in Tsinober, 1998ab, 2000).

The nonlinear terms $\omega_j s_{ij}$ and $\omega_i \omega_j s_{ij}$ are responsible for the vortex stretching (**VS**) and enstrophy production. In other words the essential dynamics of 3D-turbulence is contained in the *interaction* between vorticity, ω, and the rate of strain tensor, s_{ij}. Both $\omega_j s_{ij}$ and $\omega_i \omega_j s_{ij}$ vanish identically for 2-D flows.

So far, no theoretical arguments in favor of positiveness of $\langle \omega_i \omega_j s_{ij} \rangle$ have been given. The argument that the reason is the (approximate) balance between the enstrophy production and enstrophy dissipation is misleading and puts the consequences before the reasons, since it is known that, for Euler equations, the enstrophy production increases with time very rapdly, apparently without limit (see references in Tsinober, 1998ab, 2000). Another rather common view that the prevalence of vortex stretching is due to the predominance of stretching of material lines[5] is - at best - only partially true, since there exist several *qualitative* differences between the two processes. We bring them already here (see box at page 86) though many of these differences will become clear at later stages.

The contents of the box does not exhaust the differences between vorticity and material elements. For example, vorticity is divergence-free, whereas material elements are not: Lagrangian trajectories are compressible (for more details see Elperin et al., 2000 and references therein. Their paper contains interesting discussion of situations involving the presence of particles – the fluid is effectively also compressible). In presence of diffusivity the energy of a passive vector under certain conditions grows without limit, i.e. not balanced by diffusivity (Childress and Gilbert, 1995), whereas growth of enstrophy is balanced by viscous effects.

As mentioned above, $\langle \omega_i \omega_j s_{ij} \rangle$ is an essentially positive quantity in 3-D turbulence - the PDF of $\omega_i \omega_j s_{ij}$ is strongly positively skewed (see figure 6.4). This fact reflects one of the most basic *specific* properties of three-dimensional turbulent flows - the prevalence of the vortex stretching process. The enstrophy production $\omega_i \omega_j s_{ij}$ is an oustanding nonzero *odd* moment of utmost dynamical importance in turbulence. Indeed, in the hypothetical case of absence of vortex stretching and enstrophy generation

---

[5]This view originates with Taylor (1938): *Turbulent motion is found to be diffusive, so that particles which were originally neighbours move apart as motion proceeds. In a diffusive motion the average value of d^2/d_0^2 continually increases. It will be seen therefore from (1) (i.e. from $\omega/\omega_0 = d/d_0$, where d_0 is the initial distance between two neighbouring particles, and d is the distance at a subsequent time) that the average value of ω^2/ω_0^2 continually increases.* This view is widely accepted, e.g. ... *the interesting physical argument that $\langle \omega_i \omega_j s_{ij} \rangle$ is positive because two particles on average move apart from each other and therefore vortex lines are on average stretched rather than compressed* (Hunt (1973), JFM, **58**, 817). Another misconception is that ...*vorticity amplification is a result of the kinematics of turbulence*, Tennekes and Lumley (1972), p. 92.

VORTEX STRETCHING VERSUS STRETCHING OF MATERIAL LINES

• - The equation for a material line element l is a linear one and the vector l is passive, i.e. the fluid flow does not 'know' anything whatsoever about l: the vector l (as any passive vector) does not exert any influence on the fluid flow. The material element is stretched (compressed) locally at an exponential rate proportional to the rate of strain along the direction of l, since the strain is independent of l.

• - On the contrary, the equation for vorticity is a nonlinear partial differential equation and the vector ω is an active one - it 'reacts back' on the fluid flow. The strain does depend in a nonlocal manner on ω and vice versa, i.e. the rate of vortex stretching is a nonlocal quantity, whereas the rate of stretching of material lines is a local one. Therefore the rate of vortex stretching (compressing) is different from the exponential one and is unknown. There are many 'fewer' vorticity lines than the material ones - at each point there is typically only one vorex line, but infinitely many material lines. This leads to differences in the statistical properties of the two fields. In the absence of viscosity vortex lines are material lines, but they are special in the sense that they are not passive as all the other passive material lines.

• - Consequently while a material element l tends to be aligned with the largest (positive) eigenvector of s_{ij}, vorticity ω tends to be aligned with the intermediate (mostly positive) eigenvector of s_{ij} : the eigenframe of s_{ij} rotates with an angular velocity Ω_s of the order of vorticity ω.

• - For a Gaussian isotropic velocity field the mean enstrophy generation vanishes identically, $\langle \omega_i \omega_j s_{ij} \rangle \equiv 0$ whereas the mean rate of stretching of material lines is essentially positive. The same is true of the mean rate of vortex stretching $\langle \omega_i \omega_j s_{ij} \rangle |\omega|^{-2}$ (for other counter-examples see below) and for purely two-dimensional flows. This means that in turbulent flows the mean growth rate of material lines is larger than that of vorticity. The nature of vortex stretching process is dynamical and not a kinematic one as the stretching of material lines is.

• - The curvature of vortex lines increases with strain and positive rate of vortex stretching, whereas the curvature of material lines decreases with strain and positive rate of material line stretching.

• - An additional difference due to viscosity becomes essential for regions with concentrated vorticity, in which there is an approximate balance between enstrophy generation and its reduction. Vortex reconnection is allowed by nonzero viscosity. No such phenomena exist for material lines.

• - Comparing vorticity with a passive vector in the presence of the same diffusivity as viscosity, the analogy is partial not just because the equation for vorticity is nonlinear, but also because in the case of vorticity the process is due to *self*-amplification of the field of velocity derivatives, whereas in case of a passive vector it is not.

or even in cases in which only $\langle \omega_i \omega_j s_{ij} \rangle = 0$ - as assumed by Karman (1938) - the three-dimensional turbulence, as we observe it, would not exist.

6.2.2. WHY STRAIN TOO?

It is to be stressed that along with vortex stretching and enstrophy production, of special interest is the production of strain. There are several reasons for this. First, though formally all the flow field is determined entirely by the field of vorticity the relation between the strain and vorticity is strongly nonlocal (Constantin, 1994; Novikov, 1967; Ohkitani, 1994). In many cases, they are only weakly correlated (statistically) or not correlated at all. Second, energy dissipation is directly associated with strain and not with vorticity. Third, vortex stretching is essentially a process of interaction of vorticity and strain. Four, strain dominated regions appear to be the most active/nonlinear in a number of aspects (see 6.4.2). Finally, the energy cascade (whatever this means) and its final result - dissipation, are associated with predominant self-amplification of the rate of strain/production and vortex compression rather than with vortex stretching. This means that another nonzero *odd* moment $s_{ij} s_{jk} s_{jk}$ (responsible for the production of strain, see below) is not less important than the enstrophy generation, Tsinober (2000a).

Production of strain. Is turbulent dissipation due to vortex stretching?
The appropriate level of dissipation moderating the growth of turbulent energy is achieved by the build up of strain of sufficient magnitude which is described by the equations (C.17, C.18). It is seen from the latter equation that in the mean the only term contributing positively to the production of strain/dissipation, s^2, is the term $-s_{ij} s_{jk} s_{ki} = -(\Lambda_1^3 + \Lambda_2^3 + \Lambda_3^3)$ $= -3\Lambda_1 \Lambda_2 \Lambda_3$, since $\langle s_{ij} s_{jk} s_{ki} \rangle = -3/4 \langle \omega_i \omega_j s_{ij} \rangle$, and $\langle s_{ij} \frac{\partial^2 p}{\partial x_i \partial x_j} \rangle = 0$ due to homogeneity and incompressibility. Moreover, since, $\Lambda_1 > 0$ and Λ_2 is positively skewed, i.e. $\langle \Lambda_2^3 \rangle > 0$,[6] the positiveness of $-\langle s_{ij} s_{jk} s_{ki} \rangle$ comes only from the term $-\langle \Lambda_3^3 \rangle$. In other words, Λ_3 is responsible for most of the 'cascade', at least, one of the final results of the 'cascade' - dissipation of energy, which is directly associated with s_{ij} and not with ω_i. An example of ratios between $\langle \Lambda_i^3 \rangle$ is given in the table 6.1.

Hence, the 'cascade' is directly associated with compressing/squeezing of fluid elements and not with (vortex) stretching. It is noteworthy that this idea is not entirely new: '*It is clear, therefore, that production of vorticity is associated essentially with Λ_3 and production of ω_1 and ω_2. This suggests*

[6]See for example figure 6.10. This was first discovered by Ashurst et al. (1987), and confirmed by She et al. (1991), Su and Dahm (1996), Tsinober et al. (1989, 1992). In a Gaussian velocity field the PDF of Λ_2 is strictly symmetric.

TABLE 6.1. The ratios between the $\langle \Lambda_i^3 \rangle$. The values for $Re\lambda = 75$ are given for a computation in box with periodic boundary conditions, but very similar results as well as many other were obtained in a grid turbulence experiment (Tsinober et al., 1997). The values for $Re\lambda = 10^4$ are given for a field experiment in the atmospheric surface layer at the height $10m$ (Kholmyansky et al., 2001a).

	Re_λ	$\langle \Lambda_1^3 \rangle$	$\langle \Lambda_2^3 \rangle$	$\langle \Lambda_3^3 \rangle$
DNS	75	1.2	0.05	-2.25
Field	10^4	1.62	0.05	-2.67

that the most of important processes associated with production of vorticity and energy transfer resemble a jet collision and not the swirling of a contracting jet, (Betchov, 1956). Betchov arrived to this conclusion analysing the means $\langle \omega_i \omega_j s_{ij} \rangle$ and $-\langle s_{ij} s_{jk} s_{ki} \rangle$, and assuming both of them positive. Looking at the equation (C.18), it is seen that the above conclusion is also true of production of strain too, which is associated with Λ_3, and with the 'jet collision' regions such as sheet-like structures as observed in the laboratory (Frederiksen et al., 1997; Schwarz, 1990) and numerical experiments (Brachet et al., 1992; Boratav and Pelz, 1997; Chen and Cao, 1997; Flohr, 1999). As for enstrophy production we shall see in the sequel that it is true in part: roughly two thirds of its positive contribution occur in the 'jet collision' regions, the remaining third happens in the 'swirling of a contraction jet' regions. Also it is noteworthy that production of ω^2 requires two partners ω_i and s_{ij}, and interaction between the two, but production of s_{ij} is in some sense (locally) less dependent on ω_i, though without vorticity it is impossible. Indeed enstrophy production is due to the term $\omega_i \omega_j s_{ij}$ containing both vorticity and strain, whereas production of strain is due to the term $s_{ij} s_{jk} s_{ki}$ containing strain only. In this sense strain production is more self-production. Again one does not have to forget that, after all, both vorticity and strain are derivatives of the same velocity field. Among other things, the difference is manifested in correlation coefficients shown in table 6.2 for the field experiment mentioned above (Kholmyansky et al., 2001a).

The main feature is that strain production is much less correlated with enstrophy than with strain, whereas enstrophy production is equally correlated with both, but its rate is more correlated with strain. This feature is better seen in joint PDFs/scatter plots (figure 6.3).

The next important point is that the enstrophy production $\omega_i \omega_j s_{ij}$ appears in the equation (C.18) with the negative sign, so that the vortex

TABLE 6.2. Correlation coefficients between production terms versus enstrophy and strain.

	$\omega_i\omega_i s_{ij}$	$-(4/3)s_{ij}s_{jk}s_{ki}$	$\omega_i\omega_i s_{ij}/\omega^2$	$-(4/3)s_{ij}s_{jk}s_{ki}/s^2$
ω^2	0.35	0.16	0.14	0.11
s^2	0.31	0.41	0.24	0.28

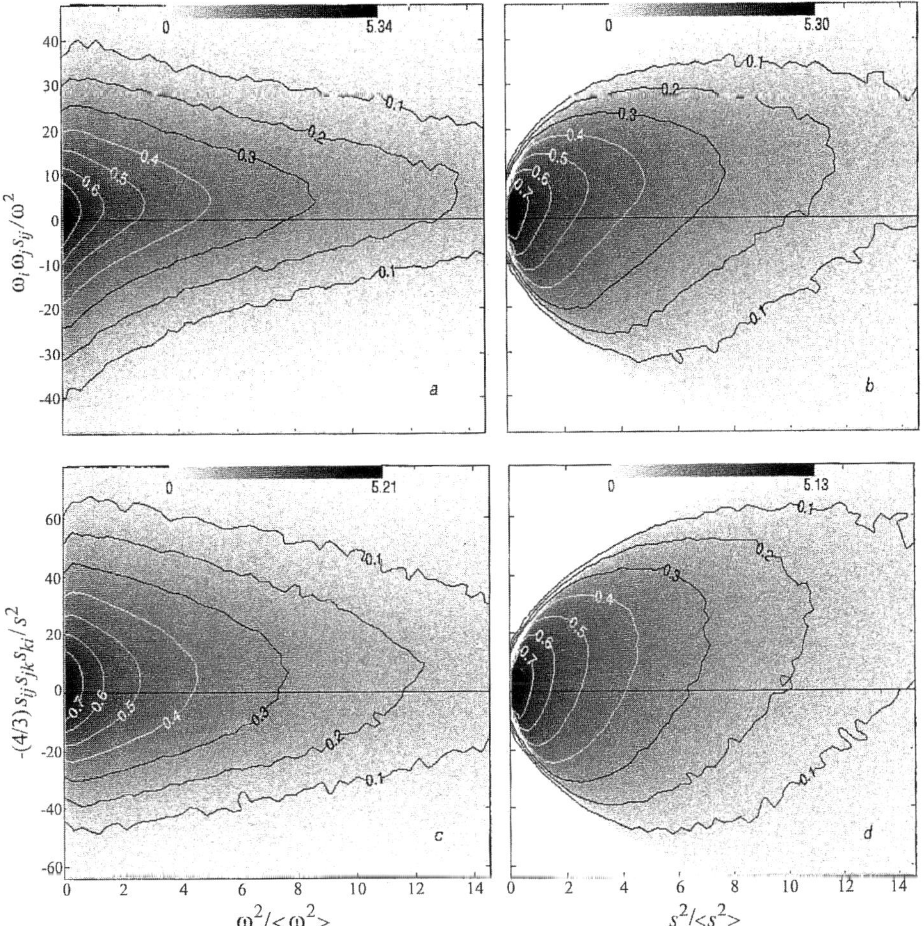

Figure 6.3. Joint PDFs/scatter plots of $\omega_i\omega_i s_{ij}/\omega^2$ and $-(4/3)s_{ij}s_{jk}s_{ki}/s^2$ versus ω^2 and s^2 from the field experiment, Kholmyansky et al. (2001a).

stretching is *opposing* the production of dissipation/strain: all instantaneous positive values of $\omega_i\omega_j s_{ij}$ make a negative contribution to the right hand side of (C.18), and since $\omega_i\omega_j s_{ij}$ is essentially a positively skewed quantity, its mean contribution to the strain production is negative. In other words, the energy cascade (whatever this means) is associated primarily with the quantity $-s_{ij}s_{jk}s_{ki}$, rather than with the enstrophy production $\omega_i\omega_j s_{ij}$ and that vortex stretching suppresses the cascade and does not aid it[7], at least in a *direct* manner (Tsinober et al., 1999; Tsinober, 2000a). On the contrary, it is the vortex *compression*, i.e. $\omega_i\omega_j s_{ij} < 0$, that aids the production of strain/dissipation and, in this sense, the 'cascade'. Negative enstrophy production is associated with strong tilting of the vorticity vector and large curvature of vortex lines (see section 6.4.2), which in turn are associated with large magnitudes of the negative eigenvalue, Λ_3, of the rate of strain tensor (Kholmyansky et al., 2001a; Tsinober, 2000a). This is in full conformity with the above mentioned fact that Λ_3 is responsible for most of the 'cascade'.

One does not have to be confused that $\langle s_{ij}s_{jk}s_{ki}\rangle = -3/4\langle\omega_i\omega_j s_{ij}\rangle$ or even by similarity of their PDFs (figure 6.4), since their pointwise relation is strongly nonlocal due to the nonlocal relation between vorticity and strain (Constantin, 1994; Novikov, 1967; Ohkitani, 1994). Consequently, locally they are very different, as can be seen from their joint PDF and scatter plots (figure 6.5): they are only weakly correlated and there are great many points with small $\omega_i\omega_j s_{ij}$ and large $-s_{ij}s_{jk}s_{ki}$ and vice versa. More details can be found in Tsinober (2000a) and Kholmyansky et al. (2001a).

6.3. Self-amplification of the field of velocity derivatives

It is commonly believed that, at least at large Reynolds numbers, the vortex stretching is a process of self- amplificaton *because ... the deformation* [i.e. rate of strain] *tensor, responsible for amplification, is expressed in terms of local characteristics* (Novikov, 1993a), or self- sustaining, since it *does not require a large scale mean flow* (Tennekes, 1989).

This belief is based on the order of magnitude estimates (Tennekes and Lumley, 1972; Novikov, 1993b), for the *mean* quantities entering the equation (C.51) for the balance of the mean enstrophy $\langle\omega^2\rangle$. In case of homogeneous turbulence with external force δ-correlated in time this estimate shows (Novikov, 1993b), that the ratio of the term associated with the external forcing to $\langle\omega_i\omega_j s_{ij}\rangle$ is of order $Re^{-3/2}$.

We shall use the example of the ABC and NH forced turbulent flows mentioned above in order to make more precise the meaning of the term *self-amplification*. Namely, we will see that the process of self-amplification refers to the whole field of velocity derivatives, i.e. both vorticity and strain,

---

[7]In contrast to the most common belief: *It seems that the stretching of vortex filaments must be regarded as the principal mechanical cause of the high rate of dissipation which is associated with turbulent motion* (Taylor, 1938a).

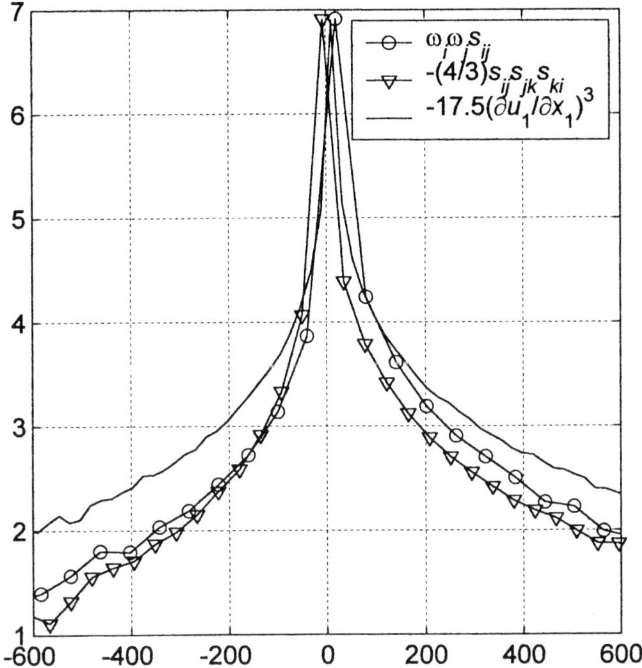

Figure 6.4. PDFs of $\frac{3}{4}\omega_i\omega_j s_{ij}$, $-s_{ij}s_{jk}s_{ki}$, and $-17.5(\partial u_1/\partial x_1)^3$ normalized on their means, $Re_\lambda = 10^4$ (Kholmyansky et al., 2001a).

since in fact there exist *two* nonlocally interconnected and weakly correlated processes: along with predominant vortex stretching/enstrophy production associated with the positiveness of $\langle \omega_i\omega_j s_{ij} \rangle > 0$, there exist a concomitant predominant self-amplification of the rate of strain/production of total strain, $s^2 \equiv s_{ij}s_{ij}$, associated with the positiveness of $\langle -s_{ij}s_{jk}s_{ki} \rangle > 0$ (Tsinober, 2000a).

The key result as obtained in the above computations relates to the comparison of the terms corresponding to the self-amplification of the field of velocity derivatives with the forcing terms in the equations (C.9) and (C.17) for the evolution of vorticity and strain, and in the equations (C.16) and (C.18) for the enstrophy, ω^2, and the total strain, s^2. Namely, it appears that the quantities $\{\omega_j s_{ij}\}^2$, $\{s_{ik}s_{kj}\}^2$, in the equations for ω_i and s_{ij}, and the quantities $\omega_i\omega_j s_{ij}$, and $-s_{ij}s_{jk}s_{ki}$ in the equations for ω^2 and s^2 are three orders of magnitude larger than the corresponding terms associated with forcing, $\{curl\mathbf{f}\}^2$ and $f_{ij}f_{ij}$, and $\boldsymbol{\omega}\cdot curl\ \mathbf{f}$ and $s_{ij}f_{ij}$ respectively. This is true not only of the mean values, but is much stronger, since the same difference in values is observed for their $max\ |\cdot|$ (see tables 6.3, 6.4 and figure 6.6).

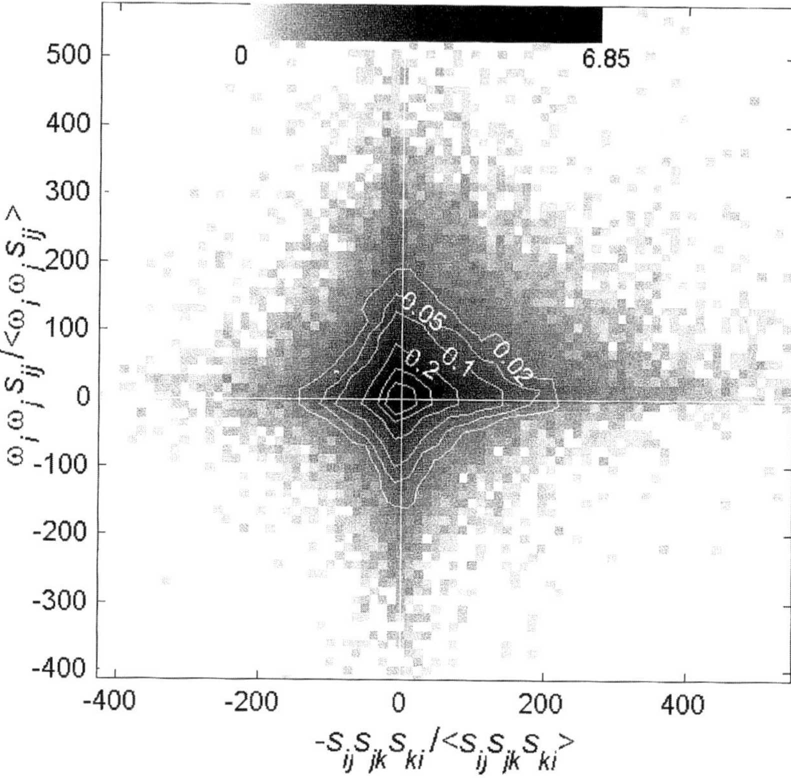

Figure 6.5. Joint PDF and scatter plot of $\frac{3}{4}\omega_i\omega_j s_{ij}$ versus $-s_{ij}s_{jk}s_{ki}$, normalized on their means, $Re_\lambda = 10^4$ (Kholmyansky et al., 2001a).

TABLE 6.3. Comparison of the mean square values and the maxima of the squares of the terms $\omega_j s_{ij}$ and $(curl\mathbf{f})_i$, and the terms $-s_{ik}s_{kj}$ and $f_{ij} \equiv \frac{1}{2}\left\{\frac{\partial f_i}{\partial x_j} + \frac{\partial f_i}{\partial x_j}\right\}$ in the equations (2) and (3).

Forcing	A B C	A B C	Nonhelical	Nonhelical
	mean	max	mean	max
$\{\omega_j s_{ij}\}^2$	8.3	$4.5\cdot 10^4$	4.6	$7.3\cdot 10^3$
$\{curl\mathbf{f}\}^2$	$4.3\cdot 10^{-4}$	$1.7\cdot 10^{-3}$	$2.3\cdot 10^{-4}$	$1.2\cdot 10^{-3}$
$\{s_{ij}s_{jk}\}^2$	18.5	$2.5\cdot 10^4$	7.7	$9.8\cdot 10^3$
$\{f_{ij}\}^2$	$8.6\cdot 10^{-4}$	$3.4\cdot 10^{-3}$	$4.7\cdot 10^{-4}$	$2.4\cdot 10^{-3}$

Moreover, the essential dominance of the self-amplification of the velocity derivatives over the forcing occurs not only in the mean and in respect

TABLE 6.4. Comparison of the mean values and the $max \mid \cdot \mid$ of the terms $\omega_i \omega_j s_{ij}$ and $\omega_i (curl\mathbf{f})_i$, and $-s_{ij} s_{jk} s_{ki}$ and $s_{ij} f_{ij}$ in the equations (4) and (5).

Forcing	A B C	A B C	Nonhelical	Nonhelical
	mean	$max\mid \cdot \mid$	mean	$max\mid \cdot \mid$
$\omega_i \omega_j s_{ij}$	3.2	$4.7 \cdot 10^3$	1.5	$1.5 \cdot 10^3$
$\omega \cdot curl\mathbf{f}$	$7.6 \cdot 10^{-3}$	1.1	$2.7 \cdot 10^{-3}$	0.64
$-s_{ij} s_{jk} s_{ki}$	2.4	$1.5 \cdot 10^3$	1.1	$7.8 \cdot 10^2$
$f_{ij} s_{ij}$	$7.6 \cdot 10^{-3}$	0.8	$2.7 \cdot 10^{-3}$	0.4

of their extremal values, but almost pointwise throughout the flow field. This was checked by looking at the volume fractions (relative number of points) of the whole flow domain, $\rho_K = V_K(|\omega \cdot curl\ \mathbf{f}|, |\omega_i \omega_j s_{ij}|)$, in which $|\omega \cdot curl\ \mathbf{f}| > K\ |\omega_i \omega_j s_{ij}|$. It appears that for $K < 100$ there are no such points at all, for $K = 100$ this fraction $\rho_K < 10^{-8}$, and $\rho_K \sim 10^{-4}$ even for $K = 10^3$. Similar results are true for the volume fractions for the rest of the quantities discussed above. In other words the process of production of the field of velocity derivatives – both vorticity and strain – is a spatially local self-amplification process in the sense that the forcing does not play any role in the production of velocity derivatives locally in space. It is noteworthy that the Reynolds number, $Re_\lambda \approx 110$, in the above simulations was very moderate. Selected runs for $Re_\lambda \approx 250$ showed that the difference between the quantities responsible for the self-amplification of velocity derivatives ($\{\omega_j s_{ij}\}^2$, $\{s_{ik} s_{kj}\}^2$, $\omega_i \omega_j s_{ij}$, and $-s_{ij} s_{jk} s_{ki}$) and the quantities associated with the external forcing ($\{curl\mathbf{f}\}^2$, $f_{ij} f_{ij}$, $\omega \cdot curl\ \mathbf{f}$ and $s_{ij} f_{ij}$) becomes much larger, so that the former are about four orders of magnitude larger than the latter. It is interesting that even at $Re_\lambda \approx 35$, the difference is still two orders of magnitude. There are clear indications that the predominance of the self-production of the velocity derivatives has a universal character (Sandham and Tsinober, 2000; Kholmyansky et al., 2001a; see Chapter 8).

There is one more important feature of the self-production of the velocity derivatives. Namely, it appears that even at Re_λ as low as ≈ 60, the intergrals over the flow domain of the enstrophy production and of its viscous destruction are approximately balanced at any time moment, $\int \omega_i \omega_j s_{ij} dV \approx -\nu \int \omega_i \nabla \omega_i dV$ (see equation [C.20]). That is, the time derivative of the overall enstrophy $d \left(\int \frac{1}{2} \omega^2 dV \right)$ is an order of magnitude smaller than both $\int \omega_i \omega_j s_{ij} dV$ and $\nu \int \omega_i \nabla \omega_i dV$,, i.e. in this respect the process is quasi-stationary. This is illustrated in figure 6.6, in which also shown

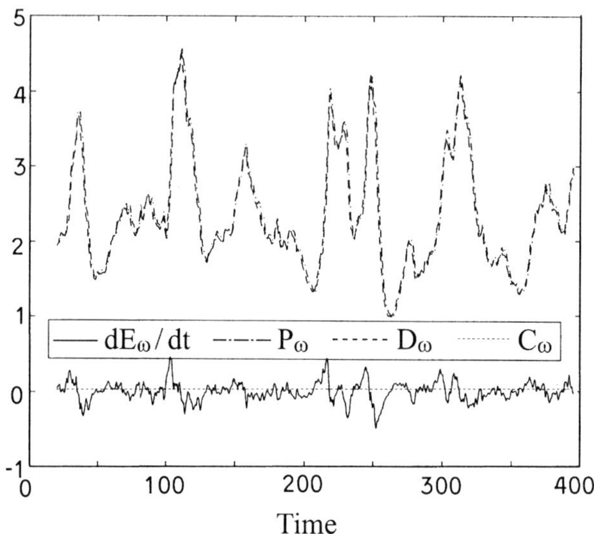

Figure 6.6. Time behaviour of dE_ω/dt, $E_\omega = V^{-1} \int \omega^2 dV$; $P_\omega = V^{-1} \int \omega_i \omega_j s_{ij} dV$; $D_\omega = \frac{\nu}{V} \int \omega_i \nabla^2 \omega_i dV$ and $C_\omega = V^{-1} \int \omega_i (curl f)_i dV$ for the case of ABC forcing at the resolution 64^3, corresponding to $Re_\lambda \approx 60$. Time is measured in turnover times. Note that the graphs for the quantities associated with the strain production, $P_s = -\frac{4}{3} V^{-1} \int s_{ij} s_{jk} s_{ki} d\mathbf{x}$,; $D_s = 2\nu V^{-1} \int s_{ij} \nabla^2 s_{ij} dV$ and $C_s = V^{-1} \int s_{ij} f_{ij} dx$, are precisely the same, since for periodical boundary conditions $P_\omega = P_s$, $D_\omega = D_s$ and $C_\omega = C_s$ (see equation [C.21]).

is the spatial integral of the corresponding forcing term $C_\omega = \int \omega_i (curl f)_i dV$, which is much smaller than all the three integrals just mentioned.

We return to figure 6.1. While there is a strong tendency for alignments between velocity, $\mathbf{u}$, and force, $\mathbf{f}$, the PDF of the cosine of the angle between vorticity and the *curl* of the force, $\cos(\boldsymbol{\omega}, curl\mathbf{f})$ is practically flat. Another feature is that these PDFs for the ABC and the NH forcings are only qualitatively similar.

On the contrary, the alignments associated with the self-amplification process are much closer quantitatively as can be seen from the two examples shown in figure 6.7.

We shall see in section 6.4 on geometrical statistics that alignments of vorticity, $\boldsymbol{\omega}$, and the vortex stretching vector, $\mathbf{W}$, $W_i = \omega_i s_{ij}$, and of vorticity, $\boldsymbol{\omega}$, and the eigenframe, $\boldsymbol{\lambda}_i$ of the rate of strain tensor, s_{ij}, are dynamically very important. This is seen from the simple relation between the enstrophy production and the above quantities $\omega_i \omega_j s_{ij} = \boldsymbol{\omega} \cdot \mathbf{W} = \omega W \cos(\boldsymbol{\omega}, \mathbf{W}) = \omega^2 \Lambda_i \cos(\boldsymbol{\omega}, \boldsymbol{\lambda}_i)$. Here Λ_i, are the eigenvalues of the rate of strain tensor, s_{ij}.

Similar behaviour is observed also for other quantities. Two groups of examples are shown in figures 6.8 and 6.9. The first group contains the

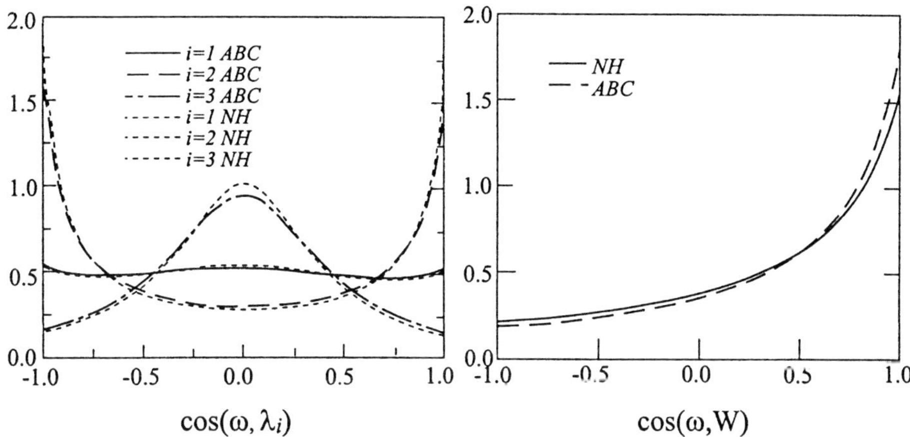

Figure 6.7. PDFs of the cosine of the angle between vorticity, ω, and the vortex stretching vector $W_i \equiv \omega_j s_{ij}$, $\cos(\omega, \mathbf{W})$ (right), and of the cosines of the angles between vorticity, ω, and the eigenframe λ_i of the rate of strain tensor, s_{ij} (left).

PDFs of quantities associated with the forcing. It is seen that they are quite different for the ABC and NH forcings (figure 6.8).

The second group shows the quantities responsible for the self amplification of the velocity derivatives. Here the PDFs for both cases, ABC and NH, are very similar (figure 6.9).

It is seen that the statistical properties as shown in figures 6.7 and 6.9 associated with the self-amplification process tend to be universal in the sense that they are weakly sensitive to the details of the forcing. Many other, such as other relevant alignments and those associated with the terms $\frac{\partial^2 p}{\partial x_i \partial x_j}$ and $s_{ij}\frac{\partial^2 p}{\partial x_i \partial x_j}$, exhibit the same tendency. The terms $\frac{\partial^2 p}{\partial x_i \partial x_j}$ in (C.17) and $s_{ij}\frac{\partial^2 p}{\partial x_i \partial x_j}$ are respectively at least order of magnitude smaller than $s_{ij}s_{jk}$ and $s_{ij}s_{jk}s_{ki}$.

Finally, we are reminded that self-amplification of strain is more 'self' and more local than that of vorticity. This is seen from the equations (C.17) and (C.18), which show that production of strain is associated with the terms $s_{ij}s_{jk}$ and $-s_{ij}s_{jk}s_{ki}$ containing strain only. On the other hand, production of vorticity is, in the first place, *interaction* of vorticity and strain, since it is associated with the terms $\omega_j s_{ij}$ and $\omega_i \omega_j s_{ij}$, as is seen from the equations (C.9) and (C.16). This does not mean that strain production is totally independent of vorticity: there is no strain production without presence of vorticity in the flow. For example, in a homogeneous field, $s_{ij}s_{jk}s_{ki} = -(4/3)\omega_i\omega_j s_{ij}$ (see also equations [C.15, C.17, C.18, C.15']).

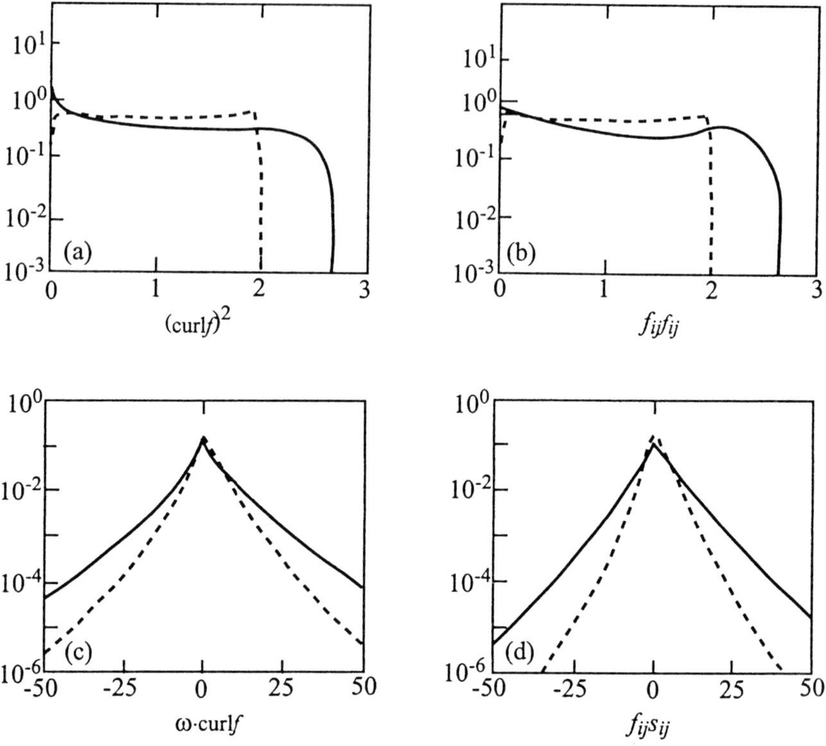

Figure 6.8.   PDFs of a) $(curl f)^2$, B) $f_{ij}f_{ij}$, C) $\omega \cdot curl f$ and d) $s_{ij}f_{ij}$.

6.4. Geometrical statistics

Geometrical invariant quantities (such as enstropy and strain production, helicity, etc., see appendix C) and relations, such as alignments between various vectors, – being independent of the frame of reference - are among the most appropriate for studying physical processes, their possible universal properties, and the characterization of the structure(s) of turbulent flows. Moreover, just like phase relations, these are *the* quantities and relations of utmost dynamical significance. A number of subtle issues related to *quantitative aspects* of structure of turbulence and other questions[8], which are beyond phenomenology, can be effectively addressed via what is denoted in the sequel by the term *geometrical statistics*[9].

[8]Such as active versus passive, weak versus strong, Gaussian versus non-Gaussian, structured versus nonstructured and some others.

[9]In a broader sense geometrical statistics has its beginning in the works of Buffon (1777). Most of the work on the subject have been summarized by Stoyan and Stoyan (1992). A number of random geometric problems suggested by turbulence were reviewed by Corrsin (1972). Some topological aspects (partially related to turbulence) are treated

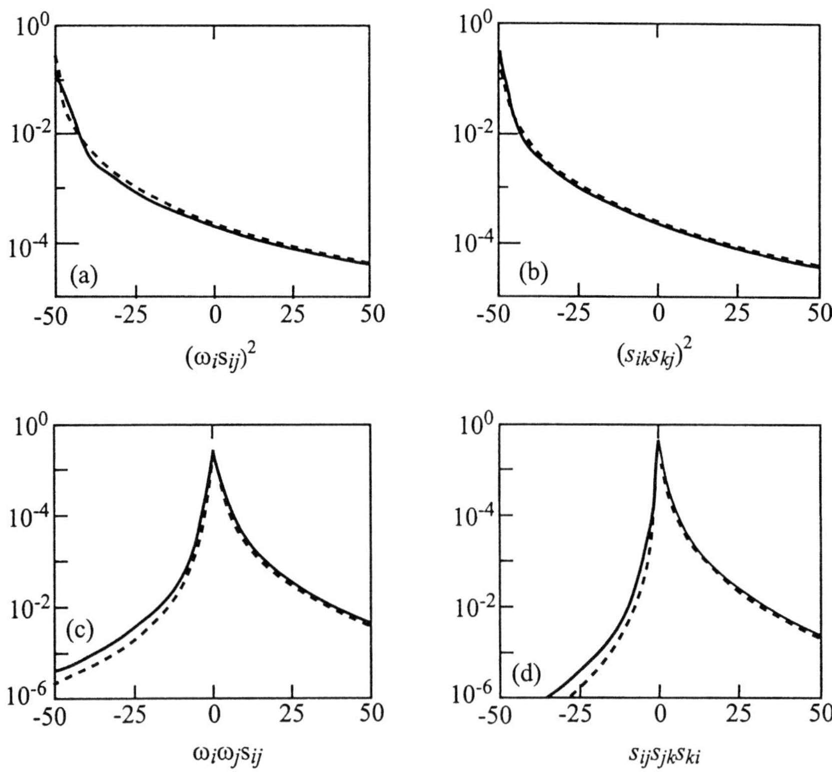

Figure 6.9.   PDFs of a) $(\omega_j s_{ij})^2$; b) $(s_{ik}s_{kj})^2$; c) $\omega_i \omega_j s_{ij}$; and d) $s_{ij}s_{jk}s_{ki}$.

Examples which belong to geometrical statistics were already given in chapter 4 and figures 6.1 and 6.7. Here we bring two more general examples.

It is rather common to use 'surrogates' of the type $(\partial u_1/\partial x_1)^n$ to represent the 'true' quantities such as dissipation, enstrophy $(n = 2)$, enstrophy production $(n = 3)$, etc. However, this is true only of their means, whereas other properties of the surrogates' and of the true quantities are generally different. Their PDFs are essentially different even in the case of a random Gaussian velocity field (Tsinober et al., 1992; and Shtilman et al., 1993). Hence most of their statistical properties are different too, and so are their spectra and fractal properties.

The widely known example of the utmost importance of geometrical relations in turbulence is the *qualitative* difference between the dynamics of 3D and 2D turbulence. In the latter $\omega_j s_{ij}$ and $\omega_i \omega_j s_{ij} \equiv 0$ because in two-dimensional flows $\boldsymbol{\omega} \perp \boldsymbol{\lambda}_i$ $(i = 1, 2)$. One of the essential aspects of

in Moffatt and Tsinober (1990, 1992), Moffatt et al. (1992) and Ricca and Berger (1996). A number of specific aspects of geometrical statistics in turbulence were raised by Constantin (1994) and Tsinober et al. (1995); see references in Tsinober (1998ab).

dynamics of 3D-turbulence – the interaction between vorticity, $\boldsymbol{\omega}$, and the rate of strain tensor, s_{ij} – depends strongly on the geometry of the field of velocity derivatives. For instance, a usual phenomenological argument results in the estimate $\omega_i\omega_j s_{ij} \sim \omega^3$, whereas in reality it is only $\omega_i\omega_j s_{ij} \sim \omega^{7/3}$ in slots of ω, but $\omega_i\omega_j s_{ij} \sim \omega^3$ in slots of s (see below), showing the importance of taking into account the mutual *orientation* of vorticity, $\boldsymbol{\omega}$, and the eigenframe, $\boldsymbol{\lambda}_i$, of the rate of strain tensor s_{ij}. In other words important dynamical aspects of 3D-turbulence contained in the interaction between vorticity, $\boldsymbol{\omega}$, and the rate of strain tensor, s_{ij}, depends strongly not only on the magnitude of vorticity and strain but also on the geometry of the field of velocity derivatives.

6.4.1. ALIGNMENTS

Various alignments comprise an important simple *geometrical* characteristics and manifestation of the *dynamics* and *structure* of turbulence. For example, there is a distinct qualitative difference between the PDFs of $\cos(\boldsymbol{\omega}, \boldsymbol{\lambda}_i)$ for a real turbulent flow and a random Gaussian velocity field. In the last case, *all* these PDFs are precisely flat. An example of special dynamical importance is the strict alignment between vorticity, ω_i, and the vortex stretching vector $W_i \equiv \omega_j s_{ij}$, since the enstrophy production is just their scalar product, $\omega_i\omega_j s_{ij} = \boldsymbol{\omega} \cdot \mathbf{W}$. In real turbulent flows, the PDF of $\cos(\boldsymbol{\omega}, \mathbf{W})$ is strongly asymmetric in full conformity with the prevalence of vortex stretching over vortex compressing, i.e. positiveness of $\langle \omega_i\omega_j s_{ij} \rangle$, whereas it is symmetric for a random Gaussian field (see figures 6.11, 6.12). Thus, the very existence of alignments such as mentioned above points to the presence of internal organization of flow at various scales, i.e. alignments belong to the rare *quantitative statistical* manifestation of the existence of structure in turbulence. They are the simplest representative of a much broader class of geometrical statistics in turbulent flows. It is noteworthy, that while the above mentioned (and some others) alignments are intimately related to the *dynamics* of turbulent flows, there are alignments which are mostly of kinematic nature, e.g. alignment between the Lamb vector $\boldsymbol{\omega} \times \mathbf{u}$ and its potential part (pressure gradient), the alignment between velocity and the eigenvectors of rate of strain tensor and some others (Tsinober, 1996a, 1998a; see also sections 6.7 and 6.8).

Alignments, by their very definition, are suitable for events of *any* magnitude, since they do not contain the amplitude of the quantities involved. Finally, alignments are *invariant* in the sense that they are independent of the system of reference and therefore, along with other invariant quantities, are the most appropriate in studying of physical processes generally and in particular for characterization of the structural nature of turbulent flows.

Due to these properties, using of alignments enables us to answer in a simple and reliable way a number of questions on turbulence structure.

6.4.2. THE GEOMETRY OF VORTEX STRETCHING

In order to address this issue let us recall some simple relations for the key quantities of turbulence dynamics – the vortex stretching vector, $W_i = \omega_j s_{ij}$, and enstrophy production, $\omega_i \omega_j s_{ij}$, and some related quantities (see also appendix C).

$$\omega_i \omega_j s_{ij} = \omega_i^2 \Lambda_i \cos^2(\boldsymbol{\omega}, \boldsymbol{\lambda}_i) = \alpha \omega^2; \quad W^2 = \omega_i^2 \Lambda_i^2 \cos^2(\boldsymbol{\omega}, \boldsymbol{\lambda}_i), \qquad (6.1)$$

Here $\alpha = \Lambda_i \cos(\boldsymbol{\omega}, \boldsymbol{\lambda}_i)$— is the rate of enstrophy production. It is seen from the relations (6.1) that indeed - as mentioned above - that part of dynamics of 3D-turbulence contained in the interaction between vorticity, $\boldsymbol{\omega}$, and the rate of strain tensor, s_{ij}, depends strongly not only on the magnitude of vorticity and strain but also on the geometry of the field of velocity derivatives, in particular on the mutual orientation of vorticity, $\boldsymbol{\omega}$, and the eigenframe, $\boldsymbol{\lambda}_i$, of the rate of strain tensor, s_{ij}. This is true especially regarding the *rate* of enstrophy production, $\alpha = \omega_i \omega_j s_{ij}/\omega^2 = \Lambda_i \cos(\boldsymbol{\omega}, \boldsymbol{\lambda}_i)$, and similar quantity for W^2, $W^2/\omega^2 = \Lambda_i^2 \cos^2(\boldsymbol{\omega}, \boldsymbol{\lambda}_i)$, both of which depend explicitly only on the orientation of vorticity and the shape of the strain tensor, but not on their magnitude.

In view of the importance of the predominant vortex stretching and positive net enstrophy production, i.e. $\langle \omega_i \omega_j s_{ij} \rangle > 0$, it is useful to introduce an angle between $\boldsymbol{\omega}$ and $\mathbf{W}$, since $\omega_i \omega_j s_{ij} \equiv \boldsymbol{\omega} \cdot \mathbf{W}$ (Tsinober et al., 1992). It is easy to see from the simple relation

$$\cos(\boldsymbol{\omega}, \mathbf{W}) = \frac{\Lambda_i \cos^2(\boldsymbol{\omega}, \boldsymbol{\lambda}_i)}{\{\Lambda_i^2 \cos^2(\boldsymbol{\omega}, \boldsymbol{\lambda}_i)\}^{1/2}}. \qquad (6.2)$$

that the alignment between $\boldsymbol{\omega}$ and $\mathbf{W}$ (i.e. positive $\omega_i \omega_j s_{ij}$) is realized in two situations: i - $\boldsymbol{\omega}$ is aligned with $\boldsymbol{\lambda}_1$ ($\Lambda_1 > 0$) and ii - $\boldsymbol{\omega}$ is aligned with $\boldsymbol{\lambda}_2$. Indeed, the contributions both to $\sigma \equiv \omega_i \omega_j s_{ij}$ and α associated with Λ_1 and Λ_2 are positive (see table 6.5). The reason that the contribution to $\langle \omega_i \omega_j s_{ij} \rangle$ associated with Λ_2 is positive is because Λ_2 is positively skewed (see figure 6.10; Ashurst et al., 1987; She et al., 1991; Su and Dahm, 1996; Tsinober et al., 1989, 1992).

At first sight it is surprising that the largest contribution to the enstrophy production and its rate comes from the regions associated with the *largest* eigenvalue, Λ_1, of the rate of strain tensor, s_{ij}, and not from the ones associated with the *intermediate,* eigenvalue Λ_2, since it is known that there exists a strong alignment tendency between $\boldsymbol{\omega}$ and $\boldsymbol{\lambda}_2$, as shown in figure 6.11. This alignment was recognized by Siggia (1981) and discovered

(Kholmyansky et al., 2001).

TABLE 6.5.   Contribution to the total mean of enstrophy production $\langle \omega^2 \Lambda_i \cos^2(\omega, \lambda_i) \rangle$ from the terms corresponding to the eigenvalues Λ_i of the rate of strain tensor s_{ij}. Grid turbulence and DNS, $Re_\lambda = 75$ and field experiment $Re_\lambda = 10^4$.

	Re_λ	$\langle \omega^2 \Lambda_1 \cos^2(\omega, \lambda_1) \rangle$	$\langle \omega^2 \Lambda_2 \cos^2(\omega, \lambda_2) \rangle$	$\langle \omega^2 \Lambda_3 \cos(\omega, \lambda_3) \rangle$
DNS	75	1.06	0.51	− 0.57
Grid	75	1.17	0.39	− 0.56
Field	10^4	1.44	0.47	− 0.97

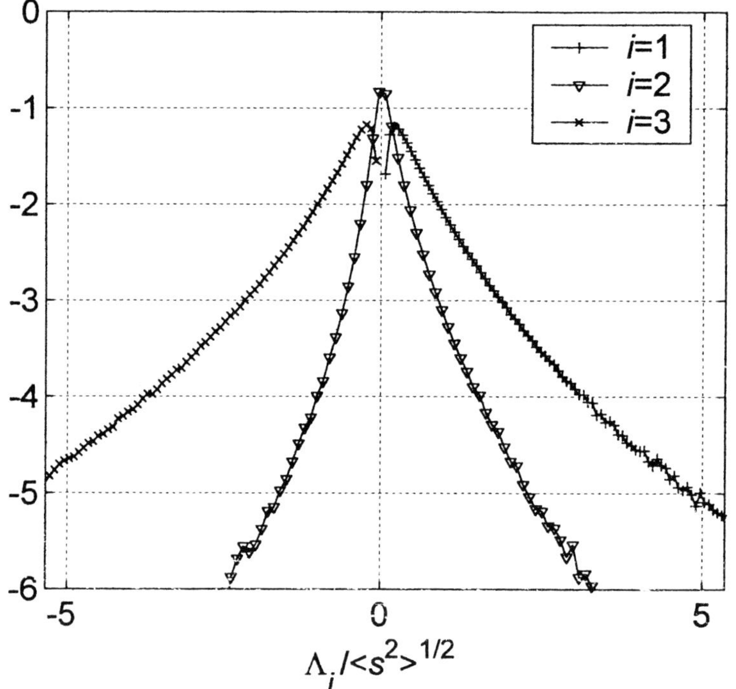

Figure 6.10.   PDFs of the eigenvalues, Λ_i, of the rate of strain tensor, s_{ij}, in the field experiment at $Re\lambda_i = 10^4$. Note the skewed PDF of the intermediate eigenvalue, Λ_2. These PDFs are similar to those in the DNS and grid turbulence at $Re_{\lambda_i} = 75$.

by Ashurst et al. (1987) (for subsequent references see Tsinober, 1998a). This apparent contradiction is resolved by noting that: i – the intermediate eigenvalue, Λ_2, assumes both positive and negative values thus reducing the terms $\omega^2 \Lambda_2 \cos^2(\boldsymbol{\omega}, \boldsymbol{\lambda}_2)$ and $\Lambda_2 \cos^2(\boldsymbol{\omega}, \boldsymbol{\lambda}_2)$, whereas Λ_1 is positive; and ii – the magnitude of Λ_1 is much larger (see table 6.6).

Meanwhile we note that the alignments between $\boldsymbol{\omega}$ and $\boldsymbol{\lambda}_1$, $\boldsymbol{\omega}$ and $\boldsymbol{\lambda}_2$

TABLE 6.6. The ratios between the $\langle\Lambda_i\rangle$ and $\langle\Lambda_i^2\rangle$ obtained in a field experiment in the atmospheric surface layer at the height $10m$, for $Re_\lambda = 10^4$, see Kholmyansky et al. (2001a).

$\langle\Lambda_1\rangle$	$\langle\Lambda_2\rangle$	$\langle\Lambda_3\rangle$
0.47	0.06	-0.53
$\langle\Lambda_1^2\rangle$	$\langle\Lambda_2^2\rangle$	$\langle\Lambda_3^2\rangle$
0.41	0.04	0.55

and between $\boldsymbol{\omega}$ and $\boldsymbol{\lambda}_3$ correspond to regions of turbulent flow that are, in several respects, qualitatively different (see below).

Turbulence background – not stuctureless random sea

Use of alignments allowed to show that - contrary to the common view - the so-called 'background' is strongly non-Gaussian, is dynamically not passive and is not structureless (figures 6.11 – 6.13). Though the strongest tendency for alignment between $\boldsymbol{\omega}$ and $\boldsymbol{\lambda}_2$ is observed for large ω^2, this alignment is still significant (see bottom of figure 6.11) in the 'background' (say $\omega^2 < \langle\omega^2\rangle$), especially taking into account that the background is occupying about 70% of the flow volume (*cf.* with the volume occupied by strong vorticity, say $\omega^2 > 3\langle\omega^2\rangle$, which is only about 6% of the flow volume). Note that this does not contradict the mostly known result about the tendency of alignment between $\boldsymbol{\omega}$ and $\boldsymbol{\lambda}_2$ in regions of concentrated vorticity; the regions with such an alignment are an order of magnitude larger than those with concentrated vorticity only.

Similar results are valid for the normalized enstrophy production $\omega_i\omega_j s_{ij}$ $\omega^{-1}W^{-1} = \cos(\boldsymbol{\omega}, \mathbf{W})$, figure 6.12. Just like in the case of $\cos(\boldsymbol{\omega}, \boldsymbol{\lambda}_2)$ the tendency for alignment between $\boldsymbol{\omega}$ and $\mathbf{W}$ exists *both* in regions of large ω^2 and s^2. However, it is much stronger for large strain s^2, as are all the nonlinearities in these regions (see section 6.5). It has been seen that the maximum of joint PDF of $\cos(\boldsymbol{\omega}, \mathbf{W})$ and ω^2 (and of $\cos(\boldsymbol{\omega}, \mathbf{W})$ and s^2) takes place at $\cos(\boldsymbol{\omega}, \mathbf{W}) \approx 1$ and $\omega^2 \approx 0$, i.e. at the points with *weakest* vorticity and *strongest* alignment between $\boldsymbol{\omega}$ and $\mathbf{W}$.

Note also the strong asymmetry of the PDF of $\cos(\boldsymbol{\omega}, \mathbf{W})$ for the background $\omega^2 < \langle\omega^2\rangle$, which is almost the same as for the whole field. This asymmetry remains significant, even for $\omega^2 < 0.1\langle\omega^2\rangle$, and becomes *stronger* for $\omega^2 < \langle\omega^2\rangle$ and $\cos(\boldsymbol{\omega}, \boldsymbol{\lambda}_2) > 0.9$ (not shown). Moreover, this asymmetry remains significant for *both* small ω^2 and s^2 (see figure 6.13) showing the

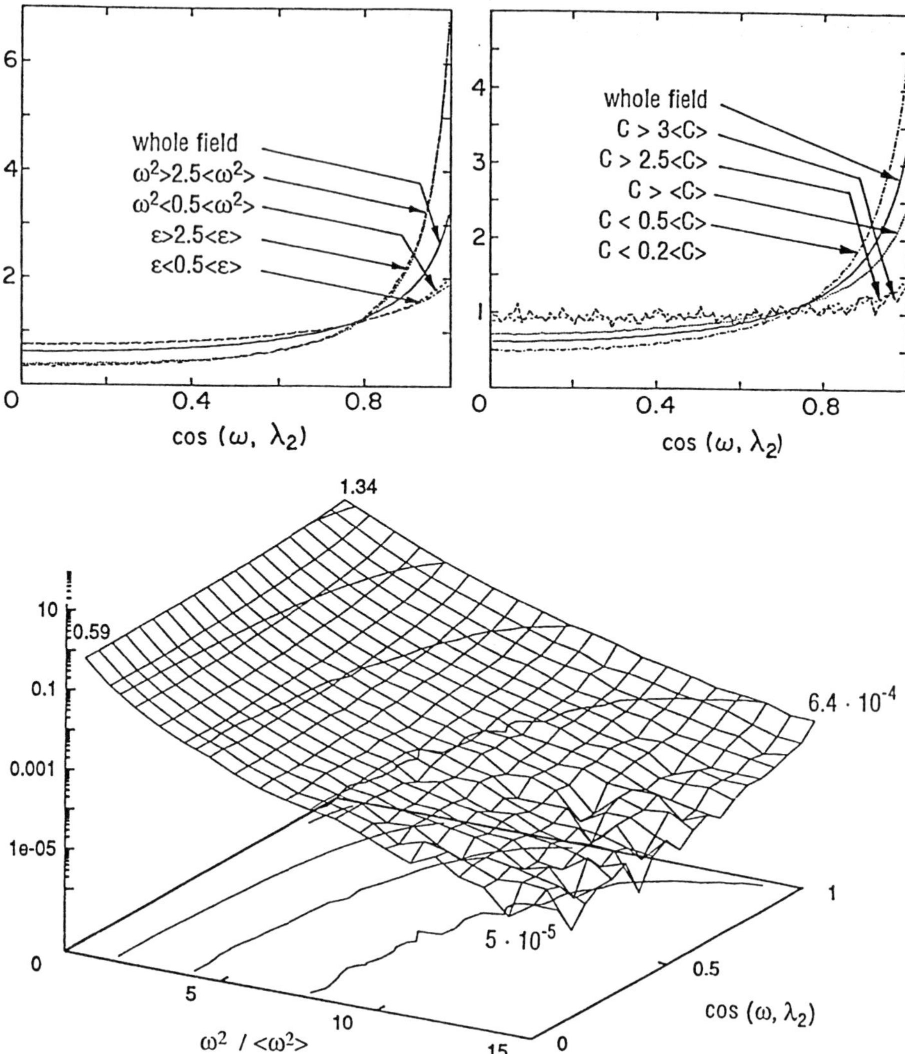

Figure 6.11. PDFs of $\cos(\omega, \lambda_2)$, DNS, $Re_\lambda = 75$. Top left - conditioned on enstrophy ω^2 and s^2, top right - conditioned on curvature C of vortex lines. Note that the tendency for alignment between ω and λ_2 exists both in regions of large ω^2 and large s^2. For a Gaussian velocity field these PDFs are precisely flat. Bottom - joint PDF of $\cos(\omega, \lambda_2)$ and ω^2. The joint PDF of $\cos(\omega, \lambda_2)$ and s^2 is similar to the one shown in this figure. It is seen that the maximum of joint PDF of $\cos(\omega, \lambda_2)$ and ω^2 (and similarly of $\cos(\omega, \lambda_2)$ and s^2) takes place at $\cos(\omega, \lambda_2) \approx 1$ and $\omega^2 \approx 0$, i.e. at the points with *weakest* vorticity and *strongest* alignment between ω and λ_2.

significance of the background. We are reminded that, for Gaussian velocity field, the PDF of $\cos(\omega, \mathbf{W})$ is symmetric.

One can see from figures 6.11 and 6.12 that the maxima of the joint

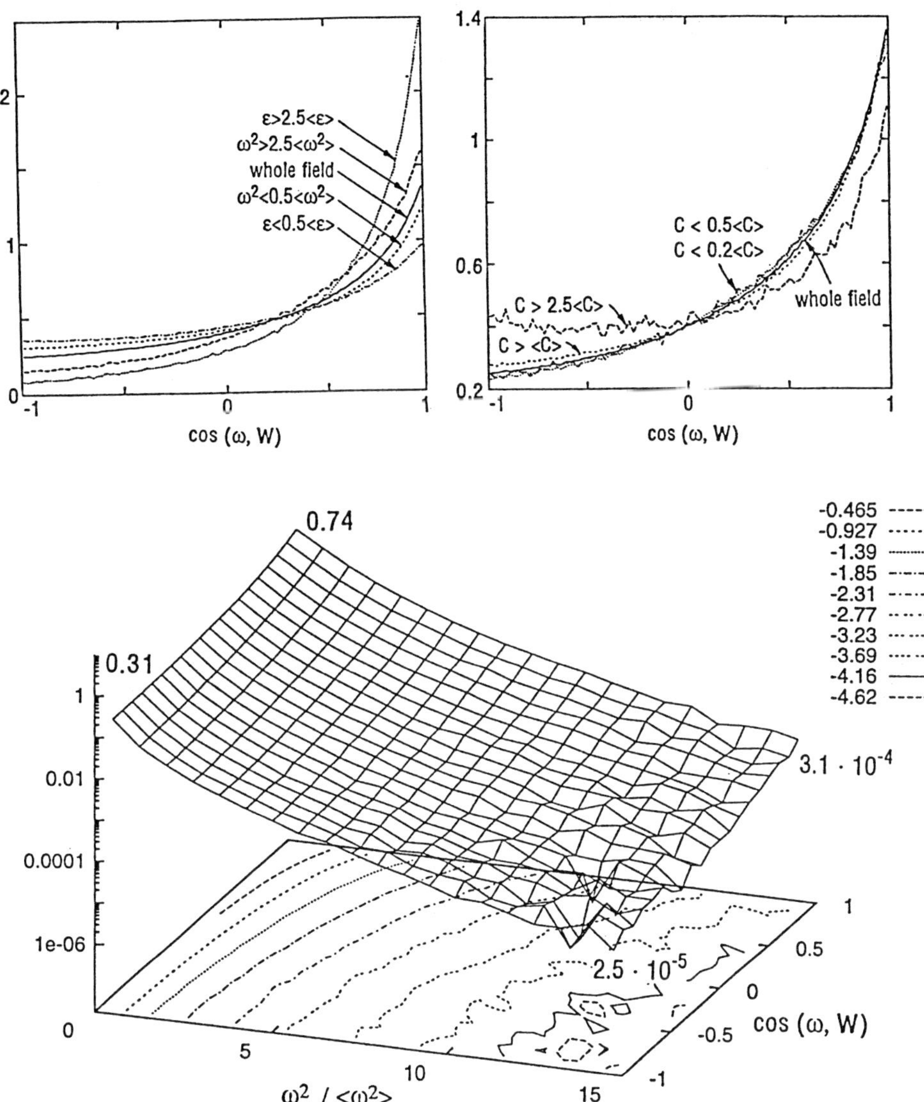

Figure 6.12.   PDFs of $\cos(\omega, W)$, $W_i = \omega_i s_{ij}$; DNS, $Re_\lambda = 75$ Top left - conditioned on enstrophy ω^2 and s^2, top right - conditioned on curvature of vortex lines; bottom - joint PDF of $\cos(\omega, W)$ and ω^2. The joint PDF of $\cos(\omega, W)$ and s^2 is similar to the one shown in this figure.

PDFs of both $\cos(\omega, \lambda_2)$ and $\cos(\omega, W)$ are located at *weakest* enstrophy and *strongest* alignment between ω and λ_2, and ω and W. The same is true for a variety of joint PDFs of other quantities (see references in Tsinober, 1998a).

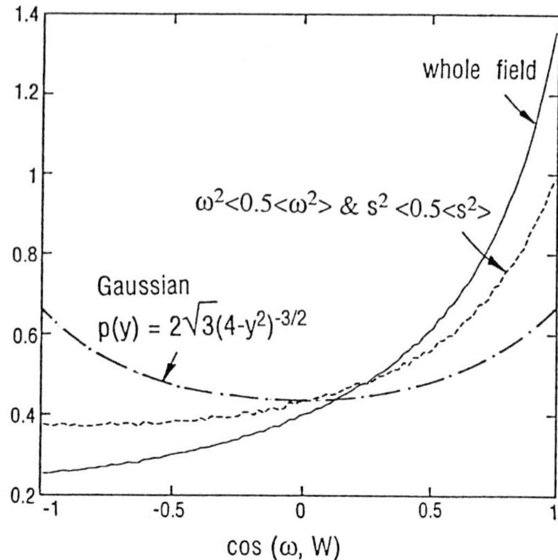

Figure 6.13. PDFs of $\cos(\omega, W)$ for the 'weakest' part of turbulent flow; DNS, $Re_\lambda = 75$.

The above results show clearly that the background is strongly non-Gaussian, not structureless and not passive.

Strained vortical (Burgers-like) objects
Regions with concentrated vorticity constitute a subset of much larger regions in which there is a tendency for alignment between $\boldsymbol{\omega}$ and $\boldsymbol{\lambda}_2$. This is clearly seen from the figure 6.11. Indeed, regions corresponding to $\cos(\boldsymbol{\omega}, \boldsymbol{\lambda}_2) > 0.9$ occupy about 20% of the total flow volume, whereas the set of points with concentrated vorticity, say, $\omega^2 > 3\langle\omega^2\rangle$, is comprised of less than 6% of the total flow volume.

The main feature and shortcoming of these objects (straight strained vortices) is that they possess *one*-dimensional vorticity and therefore zero curvature of vortex lines. Though the relation between vorticity and strain is essentially nonlocal, '*the presence of a strained vortex itself modifies the local strain field*' (Le Dizes et al., 1996) – after all both are composed of derivatives of the same velocity field. However, the special feature of the *straight* strained vortices is that they are impotent in the sense that they do not change that part of the strain by which they are strained themselves: this part of strain is prescribed *a priori*, i.e. it is independent decoupled from their vorticity. These vortices do change only that part of the strain field which is not reacting back on their vorticity. In other words, there is only one way interaction: the vorticity is strained by that part of strain which does not 'know' anything about the vorticity. In this sense such vortices are passive: the essential ingredient of nonlinearity, the

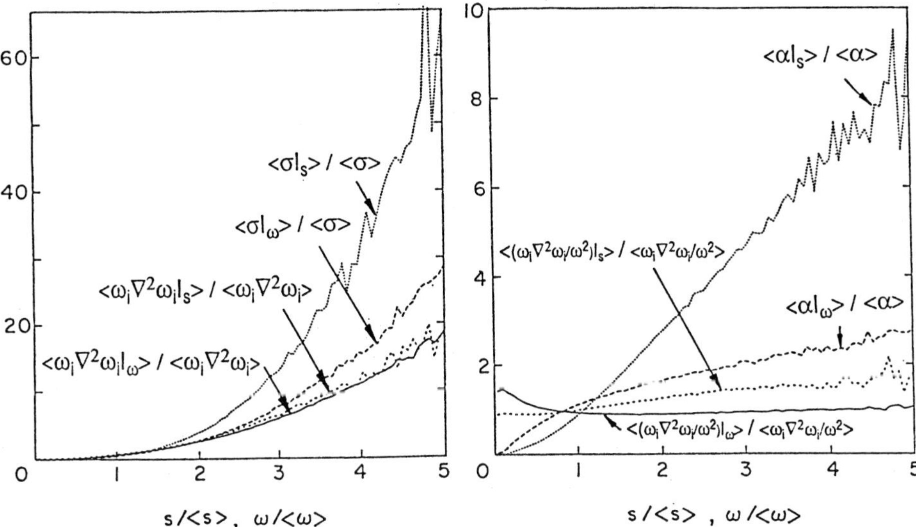

Figure 6.14. Comparison of enstrophy production $\sigma \equiv \omega_i \omega_k s_{ik}$ (left) and its rate $\alpha \equiv \omega_i \omega_k s_{ik}/\omega^2$ (right) with their viscous reduction $\nu \omega_i \nabla^2 \omega_i$ and $\nu \omega_i \nabla^2 \omega_i / \omega^2$ in slots of ω and s. DNS, $Re_\lambda = 75$.

main feature of true genuine nonlinear interaction — the *self-amplification* via *interaction* with strain— is absent in these objects. In this sense, the nonlinearity is reduced in these objects. This property is directly related to zero curvature of vortex lines in straight strained vortices (see figure 6.13, top left, figure 6.18 left and also figure 7.5) — the genuine nonlinearity is present only in regions with nonvanishing curvature. This is what is observed when looking for (apparent) singularities of Euler equations and vortex reconnection (see references in Tsinober, 1998a).

In other words regions with concentrated vorticity with small curvature in real turbulent flows seem to be mostly the result, the consequence rather than dominating factor of the turbulence dynamics. Possessing (almost) maximal enstrophy they are in an approximate equilibrium in the sense that their fairly large, (*but not largest!*, see next section) enstrophy production is approximately balanced by the viscous reduction, and in this sense, they are less active than the strain dominated regions possessing much larger (apparently maximal) enstrophy production, which is considerably larger than its viscous reduction. This is seen from the comparison of the rate enstrophy production $\alpha \equiv \omega_i \omega_k s_{ik}/\omega^2$ and its viscous reduction $\nu \omega_i \nabla^2 \omega_i / \omega^2$ in slots of ω and s as shown in figure 6.14. Indeed, the imbalance between stretching and viscous terms in slots of s is much *larger* than in slots of ω. This difference is especially large at large values of ω and s. This means that the time scale estimated from the imbalance of stretching and viscous terms $\omega^2 \{D_t(\omega^2/2)\}^{-1} \approx \{\omega_i\omega_k s_{ik}/\omega^2 + \nu\omega_i\nabla^2\omega_i/\omega^2\}^{-1}$ in slots of ω is much larger than such time scale in slots of s. In other words, the life time of regions with concentrated vorticity is large compared to that of the regions with large strain, i.e. large rate of energy dissipation. This explains – at least in part – the observability of the regions with concentrated vorticity and the difficulties in observing the regions with large dissipation. It also points to the importance of studying more carefully such regions of turbulent flows as those with strong imbalance between vortex stretching and viscous destruction of vorticity. It is noteworthy that the Burgers-like objects in real turbulent flows possess small but not vanishing curvature, so that the self-amplification of their vorticity is not vanishing, as in the perfectly straight ones used in a great variety of models[10].

Regions of strongest vorticity/strain interaction
As mentioned above, the important point is that at least in quasi-isotropic flows the largest contribution to the enstrophy production $\omega_i\omega_j s_{ij} = \omega_i^2\Lambda_i\cos^2(\boldsymbol{\omega},\boldsymbol{\lambda}_i)$ comes from the regions associated with the *largest* eigenvalue Λ_1 of the rate of strain tensor s_{ij} and not from the ones associated with the *intermediate* eigenvalue Λ_2 to which mainly belong the regions of concentrated vorticity. Namely the ratio of $\langle\omega^2\Lambda_1\cos^2(\boldsymbol{\omega},\boldsymbol{\lambda}_1)\rangle$ to $\langle\omega^2\Lambda_2\cos^2(\boldsymbol{\omega},\boldsymbol{\lambda}_2)\rangle$ is roughly $2:1$ or even more as in the field experiment mentioned several times before (see table 6.5). The same is true of other nonlinearities (see section 6.5).

This shows that there exist regions (intense and weak — both structured and dynamically active) other than concentrated vorticity regions, which at least in the above sense are dynamically more important. These regions are associated mainly with largest strain rather than enstrophy, strong tendency of alignment between $\boldsymbol{\omega}$ and $\boldsymbol{\lambda}_1$ (see table 6.5 and figure 6.16) and fairly large curvature of vorticity lines. These regions are characterized by the largest, apparently maximal, enstrophy production and its rate (as shown in figure 6.14), which are much larger than their viscous reduction as discussed above. This is consistent with the PDFs of α conditioned on ω and s (see figure 6.15) and with the results of Constantin et al. (1996) that the dominating contribution to $\alpha = \omega_i\omega_j s_{ij}/\omega^2$ comes from the local (self) interaction of vorticity $\boldsymbol{\omega}$ and strain s_{ij}, which is absent in Burgers-like objects. The behaviour of W^2 and W^2/ω^2 in slots of ω and s is essentially the same (see figure 6.19 below). These results also show that

---

[10]See, for example, the review by Pullin and Saffman (1998) and the papers by Hosokawa (2000) and Kambe and Hatakeyama (2000). A number of arguments and facts were given in Tsinober (1998a) as to why regions of concentrated vorticity in turbulent flows are not as important as previously thought. Here we mention in addition some latest references supporting various aspects of this view: Chavanis and Sire (2000), Dernoncourt et al. (1998), Min et al. (1996), Roux et al. (1998), and Sain et al. (1998).

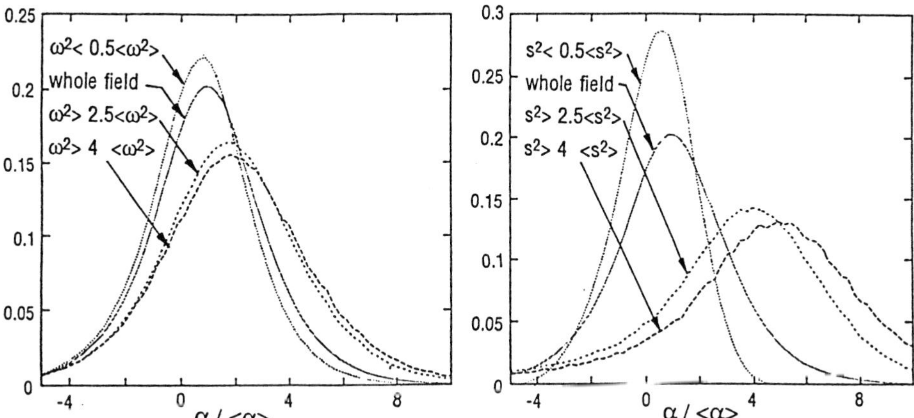

Figure 6.15.   PDF's of the enstrophy production rate α for the whole field and conditioned on ω^2 (left) and s^2 (right). DNS, $\Re_\lambda = 75$. Considerable regions with vortex compression (i.e. $\alpha < 0$) exist also for large enstrophy (see also Jimenez and Wray, 1998)), whereas in regions with large strain the rate of enstrophy generation α is mostly positive.

there is much vortex compression in regions with concentrated vorticity (see also Jimenez and Wray, 1998).

Similarly the dependence of enstrophy production $\sigma \equiv \omega_i \omega_j s_{ij}$ and its rate $\alpha \equiv \Lambda_i \cos^2(\boldsymbol{\omega}, \boldsymbol{\lambda}_i)$ on ω and on $s \equiv (s_{ij} s_{ij})^{1/2}$ is qualitatively different for small and large curvature of vortex lines in such a way that the nonlinearity is manifested stronger in regions of large curvature. In particular, the disparity in the behaviour of σ and α in slots of ω and s becomes larger at small curvature, whereas at large curvature the dependence of σ and α on ω and s is very similar. This last fact is a reflection of stronger interaction of vorticity and strain in regions with *large* curvature and *positive* α and, consequently, with non-negligible vortex folding and tilting (see the next section).

The regions just discussed comprise a *subset* of larger regions dominated by strain. Namely, these are the regions with large vortex lines curvature. There exist at least two other kinds of strain dominated regions: those with small curvature of vortex lines, which wrap around the vorticity dominated regions (tubes/worms), and which contribute mostly to the alignment of $\boldsymbol{\omega}$ and $\boldsymbol{\lambda}_2$ as shown in figure 6.11. There are also regions with large magnitude of Λ_3 and large negative α, in which most of vortex *compressing, tilting and folding* occur.

Vortex compression, tilting, folding and curvature
The most basic phenomenon in turbulence – the predominant vortex stretching, i.e. predominant enstrophy production, $\sigma \equiv \omega_i \omega_j s_{ij}$, so that $\langle \omega_i \omega_j s_{ij} \rangle >$

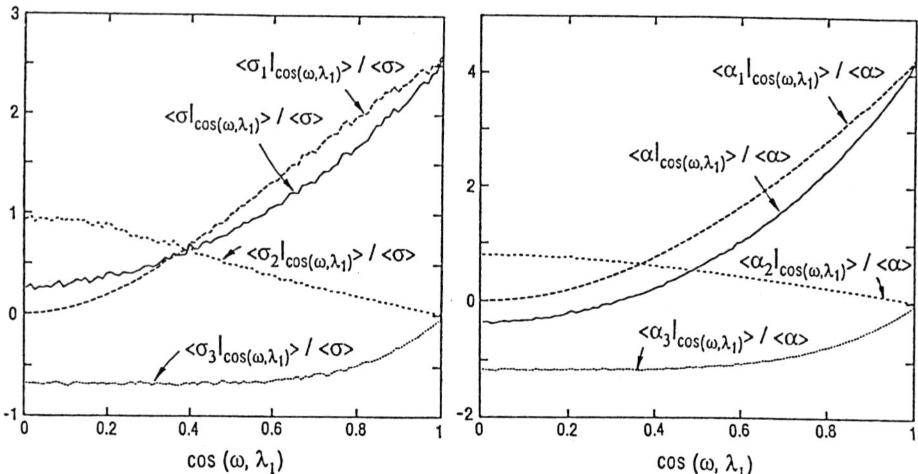

Figure 6.16. Conditional averages of enstrophy production σ (left) and its rate α (right) in slots of $\cos(\omega, \lambda_1)$. DNS, $Re_\lambda = 75$.

0 – cannot occur in a finite volume and finite energy without its concomittants – vortex compressing ($\sigma < 0$) and folding (Chorin, 1982, 1994)[11]. Hence, the importance of looking at properties of turbulent flow in regions with large curvature and $\sigma < 0$, which typically occupy about 1/3 of the whole flow volume, and for the evidence and characterization of the vortex folding in three-dimensional turbulence. These regions play an important role in the dynamics of turbulence. For example, these regions make a positive contribition to the magnitude of the vortex stretching vector $W_i \equiv \omega_j s_{ij}$ in (C.9). Indeed, $W^2 = \omega^2 \Lambda_i^2 \cos^2(\boldsymbol{\omega}, \boldsymbol{\lambda}_i)$ and $W^2/\omega^2 \equiv \Lambda_i^2 \cos^2(\boldsymbol{\omega}, \boldsymbol{\lambda}_i)$ are large for large $\Lambda_3^2 \cos^2(\boldsymbol{\omega}, \boldsymbol{\lambda}_3)$, for which the enstrophy production $\sigma \equiv \omega_i \omega_j s_{ij} = \omega^2 \Lambda_i \cos^2(\boldsymbol{\omega}, \boldsymbol{\lambda}_i)$ and its rate $\alpha = \sigma \omega^{-2} = \Lambda_i \cos^2(\boldsymbol{\omega}, \boldsymbol{\lambda}_i)$ are *negative* (see table 6.7).

TABLE 6.7. Contribution to the total mean of the magnitude of vortex stretching vector $\langle W^2 \rangle \equiv \langle \omega^2 \Lambda_i^2 \cos^2(\omega, \lambda_i) \rangle$ from the terms corresponding to the eigenvalues Λ_i of the rate of strain tensor s_{ij}. DNS, $Re_\lambda = 75$. Field experiment $Re_\lambda = 10^4$.

	Re_λ	$\langle \omega^2 \Lambda_1^2 \cos^2(\omega, \lambda_1) \rangle$	$\langle \omega^2 \Lambda_2^2 \cos^2(\omega, \lambda_2) \rangle$	$\langle \omega^2 \Lambda_3^2 \cos^2(\omega, \lambda_3) \rangle$
DNS	75	0.53	0.15	0.32
Field	10^4	0.52	0.12	0.36

Similarly, enstrophy production (and α) can be small, whereas W^2 (and

---

[11]The term folding was introduced by Reynolds in 1894 in the context of folding of material lines.

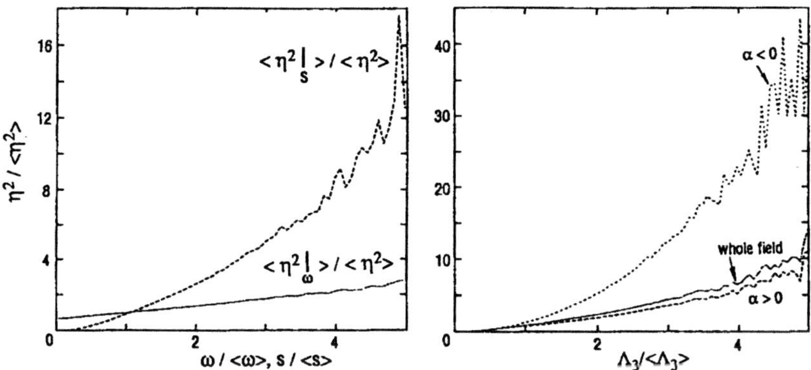

Figure 6.17. Conditional averages of the magnitude of the rate of change of vorticity direction $\eta^2 = W^2/\omega^2 - \alpha^2 = \Lambda_i^2 \cos^2(\omega, \lambda_i) - \{\Lambda_i \cos^2(\omega, \lambda_i)\}^2$. Left - in slots of ω and s, from which it is seen that the direction of vorticity is changing much stronger in strain dominated regions. Right - in slots of Λ_3, showing that this rate of change is (apparently) largest in (sub)regions of vortex compression with large magnitude of Λ_3. Similar increase of η^2 is observed in slots of Λ_1 and Λ_2 too, but at slower rates (not shown). Note that it is not so simple to separate the contributions to η^2 associated with the eigenvalues Λ_i.

$W^2/\omega^2)$ can be large.

A closely related process is the vortex tilting, which is characterized by the rate of change of direction of vorticty. This rate is obtained from the equations (6.3, 6.4) for the magnitude of vorticity ω and its unit vector $\varpi_i = \omega_i/\omega$, which are equivalent to the equations (C.9, C.16) without the forcing terms

$$D_t\omega = \alpha\omega + vt, \quad D_t\varpi_i = s_{ij}\varpi_j - \alpha\varpi_i + vt, \qquad (6.3, 6.4)$$

where and vt stands for viscous terms.

The vector $\eta_i = s_{ij}\varpi_j - \alpha\varpi_i = W_i/\omega - \alpha\omega_i/\omega$ is the inviscid rate of change of the unit vector ϖ along the direction of vorticity ω, and is responsible for the rate of change of its *direction*, Constantin (1994), and $\eta \perp \omega$, i.e. vector η is associated with vorticity *tilting*. Its magnitude is $\eta^2 = W^2/\omega^2 - \alpha^2 = \Lambda_i^2 \cos^2(\omega, \lambda_i) - \{\Lambda_i \cos^2(\omega, \lambda_i)\}^2$. From this it is seen that in regions with negative enstrophy production the rate of change η of the unit vector ϖ can be large, since, as mentioned, these regions make a positive contribution to the magnitude of the vortex stretching vector $W_i \equiv \omega_j s_{ij}$, so that $W^2/\omega^2 = \Lambda_i^2 \cos^2(\omega, \lambda_i)$ can be large and $\alpha^2 = \{\Lambda_i \cos^2(\omega, \lambda_i)\}^2$ can be small. This happens in regions associated with large magnitudes of Λ_3 as is seen from figure 6.17.

It is reasonable to associate the above process with large curvature of vortex lines and similar quantities, which should reflect their folding and tilting - at least the resulting aspect of these processes. Hence among the

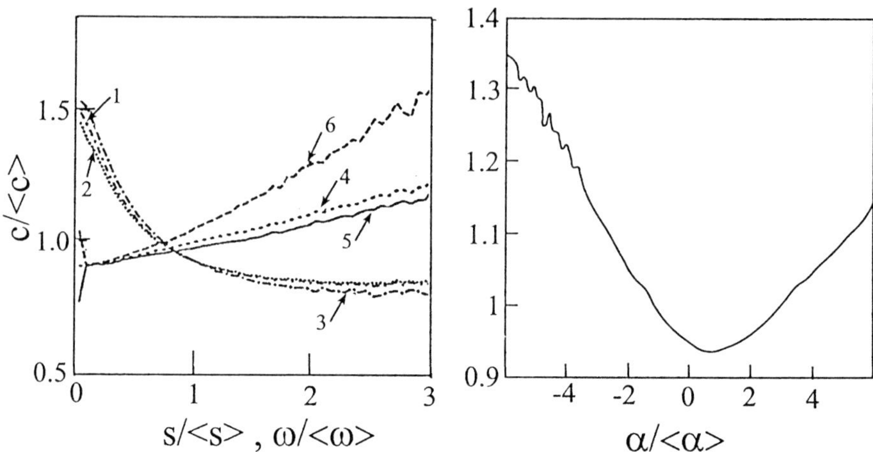

Figure 6.18. Conditional averages of curvature C of vortex lines, DNS, $Re_\lambda = 75$. Left - in slots of ω and s for the whole field and for positive and negative rate of enstrophy generation α: 1 – $\langle c|_\omega\rangle/\langle c\rangle$, 2 – $\langle c|_\omega\rangle/\langle c\rangle$ and $\alpha > 0$, 3 – $\langle c|_\omega\rangle/\langle c\rangle$ and $\alpha < 0$, 4 – $\langle c|_s\rangle/\langle c\rangle$, 5 – $\langle c|_s\rangle/\langle c\rangle$ and $\alpha > 0$, 6 – $\langle c|_\omega\rangle/\langle c\rangle$ and $\alpha < 0$. Note the *qualitatively* different behaviour of curvature in slots of ω (decreasing) and in slots of s (increasing). Right - in slots of rate of enstrophy generation α. $Re_\lambda = 75$. Note that the curvature of vortex lines *increases* with $|\alpha|$ *both* for negative and positive α.

questions of interest are those about the properties of curvature and the relation between curvature and dynamically relevant quantities such as enstrophy ω^2, enstrophy production σ, rate of enstrophy generation $\alpha \equiv \sigma/\omega^2$ and relations such as various alignments. Of course, the ultimate clarification of such relations can be obtained from looking at global properties. One can hope that some insights can be gained from local analysis, i.e. from working with point quantities at a particular time moment.

In a simplified form, the logic is that strong stretching results in strong vorticity: indeed regions with strong vorticity are known to be tube-like with a small curvature, as observed visually in a number of numerical simulations. However, closer inspection shows that the matters are much more complicated (figure 6.18) due to a number qualitative differences between material and vortex lines (see the box at page 86). One can see that, indeed, the curvature decreases in slots of ω. However, this behaviour is practically the same for the whole field, for positive and for *negative* rate of enstrophy production α (the reader is reminded again that typically regions with $\alpha > 0$ occupy about 2/3 of the turbulent flow field, and regions with $\alpha < 0$ comprise about a 1/3 of the whole flow volume). This last fact, i.e. the behaviour of curvature C versus ω for *negative* rate of enstrophy production ($\alpha < 0$) and strong increase of curvature with strain[12] (see figure

---

[12]Again for the whole field and both for $\alpha > 0$ and for $\alpha < 0$.

6.18) undermines the simple analogy with the behaviour of material lines in turbulent flows. Similarly, as is expected the curvature of vortex lines is *increasing* with $|\alpha|$ for $\alpha < 0$ due to folding of vortex lines, but again, most interestingly the same behaviour of C is observed for $\alpha > 0$ due to self-induction unlike the case of material lines. This is consistent with the results on the comparison of dependence of enstrophy generation $\omega_i \omega_k s_{ik}$ and its viscous reduction $\nu \omega_i \nabla^2 \omega_i$ on ω and s (figure 6.14) and curvature C. Namely, the preferential alignment between ω and λ_2 is correlated with small curvature and there is no preferential alignment between ω and λ_2 at large curvature (figure 6.11, top right).

The above shows that the 'most nonlinear' are the regions with large curvature, dissipation, i.e. strain, and preferable alignment between ω and λ_1, and not the regions of concentrated vorticity with small curvature and preferable alignment between ω and λ_2, such as the filaments observed in direct numerical simulations of the Navier-Stokes equations and laboratory experiments[13]. This brings us to next issue.

6.5. Depression of nonlinearity

The notion known as *depression of nonlinearity* was introduced by Kraichnan and Panda (1988). Since this paper, several aspects of this problem have been addressed (see references in Tsinober et al., 1999).

Kraichnan and Panda suggested comparing the nonlinearities in real turbulent flows with their Gaussian counterparts. This is meaningful for even moments only, for example, $\langle |\mathbf{u} \times \boldsymbol{\omega}|^2 \rangle / \langle |\mathbf{u} \times \boldsymbol{\omega}|^2 \rangle_G < 1 \ (\sim 0.8)$;
$\langle |\mathbf{u} \times \boldsymbol{\omega} - \boldsymbol{\nabla}(p + \frac{1}{2}u^2)| \rangle / \langle |\mathbf{u} \times \boldsymbol{\omega} - \boldsymbol{\nabla}(p + \frac{1}{2}u^2)| \rangle_G < 1 \ (\sim 0.5 \div 0.6)$;
$\langle W^2 \rangle / \langle W^2 \rangle_G < 1 \ (\sim 0.7 \div 0.8)$; $W_i \equiv \omega_j s_{ij}$. In this sense nonlinearity is reduced. However, as measured by odd moments the real nonlinearity is 'infinitely' larger, since for a Gaussian velocity field the odd moments vanish identically. This includes the longitudinal velocity structure functions of odd order $S_{2n+1}(r) = \langle \{ [\mathbf{u}(\mathbf{x} + \mathbf{r}) - \mathbf{u}(\mathbf{x})] \cdot \mathbf{r}/r \}^{2n+1} \rangle$, enstrophy generation $\langle \omega_i \omega_k s_{ik} \rangle$ and its rate $\langle \omega_i \omega_k s_{ik}/\omega^2 \rangle$, $\langle s_{ij} s_{jk} s_{ki} \rangle$ and many others. In other words, the build up of *odd* moments is an important *specific* manifestation of the nonlinearity of turbulence, and is, as well, a manifestation of its structure – there is no turbulence without odd moments; the nonzero $\langle \omega_i \omega_k s_{ik} \rangle$ is associated with the strict alignment between ω and $\mathbf{W}$. In this sense,m this alignment is enhancing the nonlinearity. We have seen from figures 6.12, 6.13 that this alignment is significant throughout *all* the regions of turbulent flow. On the other hand, the alignment between $\mathbf{u}$

---

[13]It is noteworthy that regions of concentrated vorticity are not free of vortex compression in the same proportion as in the whole turbulent field (Jimenez and Wray, 1998; and figures 6.14, 6.18.)

and $rot\ \boldsymbol{\omega}$ is reducing $\langle\omega_i\omega_k s_{ik}\rangle$. Indeed, since $\langle\omega_i\omega_k s_{ik}\rangle = \langle\boldsymbol{\omega}{\cdot}rot(\mathbf{u}\times\omega)\rangle$ $= \langle rot\boldsymbol{\omega}\cdot(\mathbf{u}\times\omega)\rangle = -\langle\boldsymbol{\omega}\cdot(\mathbf{u}{\times}rot\boldsymbol{\omega})\rangle$, and since $-\langle\mathbf{u}{\cdot}rot\ \boldsymbol{\omega}\rangle \equiv \langle\omega\rangle^2 > 0$ there is a tendency of (anti-)alignment between $\mathbf{u}$ and $rot\ \boldsymbol{\omega}$ reducing the magnitude of $\mathbf{u}{\times}rot\ \boldsymbol{\omega}$ and thereby of $\langle\omega_i\omega_k s_{ik}\rangle$. Though this is a purely kinematic effect it is directly related to the dissipative and rotational nature of turbulent flows, since the mean dissipation $\langle\epsilon\rangle \simeq \nu\langle\omega\rangle^2$.

One more aspect of reduction of nonlinearity is related to the tendency of alignment between $\mathbf{u}$ and $\boldsymbol{\omega}$, and the so-called beltramization (see Moffatt and Tsinober, 1992; Tsinober, 1998a; and references therein). The alignment between $\mathbf{u}$ and $\boldsymbol{\omega}$ implies reduction of the magnitude of the Lamb vector, $\boldsymbol{\omega}\times\mathbf{u}$, (e.g. $\langle|\boldsymbol{\omega}\times\mathbf{u}|^2\rangle$) and corresponding increase of a quantity called helicity density $h = \mathbf{u}\cdot\boldsymbol{\omega}$ (e.g. $\langle|\mathbf{u}\cdot\boldsymbol{\omega}|^2\rangle$). This is a very difficult and controversial issue for many reasons. First, $\mathbf{u}$ and $\boldsymbol{\omega}$ are weakly correlated by their very nature - $\mathbf{u}$ is a large scale quantity, and $\boldsymbol{\omega}$ is a small scale quantity. Second, the Lamb vector is neither potential nor solenoidal, and it has a large potential part (see section 6.6). The latter can be considered as a kind of reduction of nonlinearity, since only the solenoidal part of $\boldsymbol{\omega}\times\mathbf{u}$ matters in the dynamics of vorticity.

6.5.1. RELATIVE DEPRESSION OF NONLINEARITY IN REGIONS WITH CONCENTRATED VORTICITY

We are interested here in the behaviour of the key nonlinearities related to velocity gradients in flow regions dominated by enstrophy and strain. Typical examples are the magnitude of the vortex stretching vector $W \equiv |\omega_j s_{ij}|$, the enstrophy generation $\omega_i\omega_j s_{ij}$ and its rate $\omega_i\omega_j s_{ij}/\omega^2$, the production of dissipation, i.e. the inviscid terms in the equation (C. 18), the inviscid terms in the equations (C.23, C.24) and all physically meaningful nonlinearities involving velocity derivatives. The central result is that all of them - though increasing with ω (e.g. $\omega_i\omega_j s_{ij}$ increases as $\omega^{7/3}$, i.e. faster than ω^2, but slower than ω^3), are essentially reduced in regions dominated by enstrophy as compared to the strain dominated regions. Another manifestation of depression of nonlinearity is the decrease in the curvature of the vortex lines in the regions with concentrated vorticity, and enhanced rate of change of vorticity direction in the strain dominated regions (see figures 6.17, 6.18).

Indeed, we have seen that enstrophy generation and its rate are much larger in strain dominated regions (than that in enstrophy dominated ones) with finite curvature of vortex lines. Both are associated with the largest eigenvalue Λ_1 of the rate of strain tensor and alignment between $\boldsymbol{\omega}$ and $\boldsymbol{\lambda}_1$. The main contribution to vortex stretching in these regions comes from local effects associated with the (self) interaction of $\boldsymbol{\omega}$ and s_{ij} (Constantin et al., 1996) in contrast with the enstrophy dominated regions in which the vortex

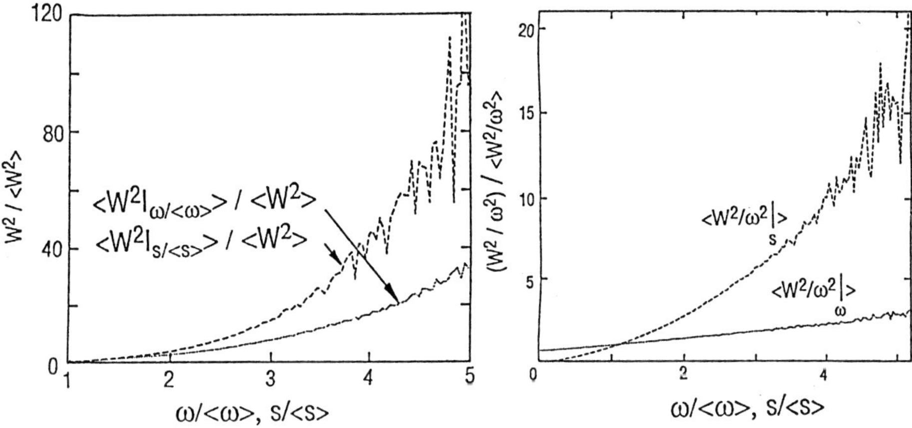

Figure 6.19. Conditional averages of squared magnitude of the vortex stretching vector W^2 (left) and its rate W^2/ω^2 (right) in slots of ω and s. DNS, $Re_\lambda = 75$.

stretching is sustained mostly by nonlocal effects. Similarly, other nonlinear dynamically relevant quantities, e.g. magnitude of the vortex stretching vector W^2 and its rate W^2/ω^2 as shown in figure 6.19, and the quantity η^2 responsible for the vorticity tilting (see figure 6.17 left) are also strongly reduced in regions of concentrated vorticity, as compared to their values in strain dominated regions.[14]

Another aspect of reduction of nonlinearity is seen clearly from figure 6.20. Namely, the nonlinearity associated with the production of $\omega_i\omega_j s_{ij}$ (see equation [C.23]) is strongly reduced in the enstrophy dominated regions (filaments/worms), whereas it is enhanced in strain dominated regions. Note the *qualitatively* different behaviour of $\omega_i\omega_j \frac{\partial^2 p}{\partial x_i \partial x_j}$ (figure 6.20 top) in enstrophy dominated regions (decreasing with ω) as compared to that in strain dominated regions (increasing with s). More results of this kind that are also related to quantities like $s_{ij}s_{jk}s_{ki}$ are presented in Tsinober et al. (1999).

6.5.2. ARE REGIONS OF CONCENTRATED VORTICITY QUASI-ONE-DIMENSIONAL?

One of the reasons for the reduced of

nonlienarity in the regions of concentrated vorticity is that the long, thin tubes-filaments-worms are believed to be in some sense locally quasi-

---

[14]It should be emphasized that, though the above results are likely to be true at large Reynolds numbers, they cannot be seen as an indication that NSE may not develop a singularity in finite time, since these results reflect statistical tendencies. For example, there exist small regions with very large enstrophy, enstrophy generation and alignment between ω and λ_1.

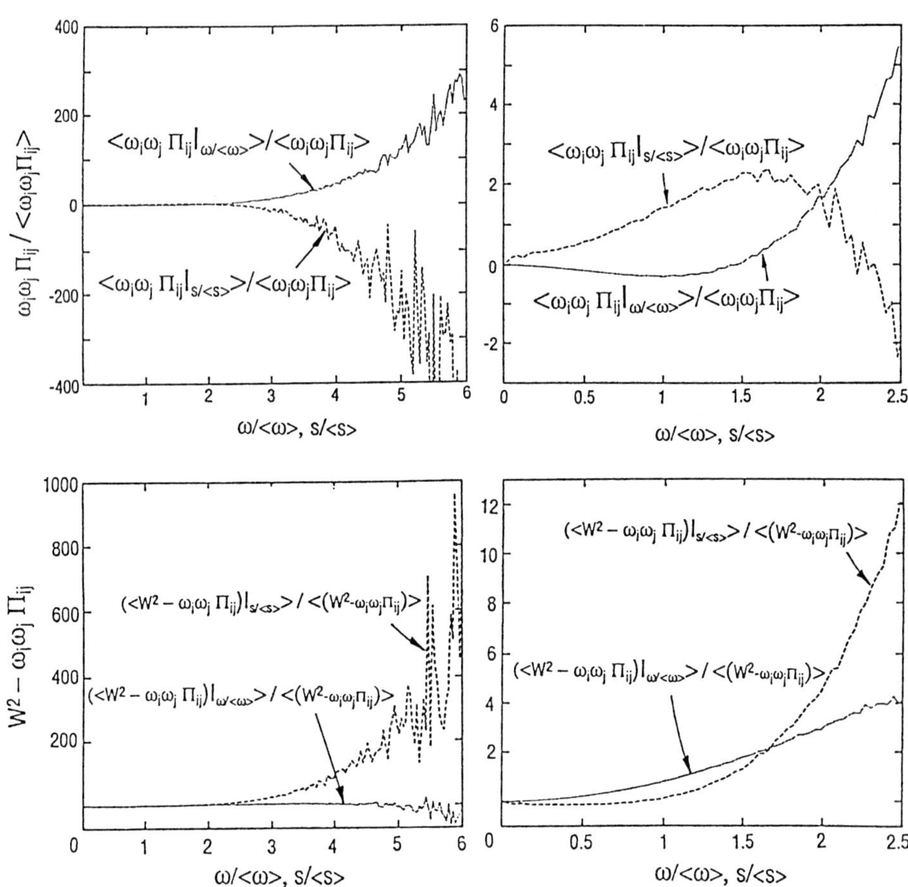

Figure 6.20. Conditional averages of the nonlocal term $\omega_i\omega_j\frac{\partial^2 p}{\partial x_i \partial x_j}$ (top) and of the inviscid rate of change of enstrophy generation $W^2 - \omega_i\omega_j\frac{\partial^2 p}{\partial x_i \partial x_j}$ (bottom) in slots of ω and s, DNS. $Re_\lambda \approx 75$. (see equation [C.23]).

one-dimensional (Frisch, 1995)[15], i.e. that nonlinearity is stronger *outside* of these structures. Hence the term depletion (expulsion) of nonlinearity. Following this line one would expect that in regions with strong alignment between vorticity ω and the intermediate eigenvector λ_2 vortex stretching and enstrophy generation should decrease as $|\cos(\omega, \lambda_2)|$ increases. Indeed W and its rate are decreasing but remain essentially finite. However, contrary to the above expectation the enstrophy generation and its rate *increase* in slots of $|\cos(\omega, \lambda_2)|$ and become *maximal* at $|\cos(\omega, \lambda_2)| \sim 1$

---

[15]Though observing the figure 13 in Jimenez et. al (1993), it is easily seen that the regions of concentrated vorticity are far from being quasi-one-dimensional objects. It is noteworthy that these authors arrived at the conclusion that *self-stretching is not important in their* (i.e. 'worms') *evolution.*

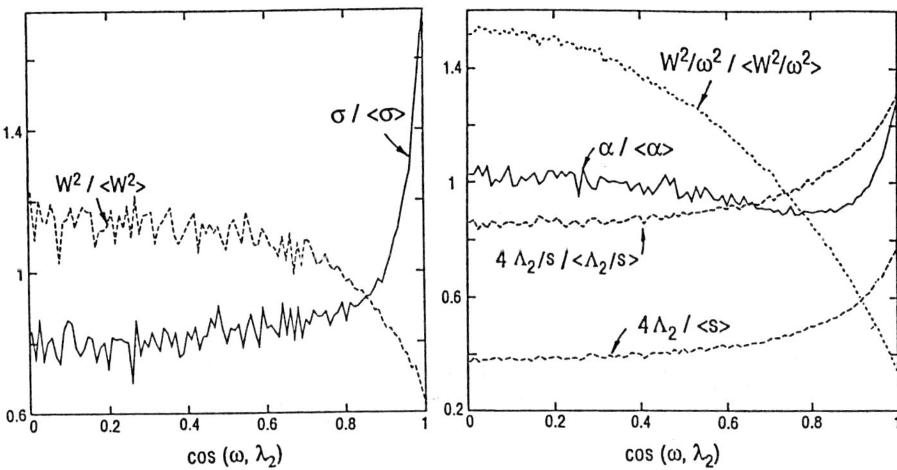

Figure 6.21. Conditional averages of: left - enstrophy generation $\sigma \equiv \omega_i \omega_j s_{ij}$, vortex stretching W^2, right - rates of enstrophy generation α, vortex stretching W^2/ω^2, intermediate eigenvalue of the rate of strain tensor Λ_2 and the ratio Λ_2/s in slots of $\cos(\omega, \lambda_2)$. DNS, $Re_\lambda = 75$.

(figure 6.21). In other words in these regions the rate of creation of enstrophy ω^2 is the largest and in this sense the nonlinearity is *stronger* and not weaker than in, at least, some of their background using $|\cos(\omega, \lambda_2)|$ as a criterion. The above tendencies are stronger in regions with strong vorticity and survive in the background, e.g. regions of weak enstrophy.

Note that none of the quantities $\omega_i \omega_j s_{ij}$, W, Λ_2 and Λ_2/s become small for $|\cos(\omega, \lambda_2)| \sim 1$ indicating that *the flow does not become locally two-dimensional*. In particular, it is important that in these regions the intermediate strain (i.e. Λ_2) is positive and is increasing along with $|\cos(\omega, \lambda_2)|$, which corresponds to strong straining in these regions (cf. with pure two-dimensional flow in which $\Lambda_2 \equiv 0$). Thus one can speculate that there is a tendency to 'localization of nonlinearity' in space which, somewhat paradoxically, is sustained by nonlocal effects due to the nonlocal relation between strain and vorticity and due to pressure ('nonlocal localization'; see next section on nonlocality of turbulence). Note, that the claim to 'localization of nonlinearity' is supported by the behaviour of Λ_2/s in slots of $|\cos(\omega, \lambda_2)|$, which is similar to the one of $\Lambda_2/\langle s \rangle$ as shown in figure 6.21. In order to get more insight it is necessary to look into more subtle aspects of geometrical statistics than just single space/time point alignments. For the moment, is it is clear that 'simple' structures in three-dimensional turbulence are qualitatively different from those in pure two-dimensional turbulence in which the nonlinearity is really depleted in such structures — however, in three-dimensional turbulence such structures do not seem to be the best candidates to look for depletion of nonlinearity in the absolute sense. Nevertheless, we have seen above that taking the enstrophy generation $\omega_i \omega_j s_{ij}$ as a measure of nonlinearity the objects with strong alignment between $\boldsymbol{\omega}$ and $\boldsymbol{\lambda}_2$ appear to be not the most nonlinear, since their enstrophy generation $\omega_i \omega_j s_{ij}$ comes mostly from the nonlocal effects and not from self-stretching.

6.6. Nonlocality

> *...the energy transfer is rather nonlocal (Deissler, 1978).*
>
> *The small and large scales are strongly coupled. (Chorin, 1994).*
>
> *The experimental evidence shows that the large and the small scales are strongly coupled and that traditional cascade picture, which promotes universality, is a crude representation (Warhaft, 2000).*
>
> *The large scales of turbulence are insensitive to viscosity at high enough Reynolds number (Mathieu and Scott, 2000).*

6.6.1. INTRODUCTION AND SIMPLE EXAMPLES

As mentioned in Chapter 1 nonlocality is among the three main reasons [16] the problem of turbulence is so difficult.

The term *nonlocality* is used here in several related meanings which will become clear in the course of the discussion of the issues throughout this section (see also Tsinober, 2000b).

We start from a simple example. Taking the position that velocity fluctuations represent the large scales and the velocity derivatives represent the small scales, one can state that, in homogeneous (not necessarily isotropic) the large and the small scales do not correlate. This can be expressed quantitatively by a correlation between velocity and vorticity. For example, in a homogeneous turbulent flow the Lamb vector $\langle \boldsymbol{\omega} \times \mathbf{u} \rangle = 0$ and also $\langle (\mathbf{u} \cdot \boldsymbol{\nabla})\mathbf{u} \rangle = 0$. If the flow is statistically reflexionally symmetric, then $\langle \boldsymbol{\omega} \cdot \mathbf{u} \rangle = 0$ too. However, as mentioned in Chapter 1, vanishing correlations do not necessarily mean absence of dynamically important relations. Indeed, the quantities $(\mathbf{u} \cdot \boldsymbol{\nabla})\mathbf{u} \equiv \boldsymbol{\omega} \times \mathbf{u} + \boldsymbol{\nabla}(u^2/2)$ and $\boldsymbol{\omega} \times \mathbf{u}$, are

---

[16]The three N's: nonlinearity, nonintegrability and nonlocality.

the main 'guilty parties' responsible for all we call turbulence. Both contain the large scales (velocity) and small scales (velocity derivatives, vorticity). So some kind of coupling between the two is unavoidable. Let us begin with the *kinematic* relation between velocity and vorticity, which is a mere consequence of the relation $\boldsymbol{\omega} = curl\ \mathbf{u}$. Therefore any altering of $\boldsymbol{\omega}$ results in its 'reacting back' on the velocity field. This point is not as trivial as may seem. Indeed, take a Helmholz decomposition of the most significant part of the nonlinear term in NSE the Lamb vector, $\boldsymbol{\omega} \times \mathbf{u}$ (Tsinober, 1990a)

$$\boldsymbol{\omega} \times \mathbf{u} = \boldsymbol{\nabla}\alpha + \boldsymbol{\nabla} \times \boldsymbol{\beta},$$

Assuming that $\boldsymbol{\omega}$ and $\mathbf{u}$ are random Gaussian and *unrelated*, i.e. $\boldsymbol{\omega} \neq rot\mathbf{u}$, the result is that

$$\left\langle (\boldsymbol{\nabla}\alpha)^2 \right\rangle = \left\langle (\boldsymbol{\nabla} \times \boldsymbol{\beta})^2 \right\rangle$$

However, if $\boldsymbol{\omega} = rot\mathbf{u}$ and $\mathbf{u}$ is quasi-Gaussian, i.e. obeys the zero-forth-cumulant relation[17] then (Tsinober, 1990a)

$$\left\langle (\boldsymbol{\nabla}\alpha)^2 \right\rangle \sim 2 \left\langle (\boldsymbol{\nabla} \times \boldsymbol{\beta})^2 \right\rangle$$

i.e. in this case the *rms* of the potential part of the Lamb vector is twice as large as its solenoidal part[18]. In other words $\boldsymbol{\omega}$ 'reacts back' on velocity, and consequently on $\boldsymbol{\omega} \times \mathbf{u}$, even for purely kinematic reasons.

More generally, vorticity does not just involve small scales. It is of special importance, since together with boundary conditions the whole flow field is determined entirely by the field of vorticity[19]. This, of course, includes the velocity field itself, and therefore the large scales are determined by the small scales. This is the simplest indication not only for direct interaction/coupling between large and small scales, but also that this interaction is bidirectional.

A natural question is what about the correlation(s) and coupling between velocity, u_i, and the strain, s_{ij}. First we recall that, just like in the case of vorticity the whole flow field is determined entirely by the field of strain. This is seen from the equation $\nabla^2 u_i = 2\partial s_{ik}/\partial x_k$, which together with boundary conditions defines uniquely the velocity field[20]. Again in a

---

[17]The so-called Millionschikov hypothesis (Monin and Yaglom, 1975), suggested by A.N. Kolmogorov.

[18]In real turbulent flows this difference is even larger (Shtilman et al., 1993).

[19]As mentioned before this is is seen from the equation $\nabla^2\mathbf{u} = -curl\boldsymbol{\omega}$, which together with boundary conditions defines uniquely the velocity field.

[20]This undermines the most common belief that turbulence is an inertial phenomenon in the sense that the precise nature of the dissipation mechanism does not affect the structure of the large (energy containing) scales, so that turbulence is statistically indistinguishable on energy-containing scales in gases, liquids, slurries, foams, and many non-Newtonian media.

homogeneous turbulent flow velocity, u_i, and the strain, s_{ij} do not correlate, $\langle u_i s_{ij} \rangle = 0$. However, velocity is correlated with small scales of 'higher order'. Namely, the correlation, $\langle u_i \nabla^2 u_i \rangle = -2 \langle s_{ij} s_{ij} \rangle = - \langle \omega^2 \rangle$, is essentially nonvanishing again for purely kinematic reasons. Nevertheless, it is of special dynamical significance directly related to the dissipation of turbulent energy, $\langle \epsilon \rangle = 2\nu \langle s_{ij} s_{ij} \rangle$, which for dynamical reasons remains finite for very small ν. Therefore the correlation $\langle u_i \nabla^2 u_i \rangle$ becomes very large at small ν. However, the corresponding correlation coefficient $\dfrac{\langle u_i \nabla^2 u_i \rangle}{\langle \mathbf{u}^2 \rangle^{1/2} \langle (\nabla^2 \mathbf{u})^2 \rangle^{1/2}} = \dfrac{\langle \omega^2 \rangle}{\langle \mathbf{u}^2 \rangle^{1/2} \langle (rot\mathbf{u})^2 \rangle^{1/2}}$ at large Reynolds numbers is roughly of the order $Re^{-1/4}$, i.e. becomes very small[21]. This, of course, does not mean that the coupling between $\mathbf{u}$ and $\nabla^2 \mathbf{u}$ becomes unimportant at large Reynolds numbers. The correlation between $\mathbf{u}$ and $\nabla^2 \mathbf{u}$ is directly related to the correlation between velocity, $\mathbf{u}$, and acceleration, $\mathbf{a}$, since, for example, in a homogeneous turbulent flow $\langle \mathbf{u} \cdot \mathbf{a} \rangle = \nu \langle \mathbf{u} \cdot \nabla^2 \mathbf{u} \rangle$ (Mann et al., 1999). Hence again coupling between large ($\mathbf{u}$) and small ($\mathbf{a}$) scales.

One can use the above example with the Lamb vector to illustrate the dynamical nature of this coupling. For this we retreat from (quasi) homogeneous isotropic flows and consider a unidirectional *in the mean* fully developed turbulent shear flow, such as the flow in a plane channel in which all statistical properties depend on the coordinate normal to the channel boundary, x_2, only. In such a flow, a simple precise kinematic relation is valid

$$d\langle u_1 u_2 \rangle / dx_2 \equiv \langle \boldsymbol{\omega} \times \mathbf{u} \rangle_1 = \langle \omega_2 u_3 - \omega_3 u_2 \rangle \neq 0, \qquad (6.5)$$

which is just a consequence of the vector identity $(\mathbf{u} \cdot \boldsymbol{\nabla})\mathbf{u} \equiv \boldsymbol{\omega} \times \mathbf{u} + \boldsymbol{\nabla}(\frac{u^2}{2})$ in which incompressibility and $d\langle \cdots \rangle / dx_{1,3} = 0$ where used, and $\langle \cdots \rangle$ means an average in some sense (e.g. time or/and over the planes $x_2 = const$, etc.). The *dynamic* aspect is that in turbulent channel flows $d\langle u_1 u_2 \rangle / dx_2 \neq 0$ is essentially different from zero at *any arbitrarily large* Reynolds number (see figure 6.22). Therefore one can see from (6.5) that at least some correlations between velocity and vorticity in such flows are essentially different from zero.

Since vorticity is basically a small scale quantity the relation (6.5) is a clear indication of a dynamically important statistical dependence between the large scales ($\mathbf{u}$) and small scales ($\boldsymbol{\omega}$). Without this dependence $d\langle u_1 u_2 \rangle / dx_2 \equiv 0$, which means that the mean flow would not 'know' about

---

[21]Indeed, $\dfrac{\langle u_i \nabla^2 u_i \rangle}{\langle \mathbf{u}^2 \rangle^{1/2} \langle (\nabla^2 \mathbf{u})^2 \rangle^{1/2}} = \dfrac{\langle \omega^2 \rangle}{\langle \mathbf{u}^2 \rangle^{1/2} \langle (rot\omega)^2 \rangle^{1/2}} = \dfrac{\langle \epsilon \rangle / \nu}{\langle \mathbf{u}^2 \rangle^{1/2} \langle (rot\omega)^2 \rangle^{1/2}} \sim \dfrac{\langle \epsilon \rangle / \nu}{U[(\epsilon/\nu)^{1/2}]/\eta}$ $\sim Re^{1/2}(\eta/L)^{1/2} = Re^{-1/4}$. We used here the standard order of magnitude phenomenological estimates (see Tennekes and Lumley, 1972). Namely, $\langle \mathbf{u}^2 \rangle^{1/2} \sim U$, $\langle \epsilon \rangle \sim U^3/L$, and $\langle (rot\boldsymbol{\omega})^2 \rangle^{1/2} \sim (\langle \epsilon \rangle / \nu)/\eta$ with U and L some integral scales of velocity and length.

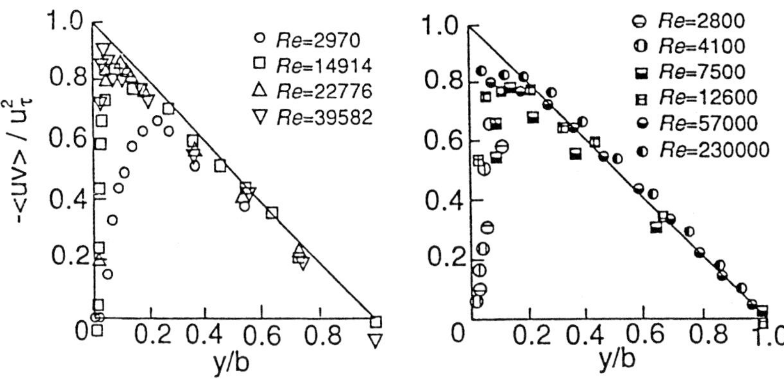

Figure 6.22. Dependence of the mean Reynolds stress $\langle u_1 u_2 \rangle$ on the distance from the wall in turbulent flows in channels of cross section with large aspect ratio. Adapted from Wei and Wilmarth (1989).

its turbulent part at all. It is noteworthy that both correlation coefficients $\frac{\langle \omega_2 u_3 \rangle}{\langle \omega_2^2 \rangle^{1/2} \langle u_3^2 \rangle^{1/2}}$, $\frac{\langle \omega_3 u_2 \rangle}{\langle \omega_3^2 \rangle^{1/2} \langle u_2^2 \rangle^{1/2}}$ (and many other statistical characteristics, e.g. some, but not all, measures of anisotropy) are of order 10^{-2} even at rather small Reynolds numbers. Nevertheless, as we have seen, in view of the dynamical importance of interaction between velocity and vorticity in turbulent shear flows[22] such 'small' correlations by no means imply absence of a dynamically important statistical dependence and a direct interaction between large and small scales. Indeed it is this interaction that results in drastic changes of the whole mean flow.

The direct interaction between large and small scales similar to the one in the above example may exist in a much broader class of turbulent flows and regions in these flows, e.g. with appropriate scale (in time and space) separation such as vorticity 'pancakes' (Brachet et al., 1992).

6.6.2. DIFFERENT ASPECTS OF NONLOCALITY

From the formal point of view a process is called local if all the terms in the governing equations are differential. If the governing equations contain integral terms, then the process is nonlocal. The Navier-Stokes equations are *integro-differential* for the velocity field in both physical and Fourier space (and any other). Therefore, generally, the Navier-Stokes equations

---

[22]The relation (6.5) is approximately valid in many important turbulent flows such as boundary layers, wakes, jets, etc. in which $d\langle \cdots \rangle/dx_{1,3} \ll d\langle \cdots \rangle/dx_2$. The argument here is based on nonvanishing *gradient* of the Reynolds stress $d\langle u_1 u_2 \rangle/dx_2$ and not the Reynolds stress itself. As noted, without this gradient the mean flow would not 'know' about its turbulent part at all. Precisely this happens when a *shear* flow is assumed to be *homogeneous*. In such a flow the gradient $d\langle u_1 u_2 \rangle/dx_2 \equiv 0$.

describe nonlocal processes[23]. The problem is intimately related to the one of decompositions/representations, which was already discussed in chapter 5 , but the relation between the two becomes more clear from what is following in the sequel.

The nonlocal nature of the Navier-Stokes equations in physical space is two-fold[24]. On one hand, it is due to pressure ('dynamic' nonlocality), since $\rho^{-1}\nabla^2 p = \omega^2 - 2s_{ij}s_{ij} = -\frac{\partial^2 u_i u_j}{\partial x_i \partial x_j}$, so that pressure is nonlocal due to the nonlocality of the operator ∇^{-2} (see appendix C). This nonlocality is strongly associated with essentially non-Lagrangian nature of pressure. For example, replacing in the Euler equations the pressure Hessian $\frac{\partial^2 p}{\partial x_i \partial x_j}$, which is both nonlocal and non-Lagrangian, by a local quantity $\frac{1}{3}\delta_{ij}\nabla^2 p = \frac{\rho}{6}\{\omega^2 - 2s_{ij}s_{ij}\}$ turns the problem into a local and integrable one and allows us to integrate the equations for the invariants of the tensor of velocity derivatives $\partial u_i/\partial x_j$ in terms of a Lagrangian system of coordinates moving with a particle (see Cantwell, 1992; and references therein. The reason for the disappearance of turbulence (and formation of singularity in finite time) in such models, called restricted Euler models, is that the eigenframe of s_{ij} in these models is *fixed* in space (Novikov, 1990b), whereas in a real turbulent flow it is oriented randomly in space and time. This means that nonlocality due to presure is essential for (self-) sustaining turbulence: no pressure Hessian - no turbulence. A related aspect is that the Lagrangian acceleration $D\mathbf{u}/Dt$ - a kind of small scale quantity - is dominated by pressure gradient, ∇p (Vedula and Yeung, 1999; see next section).

Taking the *rot* of the NSE and getting rid of the pressure does not remove the nonlocality. Indeed, the equations for vorticity (C.9) and enstrophy (C.16) are nonlocal in vorticity, $\boldsymbol{\omega}$, since they contain the rate of strain tensor, s_{ij}, due to the nonlocal relation between vorticity, $\boldsymbol{\omega}$, and the rate of strain tensor, s_{ij} ('kinematic' nonlocality)[25]. The two aspects of nonlocality are related, but are not the same. For example, in compresible flows there is no such relatively simple relation between pressure and velocity gradient tensor as above (see equation C.13), but the vorticity-strain relation remains the same.

Both aspects of nonlocality are reflected in equations (C.17) and (C.18) for the rate of strain tensor and total strain/dissipation, $s^2 \equiv s_{ij}s_{ij}$, and the equations (C.23) and (C.24) for the third order quantities. An impor-

---

[23]This does not mean that processes described by pure differential equations are local. An example is a passive object (scalar, vector) in a random velocity field.

[24]On nonlocality of turbulence in Fourier space see, for example Deissler (1979).

[25]Nonlocality of the same kind is encountered in problems dealing with the behaviour of vortex filaments in an inviscid fluid. Its importance is manifested in the breakdown of the so called localized induction approximation (LIA) as compared with the full Bio-Savart induction law, see Ricca et al. (1999).

tant aspect is that the equations (C.18), (C.23) and (C.24) contain invariant quantities $\omega_i\omega_j\frac{\partial^2 p}{\partial x_i \partial x_j}$ and $s_{ik}s_{kj}\frac{\partial^2 p}{\partial x_i \partial x_j}$ reflecting the nonlocal dynamical effects due to pressure and can be interpreted as interaction between vorticity and pressure and between strain and pressure. In particular, the equation (C.23) for the enstrophy production shows both aspects of nonlocality of vortex stretching process (see also Ohkitani and Kishiba, 1995). The first term in (C.23) (which is just the squared magnitude of the vortex stretching vector) is strictly positive $\omega_i s_{ij}\omega_k s_{ki} \equiv W^2 > 0$. This means that the nonlinear processes involving vortex stretching (or direct interaction of vorticity and strain) always tend to increase even the *instantaneous* enstrophy production. Here also the term $W_3^2 = \omega^2\lambda_3^2\cos^2(\boldsymbol{\omega}, \boldsymbol{\lambda}_3)$ associated with the negative eigenvector of the rate of strain tensor Λ_3, i.e. vortex compressing or negative enstrophy production $\omega_3^2\Lambda_3\cos^2(\boldsymbol{\omega}, \boldsymbol{\lambda}_3)$, makes a positive (!) contribution to the rate of change of enstrophy generation, $W_i^2 = \omega_i^2\Lambda_3^2\cos^2(\boldsymbol{\omega}, \boldsymbol{\lambda}_i)$. However, the inviscid rate of change of enstrophy generation contains also a second term reflecting the interaction between vorticity and the pressure Hessian $\frac{\partial^2 p}{\partial x_i \partial x_j}$. This is the term $-\omega_i\omega_j\frac{\partial^2 p}{\partial x_i \partial x_j}$. Without this term the question why $\langle\omega_i\omega_j s_{ij}\rangle > 0$ would be immediately answered. It appears (Tsinober et al., 1995) that $\langle\omega_i\omega_j\frac{\partial^2 p}{\partial x_i \partial x_j}\rangle$ is positive and is about $\langle W^2\rangle/3$, i.e. in the mean, the nonlinearity in (C.23) is reduced by this nonlocal term, since for a Gaussian velocity field $\langle\omega_i\omega_j\frac{\partial^2 p}{\partial x_i \partial x_j}\rangle \equiv 0$. The nonvanishing correlation $\langle\omega_i\omega_j\frac{\partial^2 p}{\partial x_i \partial x_j}\rangle$ (and also $\langle s_{ik}s_{kj}\frac{\partial^2 p}{\partial x_i \partial x_j}\rangle$) is also one of the manifestations of nonlocality.

Direct coupling between large and small scales
Nonlocality, in the sense as discussed above, is an indication of direct coupling between large and small scales. There exist massive evidence that this is really the case as there are many indications that this interaction is bidirectional[26]. We should first mention the well known effective use of fine honeycombs and screens in reducing large scale turbulence in various experimental facilities (Laws and Livesey, 1978; Tan-Attichat et al., 1982). The experimentally observed phenomenon of strong drag reduction in turbulent flows of dilute polymer solutions and other drag reducing additives (Gyr and Bewersdorff, 1995), is another example of such a 'reacting back' effect of small scales on the large scales, (see Chapter 8). Third, one can substantially increase the dissipation and the rate of mixing in a turbulent flow (jet) by *directly* exciting the small scales (Wiltse and Glezer, 1998)[27].

---

[26]Note that in case of passive objects, there is no such a bidirectional relation – it is only one way.

[27]These authors claim also that *forcing induces coupling (or long-range interactions) between small and large scales within the flow*, whereas this coupling does exist indepe-

Before proceeding further, we sould alos mention a related phenomenon concerning the stability of spatially periodic flows. Such flows may destabilize *directly into small-scale* three-dimensional structures (Pierrehumbert and Widnall, 1982).

Anisotropy. One of the manifestations of direct interaction between large and small scales is the anisotropy in the small scales. Though local isotropy is believed to be one of the universal properties of high Re turbulent flows, it appears that it is not so universal: in many situations the small scales do not forget the anisotropy of the large ones. There exists considerable evidence for this point, which has a long history starting somewhere in the 50s (see references in Ferchichi and Tavoularis, 2000; Shen and Warhaft, 2000; Sreenivasan and Antonia, 1997; Tsinober, 1993b, 1998b; Yeung et al., 1995; and Warhaft, 2000). Along with other manifestations of direct interaction between large and small scales, the deviations from local isotropy seem to occur due to various external constraints like boundaries, initial conditions, forcing (e.g. as in DNS), mean shear/strain, centrifugal forces (rotation), buoyancy, magnetic field, etc., which usually act as an organizing factor, favoring the formation of coherent structures of different kinds (quasi-two-dimensional, helical, hairpins, etc.). These are as a rule, large scale features which depend on the particularities of a given flow and thus are not universal. These structures, especially their edges seem to be responsible for the contamination of the small scales. This 'contamination' is unavoidable even in homogeneous and isotropic turbulence, since there are many ways to produce such a flow (i.e. many ways to produce the large scales). It is the difference in the mechanisms of large scale production which 'contaminates' the small scales. Hence, nonuniversality.

Let us turn first again to the 'simple' example above and look at the properties in the proximity of the midplane, $x_2 \approx 0$, of the turbulent channel flow. In this region $dU/dx_2 \approx 0$, but the flow is neither homogeneous nor isotropic, since though $\langle u_1 u_2 \rangle \approx 0$ in this region too, $d \langle u_1 u_2 \rangle /dx_2$ is essentially $\neq 0$ and is *finite* independently of the Reynolds number, as far as the data allow to make such a claim (see figure 6.22). This is also a clear indication of nonlocality, since in the bulk of the flow, i.e. *far from the boundaries, $dU/dx_2 \sim 0$.*

The first experimental evidence on anisotropy in small scales at large Reynolds numbers was provided by the atmospheric boundary layer experiments. It was found that the skewness of the derivative of temperature fluctuations is not small and is of order 1 (Stewart, 1969), whereas for a locally isotropic flow it should be close to zero. This and related results were obtained later in a number of laboratory flows, field observations and in numerical simulations (see Sreenivasan and Antonia, 1997; and Warhaft, 2000

dently of forcing, and in the presence of the latter is manifested in enhanced dissipation.

for further references). An important feature of these flows is the presence of a mean gradient of the passive scalar – the rest is not so important: the phenomenon is observed for a Gaussian and two dimensional velocity field (Holzer and Siggia, 1994). This is related to the weak sensitivity of the passive scalar field to the details of the velocity field and its Reynolds number (Kraichnan, 1968; Warhaft, 2000). On the other hand, passive objects exhibit chaotic behaviour and mixing in purely laminar flows - a phenomenon closely related to what is called Lagrangian chaos/chaotic advection. Therefore generally, the behaviour of a passive scalar may not reflect the structure of turbulent flows (chapter 4).

As mentioned in chapter 5, similar observations have, quite recently, been made for the velocity increments and velocity derivatives in the direction of the mean shear both in numerical and laboratory experiments (see references in Shen and Warhaft, 2000). It was found that the stastistical properties of velocity increments and velocity derivatives in the direction of the mean shear do not conform with and do not confirm the hypothesis of local isotropy. *Moreover, our results imply that the large scales are directly coupled to the small scales. (The anisotropy disappears when the large scale shear is removed;* Shen and Warhaft, 2000). More precisely these results imply that there is a *direct* influence of mean shear on the small scales due to the permanent bias of the mean shear to which the field of fluctuations is exposed on account of its large *residence* time in the mean shear. Of course, one of the *possible* explanations of the results produced by Shen and Warhaft is *that the large scales are directly coupled to the small scales.* However, this does not mean that there is no such coupling *when the large scale shear is removed.* This coupling is an intrinsic/generic property of turbulent flows and exists independently of the presence of mean shear or other external factors, but has different manifestations for different external factors.

It should be mentioned that in the experiments by Shen and Warhaft the value of the Corrsin criterion $S_C^* = (dU/dx_2)(\nu/\langle\epsilon\rangle)^{1/2} \approx 2.4 \cdot 10^{-2}$. This criterion represents the ratio of the Komlogorov time scale, $\tau_\eta = (\nu/\langle\epsilon\rangle)^{1/2}$, to the time scale, $(dU/dx_2)^{-1}$, associated with the mean shear, and it should be small enough in order to have isotropy in small scales. The main problem is how small. There is no agreement on this issue, but there is evidence that in order to have one decade of isotropic inertial range in boundary layer flows (both simple and complex) at $Re_\lambda \approx 1500$, it is necessary that $S_C^* < 10^{-2}$ (see Saddoughi, 1997; and references therein. This brings us to the next issue.

Other manifestations of direct coupling between large and small scales. An important observation was made by Praskovsky et al. (1993), Sreenivasan and Dhruva (1998), and Kholmyansky and Tsinober (2001). The specific featureof these large scale experiments is rather large Taylor microscale Reynolds number $Re_\lambda \sim 10^4$. At these large Reynolds numbers, there is also clear evidence of strong coupling between large and small scales. In view of severe limitations on statistical convergence at such Reynolds numers, the approach is different from the one undertaken by Shen and Warhaft (2000). Namely, if the large scales are *not* coupled directly to the small scales, there should be no dependence of conditional statistics of the small scale quantities (e.g. velocity increments, estrophy, total strain) conditioned on the large scale quantities (velocity). The results from Praskovsky et al. (1993), Sreenivasan and Dhruva (1998) and Kholmyansky and Tsinober (2001) show the opposite – there is such a dependence. Below we present some results from Kholmyansky and Tsinober (2001), since in their experiment the mean shear was rather small, less than $0.1s^{-1}$. This corresponds to the value of the Corrsin criterion $S_C^* = (dU/dx_2)(\nu/\langle\epsilon\rangle)^{1/2} \approx 2 \cdot 10^{-3}$, which is an order of magnitude smaller than in the experiments by Shen and Warhaft (2000) and five times lower than the value 0.01 mentioned above.

Some results similar to ones obtained by Praskovsky et al. (1993) are shown in figure 6.23 (left column) in parallel with those conditioned on the centered magnitude of the vector of velocity fluctuations $v = u - \langle u \rangle$, where $u^2 = u_1^2 + u_2^2 + u_3^2$. Similar behaviour is observed for conditonal statistics of $\langle \delta u_i^n \rangle$ for all $i = 1, 2, 3$ and $n = 2, 3, 4$. Two aspects deserve special comment.

First, there is a clear tendency that the conditional averages of the structure functions increase with the *energy* of fluctuations, as is seen from the right column of the figure 6.23. Second, such a tendency, that is the direct coupling, is observed also for the smallest distance of the order of Kolmogorov scale $\sim \eta$, which was used for estimates of the derivatives in the streamwise direction.

Analogous conditional statistics for the enstrophy ω^2 and the total strain $s_{ij}s_{ij}$ is shown in figure 6.24. The result is quite similar to the one shown in figure 6.23. for the smallest distance $\sim \eta$.

An interesting observation, which seems to be related to the coupling between large and small scales, is that, in 3-D turbulence not only $\langle \omega_i \omega_j s_{ij} \rangle$ and $-\langle s_{ij}s_{jk}s_{ki} \rangle$ are essentially positive quantities but also *all* $\int_{V_L} \omega_i \omega_j s_{ij} dV_L$, $-\int_{V_L} s_{ij}s_{jk}s_{ki} dV_L$ over volumes of the order of integral scale are essentially positive too (see figure 1 in Tsinober, 1998a; and figure 6.6).

The observations on the coupling between large and small scales and the 'reaction back' of the small scales on the large ones by no means are exausted by the references given above. As an example from the atmospheric physics we bring a quotation of the first conclusion reached at the Symposium on the nature of the so-called CAT - clear air turbulence: *The energy*

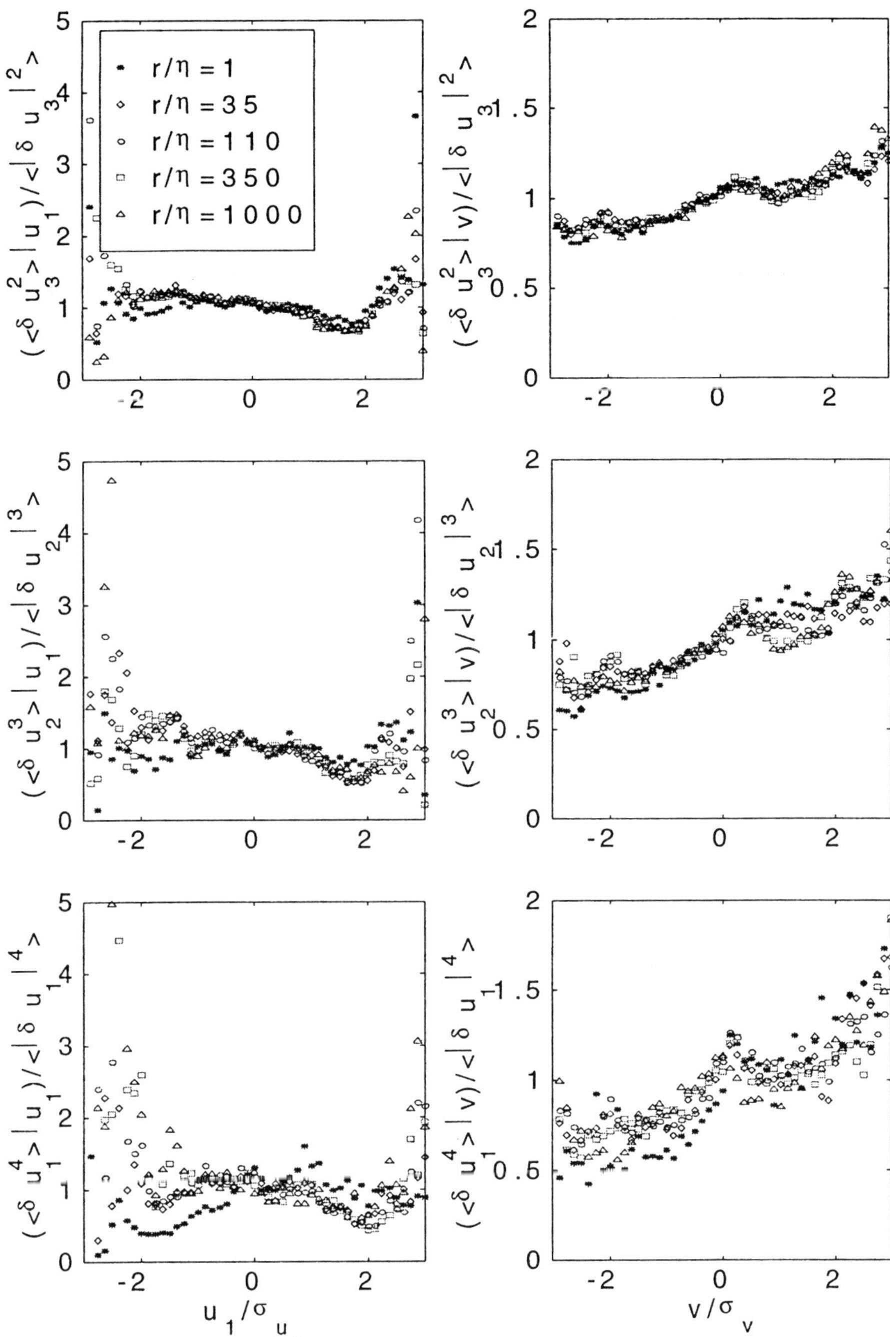

Figure 6.23. Conditional averages of velocity increments conditioned: left – on the u_1 fluctuation, right – on the centered magnitude of the vector of velocity fluctuations v (Kholmyansky and Tsinober, 2001).

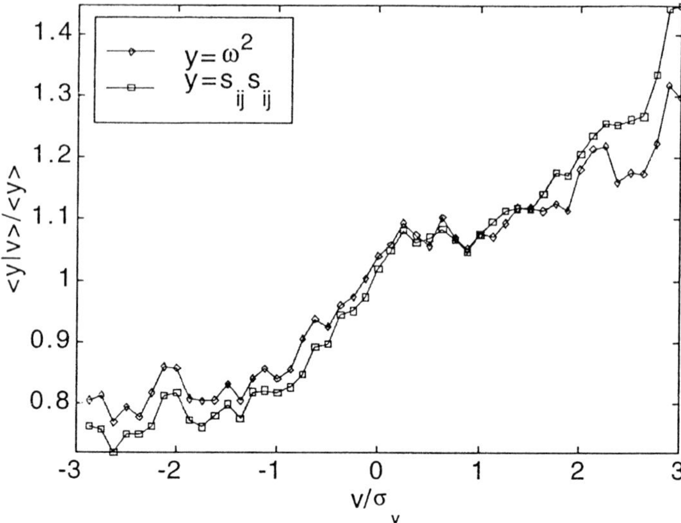

Figure 6.24. Conditional averages of enstrophy ω^2 and total strain $s_{ij}s_{ij}$ conditioned on the centered magnitude of the vector of velocity fluctuations v.

dissipated at small-scale by clear air turbulence influences the large-scale atmospheric motion. (Pao and Goldburg, 1969).

Intermittency and structure(s). Batchelor and Townsend (1949) in their studies on small scale intermittency (see Chapter 7) wrote: *All the evidence is consistent with the inference that the fluctuations are small in the region of smallest wave-numbers of equilibrium range and become increasingly large at larger wave-numbers* (p. 252) and that... *the mean separation of the visible activated regions is comparable with the integral scale of the turbuence, i.e. with the size of the energy-containing eddies.* (p. 253). This latter observation seems to be intimately related to the direct interaction/coupling of the large and small scales.

At least some of the structure(s) of turbulence reflects such a relation as well. The most popular 'structure' observed in various turbulent flows – the vortex filament/worm – has at least two essentially different scales: its length can be of the order of the integral scale, whereas its cross-section is of the order of Kolmogorov scale. Similarly, the ramp-cliff fronts in the passive scalar fields have a thickness much smaller than the two other scales.

Helicity. Helicity, $\int \boldsymbol{\omega} \cdot \mathbf{u}d\mathbf{x}$, and its density, $\boldsymbol{\omega} \cdot \mathbf{u}$, deserve here also special mention. The formal reason is that if $\langle \mathbf{u} \cdot \boldsymbol{\omega} \rangle \neq 0$, this is a clear indication of direct coupling of large and small scales[28]. A more subtle

---

[28]So it is not surprising that, in flows with nonzero mean helicity, the direct coupling between small and large scales is stronger than otherwise. The stronger coupling between the large and small scales in flows with nonzero mean helicity $\langle \mathbf{u} \cdot \boldsymbol{\omega} \rangle$ aids creation of large

E R R A T A

A. Tsinober (2001) An informal introduction to turbulence

ISBN 1-4020-0110-X

In the process of production

1. The page 150 was skipped.

2. The figures 9.3 (page 232) and 9.4 (page 234) were interchanged.

Following is the page 150 and the pages 232 and 234 with correctly placed figures.

Further corrections, if found, will be placed at the web
http://www.eng.tau.ac.il/~tsinober/book/

Suggestions and critisisms are very welcomed at
tsinober@eng.tau.ac.il

Addendum. Right after the manuscript of this book was finished, a paper by Laval et al. (2001) was published[38], which deserves a special comment. In this paper the authors have convincingly demonstrated the nonlocal nature of a particular decomposition of the flow field[39], which reflects one of the several facets of nonlocality of turbulence as described in 6.6 above (see also Kholmyansky and Tsinober, 2000; Tsinober, 2001 and references therein). Laval et al. interpret their results as evidence *that nonlocal interactions are responsible for intermittency corrections in the statistical behaviour of 3-D turbulence* as well as for the *deviations from Gaussianity.* By *intermittency corrections* the authors mean *the anomalous corrections* in the scaling behaviour of the structure functions (and corresponding PDFs). In view of the existing evidence (see again 6.6) the claim about the importance of nonlocal interactions - understood in a broader sense as direct and bi-directional interaction/coupling of large and small scales - is definitely correct. However, *the anomalous corrections* in the scaling exponents can hardly be interpreted as necessary and reliable manifestations of intermittency of turbulence, nor such corrections are necessarily associated with or are due to *thin vortices* as explained in the next chapter (see also 6.4.2 and Tsinober, 1998). Likewise, the non-Gaussian nature of turbulence - which is among the intrinsic/generic properties of turbulence - is far more than just a consequence of the nonlocality of turbulence and even it's intermittency, as follows, for example, from the four fifth law (see 6.8 for more).

---

[38]Laval, J.-P., Dubrulle, B. and Nazarenko, S. (2001) Nonlocality and intemittency in three-dimensional turbulence, *Phys. Fluids,* **13,** 995-2012.

[39]Laval et al. use a cutoff in Fourier space and define the large scales, U_i, and small scales, u_i, correspondingly as those below and above the cutoff; which results in essentially the same equations for U_i and u_i as (C.43, C.44) or (C.55, C.57). The authors interpret as *nonlocal terms, involving the product of a large scale and a small scale component, and a local term, involving two small-scale components.* This is not the same what the authors claim several lines before that *by nonlocal interactions,* they *mean interaction between well-separated scales (or highly elongated wave number triads),* since U_i and u_i are not that well separated, they are 'neighbours', since the smallest 'scales' of U_i are just of the same order as the largest 'scales' of u_i. In order *to study the dynamical effect of these contributions* (i.e. the terms of the type uu called local, and terms of the type Uu called nonlocal) *at small scales* the authors performed two types of numerical experiments along with the full DNS. The first type is in which the local interactions were neglected in the equations for the small scales (only), thus turning all the problem into a linear one (the RDT simulation). Among other things this resulted in stronger deviations from the Komogorov-like scaling just like in other linear problems such as for passive objects. *Replacing the removed local interactions by a simple turbulent viscosity terms allows one to restore the correct intermittency* (i.e. scaling) *and the energy characteristics.* However, this in essence empirical correction does restore the correct physics of neither the self-interaction of the small scales nor their reaction back on the large scales. The second type of simulation is in which the local interactions are the only retained. *This resulted in even higher (than in RDT) level of small scales, but exhibited in much less intermittency...* in the sence of deviations from the Komogorov-like scaling.

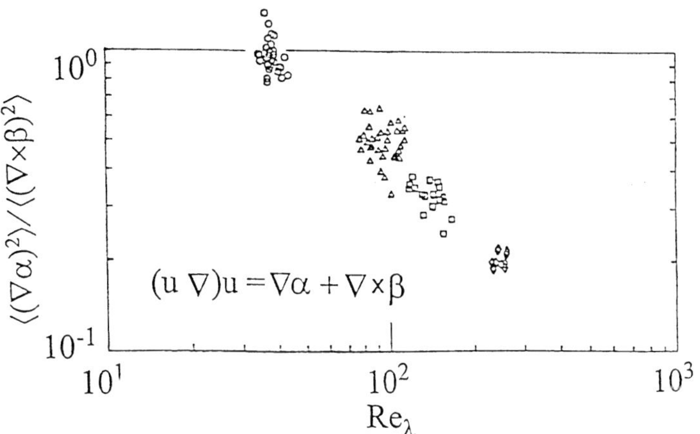

Figure 9.3. Reynolds number dependence of the ratio of the variances of the irrotational and solenoidal parts of the nonlinear term $(u \cdot \nabla)u$ in a DNS simulation of quasi-isotropic turbulence (Tsinober et al., 2001). This is a version of the bottom figure 6.29.

at such Reynolds numbers. Turbulent flows at the asymptotic regime at $Re \gg 1$ ($\to \infty$) - if such exists - do not seem to be simpler than those at very moderate Re, and are definitely much less accessible in every respect. Many will ask: but what about the (so-called) inertial range? As discussed in chapter 6, the pure inertial range is not well defined due to nonlocality of turbulence; *because of scale invariance breaking, the notion of inertial range is not well defined* (Arneodo et al., 1999). We wish to reiterate that independence of some (statistical) parameters or properties of viscosity at large Reynolds numbers does not mean that viscosity is unimportant. It means *only* that the effect of viscosity is Reynolds number independent.

9.1.3. SELF-AMPLIFICATION OF VELOCITY DERIVATIVES

As discussed in chapter 6 there is some evidence that the process of self-amplification of velocity derivatives, both vorticity and strain, is a universal phenomenon which occurs at Reynolds numbers as low as $Re_\lambda \sim 60$. This seems to be one of the key physical processes in all turbulent flows. The details of this process are not well understood, and apart from geometrical statistics and similar things, one needs much more. Since the whole flow is defined by the field of velocity derivatives (either vorticity or strain), proper understanding of the process of self-amplification of velocity derivatives in

exponents at very large values of Reynolds number. It is not at all clear why and to what extent accurate measurement and/or knowledge of exponents would aid understanding of turbulence. Moreover, the very existence of such exponents (with few exceptions) is quite problematic, e.g. Badii and Talkner (2001), Feigenbaum (1997), Tsinober (1996b).

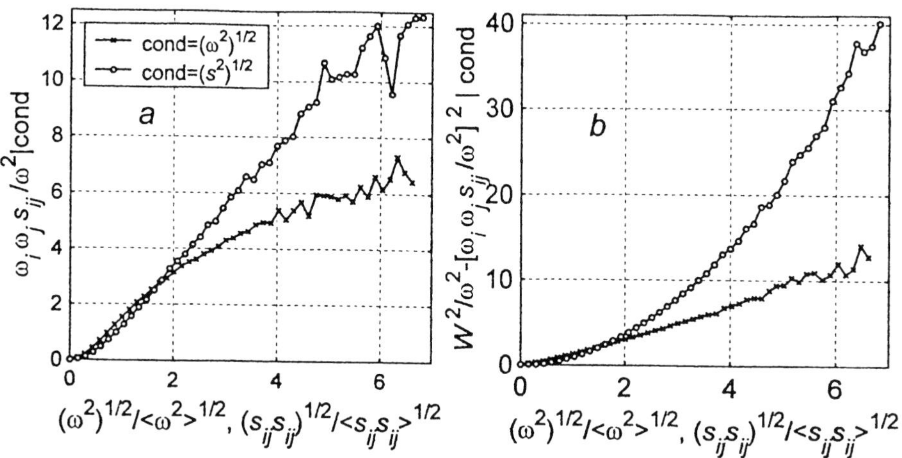

Figure 9.4. Examples of conditional averages showing the difference in the behaviour of nonlinearities in vorticity and strain dominated regions in a turbulent boundary layer at $Re_\lambda = 10^4$ (Kholmyansky et al., 2001a). The quantity $\eta^2 \equiv W^2/(\omega^2) - \{\omega_i\omega_j s_{ij}/(\omega^2)\}^2$ is responsible for vorticity tilting; see equation (6.4) and the text below.

$2\nu\langle s^2\rangle \approx \nu\langle\omega^2\rangle$ this means that at large Reynolds numbers both $\langle s^2\rangle$, $\langle\omega^2\rangle \approx \nu^{-1}$, i.e. the field of velocity derivatives is not only Reynolds number dependent, but also becomes very large. Due to the intermittent nature of the field of velocity derivatives one can expect that the maximal values of s^2, ω^2 (or $\max(|\partial u_i/\partial x_k|)$) increase even faster with the Reynolds number. This possible unboundness of the field of velocity derivatives as $Re \to \infty$ has an implication that the Newtonian approximation can break down as $\nu \to 0$, since the linear stress/strain relation is only the first term in the gradient expansion. So far, however, there seems to be no evidence that the Navier-Stokes equations are inadequate for describing turbulent flows[5]. Among the possible reasons that this (possible) violation is not so easy to detect is that, even if it happens, it will occur in rather small regions due to the strong intermittency of the field of velocity derivatives. We will not discuss here the possible breakdown of the NSE due to possible formation of singularities in finite time (see Constantin, 1995, 1996). There is a more general mathematical question: *Do the Navier-Stokes equations on a 3-dimensinal domain Ω have a unique smooth solution for all time?..* The belief is that *The solution of this problem might well be a fundamental step toward the very big problem of understanding turbulence* (Smale, 1998), and that *turbulence will be solved when well formulated mathematical statements describing the properties of Navier-Stokes systems are proven* (Sinai, 1999).

---

[5]But see Ladyzhenskaya (1975) and McComb (1990, pp. 401-403) on alternatives to NSE, and Tsinober (1993b) and references therein. It is safe to keep in mind that any equations are not Nature.

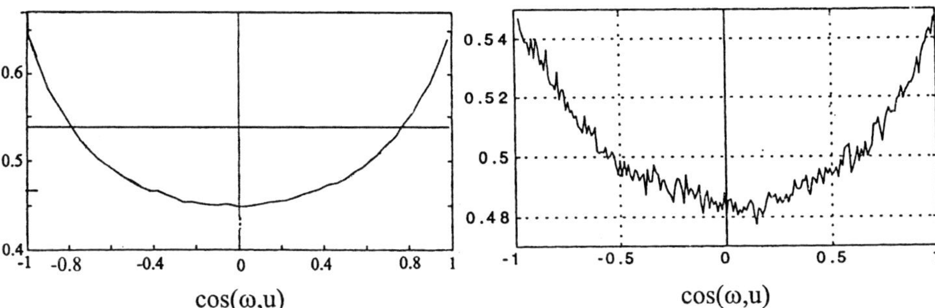

Figure 6.25. PDF of the cosine of the angle between the vectors of velocity and vorticity. Left: DNS at $Re_\lambda = 75$; right - field experiment at $Re_\lambda = 10^4$.

indication is the alignment between $\boldsymbol{\omega}$ and $\mathbf{u}$ (figure 6.25). Though the alignment is weak (because $\boldsymbol{\omega}$ is a small scale and $\mathbf{u}$ is a large scale quantity), it is significant and therefore provides another manifestation of the direct coupling between large and small scales. It is not surprising that the correlation between a large scale and a small scale quantity is small. In the particular case shown, $\langle \mathbf{u} \cdot \boldsymbol{\omega} \rangle / u\omega = 5 \cdot 10^{-4}$. However, this does not mean that the coupling between them is small as well. As mentioned above, other correlations involving large and small scales are of particular importance for shear flows. These are $\langle \omega_2 u_3 \rangle$ and $\langle \omega_3 u_2 \rangle$, which are directly related to the derivative of the Reynolds stress along the mean velocity gradient $d\langle u_1 u_2 \rangle / dx_2 \equiv \langle \omega_2 u_3 \rangle - \langle \omega_3 u_2 \rangle$. The important point is that the corresponding correlation coefficients $C_{\omega_2 u_3} = 7.7 \cdot 10^{-3}$ and $C_{\omega_3 u_2} = 1.3 \cdot 10^{-2}$ are small, but significant: if they vanished precisely the mean flow would not 'know' anything about the fluctuative part of the turbulent flow.

On closures and constitutive relations. Memory effects. Nonlocality is a concern in the problem of the relation between the fluctuations and the mean flow as in the Reynolds decomposition in the Reynolds averaged Navier-Stokes equations or resolved and unresolved scales in the large eddy simulations (LES)[29] (see appendix C for more details). Namely, this relation is a nonlinear *functional*, i.e. the field of fluctuations (the unresolved field in LES) at each time/space point depends on the mean (resolved) field in the whole time/space domain. Vice versa, the mean (resolved) flow at each time/space point depends on the field of fluctuations (unresolved scales) in the whole time/space domain. This means that, in turbulent flows, local

scale structures out small scale turbulence (see Droegemeier et al., 1993; and references therein). As mentioned, this does not mean that in case $\langle \mathbf{u} \cdot \boldsymbol{\omega} \rangle = 0$, or even $\mathbf{u} \cdot \boldsymbol{\omega} = 0$ as in two-dimensional flows, such a coupling does not exist. We return to these matters below.

[29]Or other low-dimensional representations with *modelling to account for neglected modes* (Holmes et al., 1996).

'constitutive' relations analogous to real material constitutive relations for fluids (such as stress/strain relations) do not exist. Hence, in principle any local model for the Reynolds stress (and a similar one for the subgrid stress in LES), e.g. of the kind

$$\tau_{ij}^{\text{Reynolds}} = 2\rho\nu_{ijkl}^{\text{eddy}}(Q_S,\ R_S)S_{ij} \tag{6.6}$$

is incorrect. Here Q_S, R_S are the second and the third invariants of the mean (resolved) rate of strain tensor, S_{ij}. Nevertheless, there is considerable literature on modelling (both RANS and LES) employing such local models. Therefore, an important question is what is the 'error' due nonlocal space/time contributions. In part, this problem is treated in the frame of the so called backscatter. This like all other issues of turbulence modelling is beyond the scope of this book, and the reader is referred for further references to Meneveau and Katz (2000). Two comments are apt here. First, it seems that there exist no rigorous estimates of the nonlocal contribution (NC) in representations such as

$$\tau_{ij}^{\text{Reynolds}} = 2\rho\nu_{ijkl}^{\text{eddy}}(Q_S,\ R_S)S_{ij} + NC, \tag{6.7}$$

and similar ones in LES. Second, all representations of the kind are based on quite general considerations (dimensional and similar) due to the lack of basic understanding and knowledge about the physics of the unresolved scales if such can be considered separately of the resolved ones[30].

6.7. Acceleration and related matters

As the material derivative of the velocity vector, the fluid particle acceleration field in turbulent motion

$$\mathbf{a} \equiv \frac{D\mathbf{u}}{Dt} = \frac{\partial\mathbf{u}}{\partial t} + (\mathbf{u}\cdot\boldsymbol{\nabla})\mathbf{u} = -\frac{1}{\rho}\nabla p + \nu\nabla^2\mathbf{u}$$

is among the most natural physical parameters of special interest in turbulence research for a variety of reasons, ranging from studies of small-scale intermittency to applications in Lagrangian modeling of dispersion (see references in Vedula and Yeung, 1999; and Tsinober et al., 2001).

One of the basic issues concerns the relations between the Lagrangian acceleration, $\mathbf{a}$, and its (Eulerian) 'components' $\mathbf{a}_l = \frac{\partial\mathbf{u}}{\partial t}$ and $\mathbf{a}_c = (\mathbf{u}\cdot\boldsymbol{\nabla})\mathbf{u}$

---

[30] *There is a serious need for high quality work to introduce more physics into the models, so that they will behave more like turbulence in a wider variety of situations. Such work is actually quite rewarding, since the effort to make a model that behaves more like turbulence in some sense can lead to understanding of basic physical mechanisms,* (Lumley, 1996). Definitely there is a need for more physics, and not only because of lack of it in the models. But extrapolating from the previous experience, this is too an optimistic view on what models can do.

on one, and $\mathbf{a}_i = -\frac{1}{\rho}\nabla p$ and $\mathbf{a}_s = \nu\nabla^2\mathbf{u}$ on the other. The local acceleration, $\mathbf{a}_l \equiv \partial\mathbf{u}/\partial t$, expresses the unsteady rate of change of the velocity vector at a fixed point, and the convective acceleration, $\mathbf{a}_c \equiv \mathbf{u}\cdot\nabla\mathbf{u}$, expresses its rate of change due to the spatial derivatives and also embodies nonlinearity effects. On the other hand $\mathbf{a}_i = -\frac{1}{\rho}\nabla p$ and $\mathbf{a}_s = \nu\nabla^2\mathbf{u}$ represent respectively the irrotational and the solenoidal parts of $\mathbf{a}$.

Here use is made of the results from Tsinober et al. (2001) obtained from a current DNS database (Vedula & Yeung, 1999) for isotropic turbulence at ensemble-averaged Taylor-scale Reynolds numbers ranging from 38 to 240 using up to 512^3 grid points.

6.7.1. THE RELATION BETWEEN THE TOTAL ACCELERATION AND ITS LOCAL AND CONVECTIVE COMPONENTS

This issue is known as the random Taylor hypothesis or the sweeping decorrelation hypothesis. It takes its origin from the so called Taylor hypothesis (Taylor, 1935) for computation of the spatial derivative in the direction of the mean flow, $\partial/\partial x_1$ via the time derivative $\partial/\partial t$ from the relation $\partial/\partial x_1 = -U^{-1}\partial/\partial t$ by assuming that (the grid) turbulence is transported by the mean velocity, U, without change. This approximation appears to be valid if the magnitude of turbulent fluctuations, say $\langle u^2\rangle^{1/2}$, is small enough comparing to U, i.e. $\langle u^2\rangle^{1/2}/U \ll 1$. This is seen from a (thought) experiment in which the observer (probe) is moving through the turbulent flow with some velocity U, which can be (assumed to be) very large, for instance in a real experiment when the probe is mounted on a research aircraft moving through a turbulent region in the atmosphere. For the references on the Taylor hypothesis see Tsinober et al. (2001).

Tennekes (1975) suggested that *in turbulence with high Reynolds numbers... the dissipative eddies flow past an Eulerian observer in a time much shorter than the time scale which characterizes their own dynamics. This suggests that Taylor's 'frozen-turbulence' approximation should be valid for the analysis of the consequences of large-scale advection of the turbulent microstructure.* In fact, Tennekes' hypothesis consists of two ingredients. First, it is proposed that the Lagrangian acceleration of fluid particles, $\mathbf{a}$, is in some sense small, and in order to obtain estimates and comparison of Lagrangian and Eulerian time scales Tennekes put just $\mathbf{a} = 0$. It is noteworthy that this assumption is local pointwise in space/time and is not a statistical one. The second assumption made by Tennnekes is of statistical nature, namely, that *the misrostructure is statistically independent of the energy containing eddies.*

The equality $\mathbf{a} = 0$ should not be understood literally[31]. The accel-

---

[31]Just like the misrostructure is *not* statistically independent of the energy containing

eration of fluid particles cannot be vanishing, since $\mathbf{a} = 0$ may be misinterpreted as the balance between the pressure gradient and the viscous term in the Navier-Stokes equations, which is obviously incorrect because $\mathbf{a}_i = -\frac{1}{\rho}\nabla p$ is irrotational and $\mathbf{a}_s = \nu\nabla^2\mathbf{u}$ is solenoidal. Also $\mathbf{a} = 0$ would mean that the flow is described by the equation $\partial\mathbf{u}/\partial t + (\mathbf{u}\cdot\nabla)\mathbf{u} = 0$.

More precisely $\mathbf{a} \approx 0$ means that in some sense $\mathbf{a}$ is small comparing both to $\mathbf{a}_l$ and $\mathbf{a}_c$, e.g. $\langle a^2\rangle/\langle a_l^2\rangle \ll 1$ and $\langle a^2\rangle/\langle a_c^2\rangle \ll 1$. This in turn is possible if there is mutual (statistical) cancellation between the local acceleration, $\mathbf{a}_l$, and convective acceleration, $\mathbf{a}_c$. Since these quantities are vectors, the degree of this mutual cancellation should be studied both in terms of their magnitude and the geometry of vector alignments.

In terms of magnitude, it is seen from figure 6.26 that the acceleration is much smaller than its local and convective components. Indeed, the two ratios $\frac{\langle\mathbf{a}^2\rangle}{\langle\mathbf{a}_l^2\rangle}, \frac{\langle\mathbf{a}^2\rangle}{\langle\mathbf{a}_c^2\rangle}$ decrease, and the ratio $\frac{\langle\mathbf{a}_l^2\rangle}{\langle\mathbf{a}_c^2\rangle}$ tends to unity with inceasing Reynolds number. At the highest accessible Reynolds number, $\mathrm{Re}_\lambda \approx 240$ the variance of the local acceleration, $\langle\mathbf{a}_l^2\rangle^{1/2}$, is about 0.9 of the magnitude of convective acceleration, $\langle\mathbf{a}_c^2\rangle^{1/2}$. At this Reynolds number the total acceleration is about $0.1\langle\mathbf{a}_c^2\rangle^{1/2}$, i.e. an order of magnitude smaller than both its components.

Similarly, the correlation coefficient between $\mathbf{a}_l$ and $\mathbf{a}_c$, is expected to be $\mathcal{O}(1)$ and tend to unity (with minus sign) with inceasing Reynolds number. This, indeed, is seen in figure 6.27 (lower frame). It is noteworthy that $\mathbf{a}$ and $\mathbf{a}_l$ are practically decorrelated.

An important point is that though the magnitude of the total acceleration becomes small compared to both its local, $\mathbf{a}_l$, and convective, $\mathbf{a}_c$, components as the Reynolds number increases, the latter do not compensate each other totally. This cannot happen at *any* Reynolds number, since $\mathbf{a}_l$ is divergence-free, whereas $\mathbf{a}_c$ is not, and its irrotational part, which is equal to $\mathbf{a}_i = -\frac{1}{\rho}\nabla p$, contributes most to the magnitude of the total acceleration (Shtilman et al., 1993; Tsinober, 1990a; Vedula and Yeung, 1999). Since the solenoidal part of the total acceleration, $\mathbf{a}_s = \nu\nabla^2\mathbf{u}$ on the one hand, and $\mathbf{a}_s = \mathbf{a}_l + \mathbf{a}_{cs}$ ($\mathbf{a}_{cs}$ is the solenoidal part of $\mathbf{a}_c$) on the other, the only way $\mathbf{a}_s$ can become small is that $\mathbf{a}_l$ and $\mathbf{a}_{cs}$ compensate each other, i.e. they should be almost the same in magnitude but almost antiparrallel. Indeed, the ratio $\frac{\langle\mathbf{a}_l^2\rangle}{\langle\mathbf{a}_{cs}^2\rangle} \approx 1$ similar to $\frac{\langle\mathbf{a}_l^2\rangle}{\langle\mathbf{a}_c^2\rangle}$ shown in figure 6.26, and their correlation coefficient is very close to -1 (figure 6.27, middle frame).

The geometrical aspect of the above results is seen from the alignments between the vectors involved. Namely, if $\mathbf{a} = \mathbf{a}_l + \mathbf{a}_c$ is small compared to $\mathbf{a}_l$ and $\mathbf{a}_c$, then the last two vectors are expected to be (anti-) aligned, i.e.

eddies (see previous section).

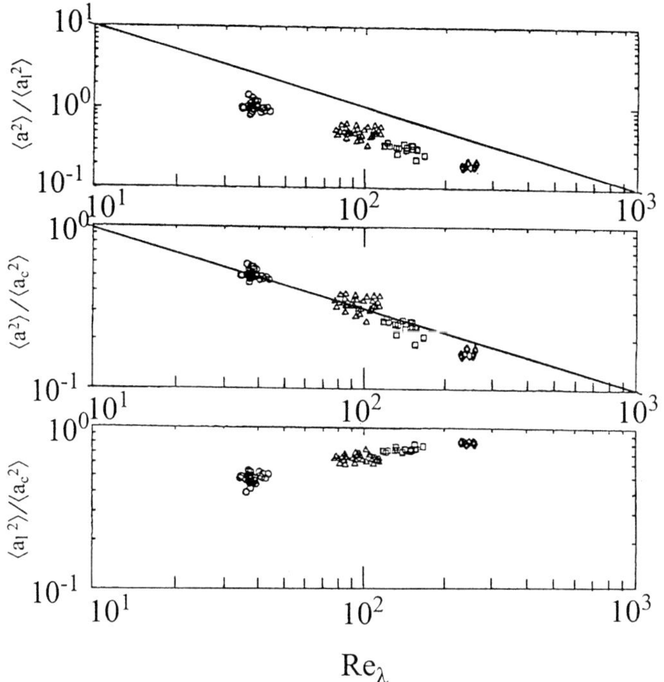

Figure 6.26. Reynolds number dependence of the ratios $\frac{\langle \mathbf{a}^2 \rangle}{\langle \mathbf{a}_l^2 \rangle}$, $\frac{\langle \mathbf{a}^2 \rangle}{\langle \mathbf{a}_c^2 \rangle}$ and $\frac{\langle \mathbf{a}_l^2 \rangle}{\langle \mathbf{a}_c^2 \rangle}$.

the cosine of the angle between $\mathbf{a}_l$ and $\mathbf{a}_c$, $\cos(\mathbf{a}_l, \mathbf{a}_c)$ should be negatively skewed, i.e. to have a maximum around -1. This, indeed, is observed in figure 6.28.

This alignment exhibits an essential dependence on Reynolds number: the tendency of alignment between $\mathbf{a}_l$ and $\mathbf{a}_c$ is strongly enhanced with increasing Reynolds number. The alignment between $\mathbf{a}_l$ and $\mathbf{a}_c$ is a consequence of even stronger alignment between $\mathbf{a}_l$ and $\mathbf{a}_{cs}$, with stronger Re-dependence as well. On the other hand both 'components' of $\mathbf{a}_s$, i.e. $\mathbf{a}_l$ and $\mathbf{a}_{cs}$, are about twenty times larger than $\mathbf{a}_s$ itself even at the smallest Reynolds number, and at the largest Re_λ, $\langle \mathbf{a}_s^2 \rangle^2 \approx 3 \cdot 10^{-3}$ of $\langle \mathbf{a}_{cs}^2 \rangle^2$ and/or $\langle \mathbf{a}_l^2 \rangle^2$ (see figure 6.29, upper frame).

In other words, there is a tendency of compensation between the local acceleration, $\mathbf{a}_l = \frac{\partial \mathbf{u}}{\partial t}$, and the solenoidal part, $\mathbf{a}_{cs}$, of the convective acceleration, $\mathbf{a}_c = (\mathbf{u} \cdot \nabla)\mathbf{u}$. It is this tendency, which is increasing with Reynolds number, that makes the solenoidal part of the acceleration, $\mathbf{a}_s$, much smaller than the irrotational part of the acceleration, $\mathbf{a}_i = \mathbf{a}_{ci}$ ($\mathbf{a}_{ci}$ is the irrotational part of $\mathbf{a}_c$). It is natural to call this tendency the reduction

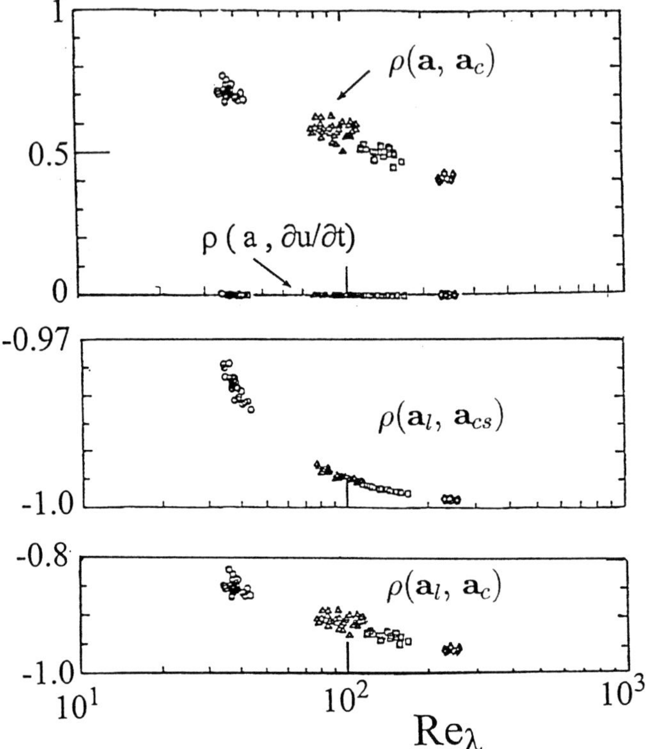

Figure 6.27. Reynolds number dependence of the correlation coefficients between $\mathbf{a}_c$ and $\mathbf{a}_l$, $\mathbf{a}_{cs}$ and $\mathbf{a}_c$, $\mathbf{a}_l$ and $\mathbf{a}_c$.

of solenoidality of the total acceleration. In fact this tendency is seen in figure 1 of Vedula and Yeung (1999). We return to this matter in the next section.

This tendency should be distinguished from an opposite tendency when comparing the solenoidal part, $\mathbf{a}_{cs}$, of the convective acceleration, i.e. of the nonlinearity $(\mathbf{u} \cdot \boldsymbol{\nabla})\mathbf{u}$, (and also $\mathbf{a}_l$), and its potential part, $\mathbf{a}_{ci}$: as the Reynolds number increases both $\mathbf{a}_l$ and $\mathbf{a}_{cs}$ become large compared to $\mathbf{a}_i$ (or $\mathbf{a}_{ci}$ which is the same). For example, the ratio $\frac{\langle \mathbf{a}_{ci}^2 \rangle}{\langle \mathbf{a}_{cs}^2 \rangle}$ decreases substantially with Reynolds number as shown in the lower frame of figure 6.29. In other words, along with the above-mentioned tendency of reduction of solenoidality of the total acceleration, there is a concomitant tendency of enhancement of solenoidality of its convective part, $\mathbf{a}_c$, i.e. the nonlinearity, $(\mathbf{u} \cdot \boldsymbol{\nabla})\mathbf{u}$, tend to become more solenoidal as the Reynolds number increases. It appears that this tendency is, to a large extent, of kinematical

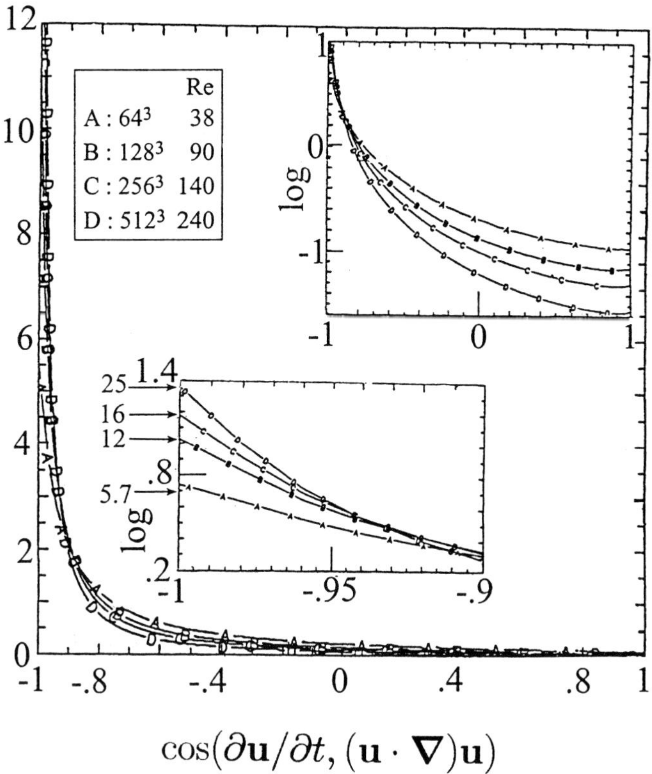

$$\cos(\partial \mathbf{u}/\partial t, (\mathbf{u} \cdot \nabla)\mathbf{u})$$

Figure 6.28. Alignment (PDFs of the cosine of the angle) between $\mathbf{a}_l$ and $\mathbf{a}_c$. The insets show this dependence with the vertical in log and in the proximity of $\cos(\mathbf{a}_l, \mathbf{a}_c) \sim -1$. It is noteworthy that the strong alignment between $\mathbf{a}_l$ and $\mathbf{a}_c$ was observed in recent laboratory experiments (Lüthi et al., 2001) and in the atmospheric surface layer (Kholmyansky et al., 2001b).

nature (see below Tsinober, 1990a; and Pinsky et al., 2000). Namely, for a Gaussian velocity field the solenoidal part of the convective acceleration $\mathbf{a}_c \equiv (\mathbf{u} \cdot \nabla)\mathbf{u}$ is larger than its irrotational part.

6.7.2. THE RELATION BETWEEN THE TOTAL ACCELERATION AND ITS IRROTATIONAL AND SOLENOIDAL COMPONENTS

The relation between $\mathbf{a}$, $\mathbf{a}_i = -\frac{1}{\rho}\nabla p$ and $\mathbf{a}_s = \nu \nabla^2 \mathbf{u}$ is qualitatively different from that considered in the previous section. Namely, as is seen from figure 6.30 (see also figure 6.31), the main contribution to the total acceleration comes from its irrotational part, $\mathbf{a}_i$ (Vedula and Yeung, 1999). Consequently, the correlation between $\mathbf{a}$ and $\mathbf{a}_i$ is close to unity and is

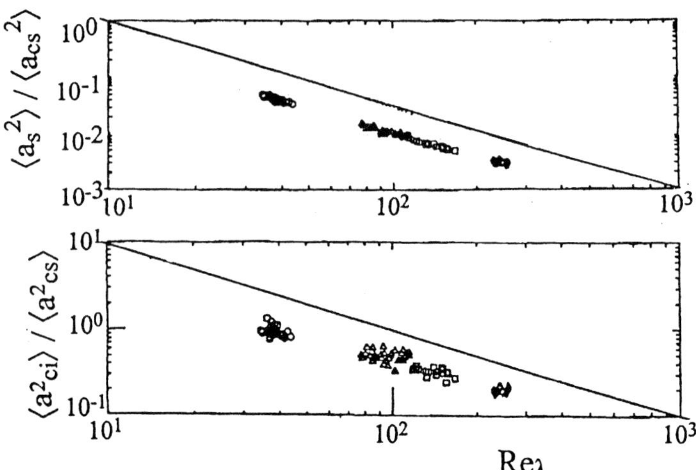

Figure 6.29. Reynolds number dependence of the ratios $\frac{\langle \mathbf{a}_s^2 \rangle}{\langle \mathbf{a}_{cs}^2 \rangle}$ and $\frac{\langle \mathbf{a}_{ci}^2 \rangle}{\langle \mathbf{a}_{cs}^2 \rangle}$; $\mathbf{a}_s = \nu \nabla^2 \mathbf{u}$, and $\mathbf{a}_{cs}$, $\mathbf{a}_{ci}$ are the solenoidal and potential parts of the convective acceleration $\mathbf{a}_c = (\mathbf{u} \cdot \nabla)\mathbf{u}$.

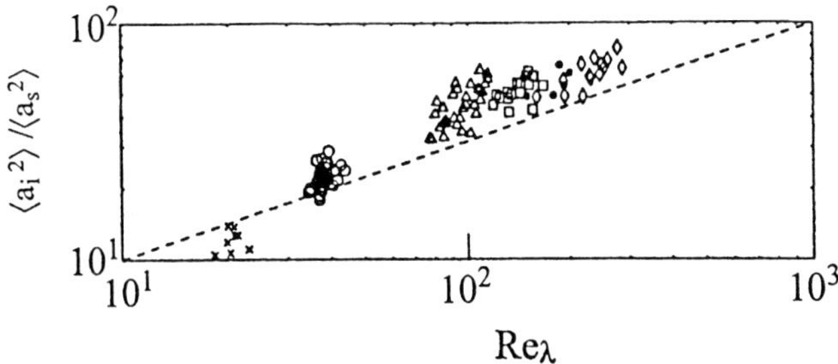

Figure 6.30. Reynolds number dependence of the ratio $\frac{\langle \mathbf{a}_i^2 \rangle}{\langle \mathbf{a}_S^2 \rangle}$. This is an updated figure equivalent to figure 1a in Vedula and Yeung (1999).

~ 0.98 at $\mathrm{Re}_\lambda \approx 240$. The variance of the solenoidal part of the acceleration $\frac{\langle \mathbf{a}_s^2 \rangle}{\langle \varepsilon \rangle^{3/2} \nu^{-1/2}}$ is independent of Reynolds number in the range investigated (Vedula and Yeung, 1999), so that the increase of the ratios $\frac{\langle \mathbf{a}^2 \rangle}{\langle \mathbf{a}_S^2 \rangle}$ and $\frac{\langle \mathbf{a}_i^2 \rangle}{\langle \mathbf{a}_S^2 \rangle}$ with Reynolds number is due to the increase of the irrotational part of the acceleration.

Again the geometrical aspect is seen from the alignments between the vectors involved. Namely, the only strongly exhibited alignment is the one between $\mathbf{a}$ and $\mathbf{a}_i$ (figure 6.32). The alignment between $\mathbf{a}$ and $\mathbf{a}_s$ is weak, and there is no alignment between $\mathbf{a}_i$ and $\mathbf{a}_s$, since they are orthogonal in the sense that $\langle \mathbf{a}_i \cdot \mathbf{a}_s \rangle = 0$. All these alignments are weakly sensitive to the Reynolds number.

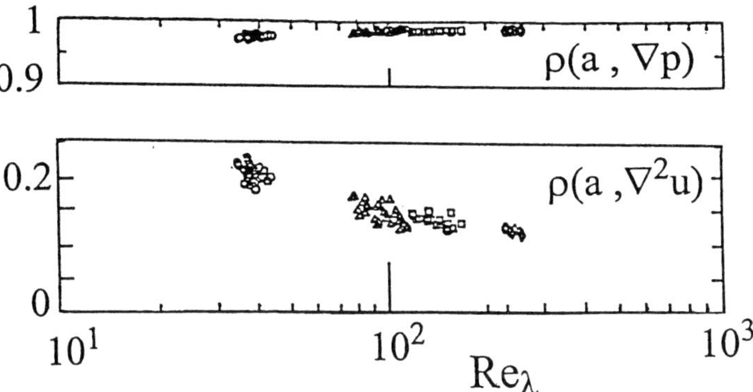

Figure 6.31. Correlation coefficients between $\mathbf{a}$ and $\mathbf{a}_i$, and $\mathbf{a}$ and $\mathbf{a}_s$ at several Reynolds numbers.

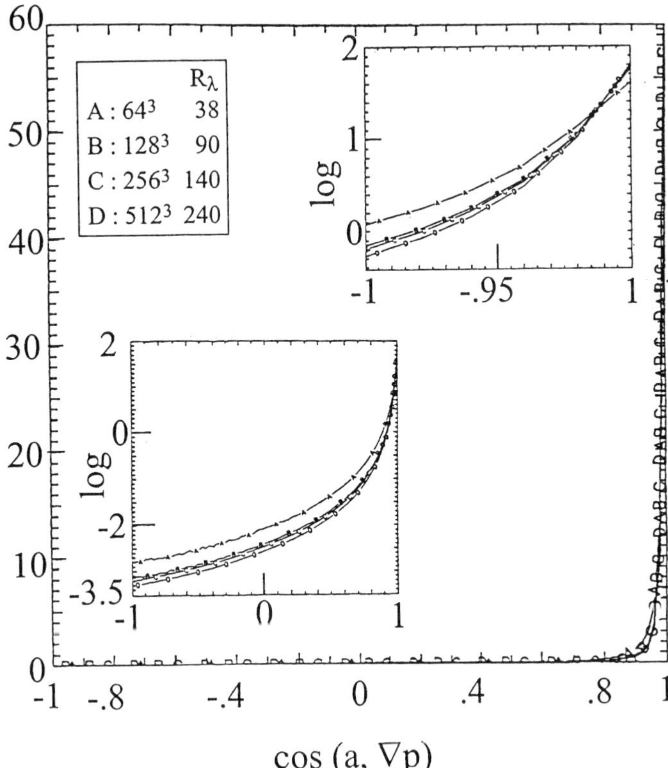

Figure 6.32. Alignments (PDF of the cosines of the angle) between $\mathbf{a}$ and $\mathbf{a}_i$ at different Reynolds numbers. The insets show this dependence with the vertical in log and in the proximity of $\cos(\mathbf{a}, \mathbf{a}_i) \sim 1$.

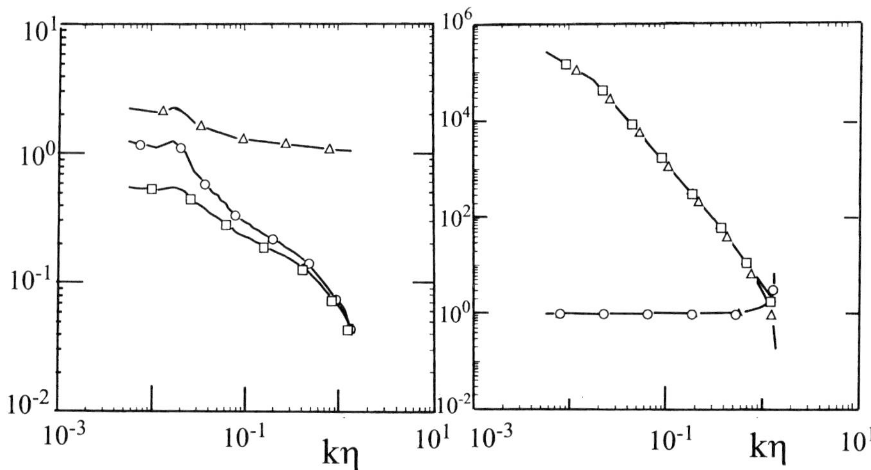

Figure 6.33. Ratios: left – $\frac{S_a}{S_{a_l}}(\square)$, $\frac{S_a}{S_{a_c}}(\circ)$ and $\frac{S_{a_c}}{S_{a_l}}(\triangle)$; right – $\frac{S_a}{S_{a_i}}(\circ)$, $\frac{S_a}{S_{a_s}}(\square)$ and $\frac{S_{a_i}}{S_{a_s}}(\triangle)$ as functions of the wave number at $Re_\lambda = 240$. Here η - is the Kolmogorov microscale.

6.7.3. SCALE DEPENDENCE

The scale dependence can be seen by looking at one dimensional spectra and/or the spectral analogues of the quantities addressed above, e.g. $\frac{S_a}{S_{a_l}}$, where S_a, S_{a_l} are the one dimensional energy spectra of $\mathbf{a}$, $\mathbf{a}_l$.

The tendencies described above regarding the relations between $\mathbf{a}, \mathbf{a}_l$ and $\mathbf{a}_c$ are expected to be strongest at smaller scales. This is clearly seen from figure 6.33, i.e. the ratio $\frac{S_{a_c}}{S_{a_l}}$ tends to unity, whereas the ratios $\frac{S_a}{S_{a_l}}, \frac{S_a}{S_{a_c}}$ decrease with $k\eta$ and become much smaller than their overall analogues $\frac{\langle \mathbf{a}^2 \rangle}{\langle \mathbf{a}_i^2 \rangle}$, $\frac{\langle \mathbf{a}^2 \rangle}{\langle \mathbf{a}_s^2 \rangle}$ at largest Reynolds numbers, i.e. these relations are quite sensitive to the Reynolds number.

A different picture is seen in respect with the ratios $\frac{S_a}{S_{a_i}}, \frac{S_a}{S_{a_s}}$ and $\frac{S_{a_i}}{S_{a_s}}$. As expected the ratio $\frac{S_a}{S_{a_i}}$ is of order 1, except at the smallest scales, where it is larger due to the contribution from $\mathbf{a}_s$. Both ratios $\frac{S_a}{S_{a_s}}$ and $\frac{S_{a_i}}{S_{a_s}}$ are much larger than $\frac{S_a}{S_{a_i}}$, especially at large scales, and are quite similar except again in the small scales due to the same reason as above. The pattern is the same qualitatively at all four Reynolds numbers, but is quite different quantitatively: at large Reynolds numbers (the only one shown in figure 6.33) the differences become orders of magnitude larger than at small Reynolds numbers.

6.7.4. KINEMATICAL VERSUS DYNAMICAL EFFECTS

By kinematical we mean the effects which are exhibited in random Gaussian analogues of the real fields. In some aspects these effects can be dominant (Shtilman et al., 1993; Tsinober, 1990a). The reference random Gaussian fields are Gaussian in the sense of the velocity gradients being (artificially) Gaussian, but have the same form of the energy spectrum and Reynolds number corresponding to each grid resolution.

It appears that the behaviour of real fields and of their Gaussan counterparts is qualitatively the same in a number of aspects. This is true, for example, of the alignments discussed above and also of the joint PDFs of $\mathbf{a}_c$ and $\mathbf{a}_l$ and of $\mathbf{a}$ and $\mathbf{a}_i$. In other words, the discussed effects are mainly (but not entirely) of kinematical nature (see also figure 6.38).

Along with the significant qualitative resemblance between the real and the random Gaussian cases, there is a quantitative difference. This difference is exhibited in several ways. First, the Reynolds number dependence of the effects described above is stronger in the case of real flow field. Second, the essential difference between the real and the Gaussian fields is seen from the PDFs of the magnitudes of $\mathbf{a}$, $\mathbf{a}_c$, and $\mathbf{a}_l$, and $\mathbf{a}$, $\mathbf{a}_c$, and $\mathbf{a}_l$. All the PDFs for the real flow field are more intermittent than those for their Gaussian counterparts, in the sense that they have much higher flaring tails. The Reynolds number dependence of this efffect is also stronger for the real field. For more details see Tsinober et. al (2001).

It is noteworthy that Gaussian fields are by definition not intermittent: the non-Gaussian shape of the PDFs of quantities like a_i for a Gaussian velocity field is because a_i is nonlinear in velocity.

Thus the quantitative meaning of the smallness of the total acceleration, $\mathbf{a}$, is that it appears to be small as *in comparison with* to its local and convective components, $\mathbf{a}_l = \frac{\partial \mathbf{u}}{\partial t}$ and $\mathbf{a}_c = (\mathbf{u} \cdot \nabla)\mathbf{u}$. Already at $Re_\lambda = 240$, the variance of $\mathbf{a}$ is more than an order of magnitude smaller than the variance of $\mathbf{a}_l$ and $\mathbf{a}_c$. At this Reynolds number the local and the convective accelerations are strongly (anti) correlated with a correlation coefficient exceeding $|-0.9|$. On the other hand, $\mathbf{a}$ is of the same order than its potential part $\mathbf{a}_i = -\frac{1}{\rho}\nabla p$, and both are much larger than the solenoidal part of the acceleration $\mathbf{a}_s = \nu\nabla^2\mathbf{u}$. The smallness of $\mathbf{a}_s$ is maintained by strong cancellation between local acceleration $\mathbf{a}_l$ (which is solenoidal), and the solenoidal part of the convective accleration $\mathbf{a}_{cs}$ ($\mathbf{a}_c = \mathbf{a}_{cs} + \mathbf{a}_{ci}$), so that the irrotational part of $\mathbf{a}_c$, which is equal to the irrotational part of the total acceleration, $\mathbf{a}_{ci} = \mathbf{a}_i$, comprises the main contribution to the total acceleration, but is much smaller than both $\mathbf{a}_l$ and/or $\mathbf{a}_c$.

We summarize the above as follows

$$\langle \mathbf{a}_s^2 \rangle \ll \langle \mathbf{a}^2 \rangle \approx \langle \mathbf{a}_i^2 \rangle = \langle \mathbf{a}_{ci}^2 \rangle \ll \langle \mathbf{a}_c^2 \rangle \approx \langle \mathbf{a}_{cs}^2 \rangle \approx \langle \mathbf{a}_l^2 \rangle, \qquad (6.8)$$

The meaning of $\ll$ in the above relation is (at least) 'an order of magnitude smaller than' at the largest accessible Reynolds numbers. The relation between various components of acceleration as expreseed by (6.8) becomes stronger in small scales. It is natural to expect that, at still larger Reynolds numbers the meaning of '$\ll$' in the above relation will become 'several orders of magnitude smaller than'. If true this fact provides some justification for the 'random sweeping decorrelation hypothesis'. We tend, however, to give a more limited interpretation to this hypothesis in the sense that the misrostructure (whatever this means) is *statistically decorrelated* of the energy containing eddies. This is different from the original assumption made by Tennekes (1975), in which he held that *the misrostructure is statistically independent of the energy containing eddies.* The large and small scales are statistically not independent, though they are practically decorrelated (see previous section).

It should be stressed that, though the 'components' $\mathbf{a}_l$ and $\mathbf{a}_{cs}$ (or $\mathbf{L}_s$) of the solenoidal part of the acceleration, $\mathbf{a}_s = \mathbf{a}_{cs} + \mathbf{a}_l$, mostly cancel each other, this fact does not mean that separately they are unimportant. Their importance is seen at the level of velocity derivatives. For example, the main contribution to the enstrophy production is associated with $\mathbf{a}_{cs}$ (or $\mathbf{L}_s$), but not with $\mathbf{a}_l$. Namely, the enstrophy production term is $curl\{\mathbf{a}_{cs} - (\mathbf{u}\nabla)\omega\}$ and is approximately balanced by the viscous term in the equation for the enstrophy, whereas $curl\{\mathbf{a}_l + (\mathbf{u}\nabla)\omega\}$ is much smaller than $curl\{\mathbf{a}_{cs} - (\mathbf{u}\nabla)\omega\}$ and is balanced by the sum of the enstrophy production and the viscous terms (see section 6.3).

6.8. Non-Gaussian nature of turbulence

> *Turbulence – being essentially non-Gaussian – is such a rich phenomenon that it can 'afford' a number of Gaussian-like manifestations, some of which are not obvious and even non-trivial.*

The non-Gaussian nature of turbulence is another 'N' contributing to the difficulty of turbulence – in addition to the three mentioned before non-linearity, non-integrability and non-locality.

A rather common view/assumption was that the field of turbulent fluctuations is - in some sense(s) - nearly Gaussian. For example, there were many attempts to treat the problem as a perturbation of a Gaussian field[32].

---

[32]Even Hopf (1962) was tempted by quasi-Gaussianity.

More recently, it was proposed that some parts of a turbulent field are nearly Gaussian, so that it can be decomposed into a (nearly) Gaussian part and a (strongly) non-Gaussian one (e.g. Chertkov et al., 1999; Farge et al., 1999); Katul et al., 1994; Lewalle et al., 2000; She et al., 1991). This is a tempting assumption, since it helps to simplify the problem. However, without entering into a comprehensive review of the whole issue, we shall argue that such an assumption is inadequate[33]. For this purpose, we give a number of (counter) examples with the emphasis on the dynamical aspects. Additional aspects are addressed in the next chapter.

6.8.1. ODD MOMENTS

First we mention that a purely Gaussian velocity field is dynamically impotent. Indeed, in such a field, all the odd moments vanish[34]. This contradicts the Kolomogorov 4/5 law, turns the Karman-Howarth equation into a linear one (see Monin and Yaglom, 1971), and prevents the production of enstrophy and strain (i.e. dissipation): in a Gaussian velocity field $\langle \omega_i \omega_j s_{ij} \rangle \equiv 0$ and $\langle s_{ij} s_{jk} s_{ki} \rangle \equiv 0$. Even if the flow field is initially Gaussian, the dynamics of turbulence makes it non-Gausssian with finite rate. This is seen by taking $\langle \ldots \rangle$ from the equation (C.23) (dropping the viscous term)

$$\frac{D}{Dt} \langle \omega_i \omega_j s_{ij} \rangle = \langle \omega_j s_{ij} \omega_k s_{ik} \rangle - \langle \omega_i \omega_j \frac{\partial^2 p}{\partial x_i \partial x_j} \rangle \tag{6.9}$$

For a Gaussian velocity field $\langle \omega_i \omega_j s_{ij} \rangle_G = 0$, $\left\langle \omega_i \omega_j \frac{\partial^2 p}{\partial x_i \partial x_j} \right\rangle_G = 0$ and $\langle \omega_j s_{ij} \omega_k s_{ik} \rangle_G = \frac{1}{6} \langle \omega^2 \rangle^2 > 0$. Since the quantity $\omega_j s_{ij} \omega_k s_{ik} \equiv W^2, W_i = \omega_j s_{ij}$, it is positive pointwise for any vector field. Hence at $t = 0$

$$\left\{ \frac{D}{Dt} \langle \omega_i \omega_j s_{ij} \rangle \right\}_{t=o} = \{ \langle \omega_j s_{ij} \omega_k s_{ik} \rangle \}_{t=0} > 0, \tag{6.10}$$

It follows from the equation (6.10) that, at least for a short time interval t, the mean enstrophy production will become positive. We remind that for later moments the vorticity–pressure hessian correlation $\left\langle \omega_i \omega_j \frac{\partial^2 p}{\partial x_i \partial x_j} \right\rangle$ becomes finite, and nothing is known rigorously. As follows from DNS of NSE in a periodic box, the corelation $\left\langle \omega_i \omega_j \frac{\partial^2 p}{\partial x_i \partial x_j} \right\rangle$ is positive, but is smaller than $\langle \omega_j s_{ij} \omega_k s_{ik} \rangle \equiv \langle W^2 \rangle$. Namely, $\left\langle \omega_i \omega_j \frac{\partial^2 p}{\partial x_i \partial x_j} \right\rangle \sim \frac{1}{3} \langle W^2 \rangle$, so that the RHS of

---

[33]Most frequently it is assumed that the 'unresolved' (i.e. mainly the small scale) part of the turbulent field is cnearly Gaussian, in contradiction to observations showing that this part of flow is the most non-Gaussian (see also 6.4).

[34]Therefore they are so convenient in assessing the 'deviations' from Gaussianity.

(6.9) remains positive (Tsinober et al., (1995). The nonzero $\left\langle \omega_i \omega_j \frac{\partial^2 p}{\partial x_i \partial x_j} \right\rangle$ is another manifestaion of non Gaussianity. It is noteworthy that the equation (6.9) with $\left\langle \omega_i \omega_j \frac{\partial^2 p}{\partial x_i \partial x_j} \right\rangle = 0$ is precisely the one arising using the quasi-Gaussian approximation $\frac{D^2}{Dt} \left\langle \omega^2 \right\rangle = \frac{1}{3} \left\langle \omega^2 \right\rangle^2$ (Proudman and Reid, 1954; Orszag, 1977), since $\frac{D}{Dt} \left\langle \omega_i \omega_j s_{ij} \right\rangle = \frac{1}{2} \frac{D^2}{Dt} \left\langle \omega^2 \right\rangle$ and under quasi-Gaussian approximation $\left\langle \omega_j s_{ij} \omega_k s_{ik} \right\rangle = \frac{1}{6} \left\langle \omega^2 \right\rangle^2$ and $\left\langle \omega_i \omega_j \frac{\partial^2 p}{\partial x_i \partial x_j} \right\rangle = 0$. The essential point is that at $t = 0$, the relation (6.10) is precise due to the freedom of the choice of the intitial condition. On the other hand, we have seen that, if the initial conditions are Gausssian, the flow ceases to be Gaussian with *finite rate*. In other words, it is seen directly from (6.10) that *turbulence cannot be Gaussian* (see also Novikov, 1967). In this sense, Gaussian initial conditions are not 'good', since no flow state existing in reality is Gaussian. In a similar way, one can see from the equation (C.24) that the mean rate of production of strain becomes positive at small times (at $t = 0$ it vanishes) for an initially Gaussian velocity field.

The next point is that the Kolmogorov 4/5 law, $S_3(r) = -\frac{4}{5} \langle \epsilon \rangle r$, $S_3(r) = \left\langle \left(\Delta u_{\parallel} \right)^3 \right\rangle$, is a clear demonstration of the non-Gaussian and dissipative nature of turbulence; $S_3(r)$ is essentially nonvanishing.

A special aspect of the the non-Gaussian nature of turbulence as manifested in the 4/5 law is seen when one looks at the analogy between the 4/5 law and Yaglom's 4/3 law for fluctuations of a passive scalar, θ. The latter has the following form $\left\langle \Delta u_{\parallel} (\Delta \theta)^2 \right\rangle = -\frac{4}{3} \langle \epsilon_\theta \rangle r$, where $\Delta u_{\parallel} \equiv [\mathbf{u}(\mathbf{x} + \mathbf{r}) - \mathbf{u}(\mathbf{x})] \cdot \mathbf{r}/r$, $\Delta \theta = \theta(\mathbf{x} + \mathbf{r}) - \theta(\mathbf{x})$, and $\epsilon_\theta = \mathcal{D} \frac{\partial \theta}{\partial x_i} \frac{\partial \theta}{\partial x_i}$ is the dissipation of the passive scalar. The analogy, though useful in some respects (Antonia et al., 1997)[35], is violated for a Gaussian velocity field. Namely, the 4/3 law remains valid for such (as any other isotropic) velocity field, whereas the 4/5 law is not, because $S_3(r) \equiv 0$ for a Gaussian velocity field. This points to serious limitations on analogies between the passive and active fields mentioned above[36].

More generally, the build-up of odd moments, such as $S_3(r)$, $\langle \omega_i \omega_k s_{ik} \rangle$, $\langle s_{ij} s_{jk} s_{ki} \rangle$ and many others is an important manifestation of the nonlinearity and non-Gaussianity of turbulence. It is one of the prominent and distinctive *specific* features of turbulent flows of utmost dynamical significance involving such processes as production of enstrophy and total strain (dissipation). Hence the particular emphasis on the *odd* moments. For this

---

[35]Antonia et al. (1997) looked at the analogy between the 4/3 law for the passive scalar and the 4/3 law for the velocity field in the form $\left\langle \Delta u_{\parallel} (\Delta \mathbf{u})^2 \right\rangle = -\frac{4}{3} \langle \epsilon \rangle r$, which turns into the 4/5 law by isotropy.

[36]This includes the *well established phenomenological parallels between the statsistical description of mixing and fluid turbulence itself* (Shraiman and Siggia, 2000).

reason, the quantity $cos(\boldsymbol{\omega}, \mathbf{W}) = \omega_i\omega_j S_{ij}|\omega|^{-1}|W|^{-1}$ proved quite so useful in the diagnostics of the non-Gaussian nature of the 'random structureless' sea in turbulent flows, which appeared to be quite the opposite, i.e. not stuctructureless, dynamically not passive and essentially non-Gaussian (see section 6.4). This is in contrast with various recent proposals that the 'weak', in some sense, part of the field is nearly Gaussian. It seems that there exist no such part(s). More work is neccessary to clarify the issue starting with the quasi-Gaussian manifestations of turbulent flows.

6.8.2. QUASI-GAUSSIAN MANIFESTATIONS

Single point moments of velocity fluctuations are known to be close to Gaussian[37]. However, this does not mean that the field of velocityfluctuations is really Gaussian. For example, non-Gaussian behaviour of the field of velocity fluctuations has been observed in experiments by Frenkiel et al. (1979) on grid turbulence. They measured correlations of the type $R^{m,n}(\tau)$ $= \frac{\langle u^m(t)u^m(t+\tau)\rangle}{\langle u^2(t)\rangle^{(m+n)/2}}$. The non-Gaussian behavior of the field of velocity fluctuations is seen clearly for $\tau \neq 0$: the odd order correlations, $R^{2,1}(\tau)$, $R^{4,1}(\tau)$, $R^{3,2}(\tau)$, are essentially nonzero especially around $U\tau/M \sim 1$ (see an example shown in figure 6.34). It is noteworthy that at $h = 0$ the $R^{2,1}(\tau)$, $R^{4,1}(\tau)$, $R^{3,2}(\tau)$ are indistinguishable from zero showing the importance of two-point correlations. The even moments up to the sixth order (not only numbers but the whole correlation functions) are very close to Gaussian too, see figure 6.35 left. This means, for example, that the fourth order correlation shown in this figure is very well approximated by the Millionschikov hypothesis (zero-fourth cumulant) *as* for a Gaussian velocity field. The velocity derivatives in the experiments of Frenkiel et al. (1979) exhibit clear non-Gaussian behaviour for all moments.

There are more intricate examples of nearly Gaussian manifestastions of turbulent flows. Two examples are shown in figures 6.35 and 6.36. The example in figure 6.36 is interesting in that the Gaussian-like behaviour is exhibited by third order quantitites. For other examples, see Tsinober (1998a).

The fourth example involves pressure. It was shown by Holzer and Siggia (1993) that for a Gaussian velocity field the PDF of pressure is strongly negatively skewed and has exponential tails. This result is not unexpected, though it is not easy in this case to demonstrate it directly due to nonlocal

---

[37]This is not true of strong fluctuations of velocity. It was observed in laboratory and numerical experiments that the single point PDF of velocity at large amplitudes of velocity fluctuations is sub-Gaussian and recently was confirmed theoretically using the instanton formalism (see references in Tsinober, 1998b; . for other quasi-Gaussian manifestations see Brun and Pumir, 2001; and references therein).

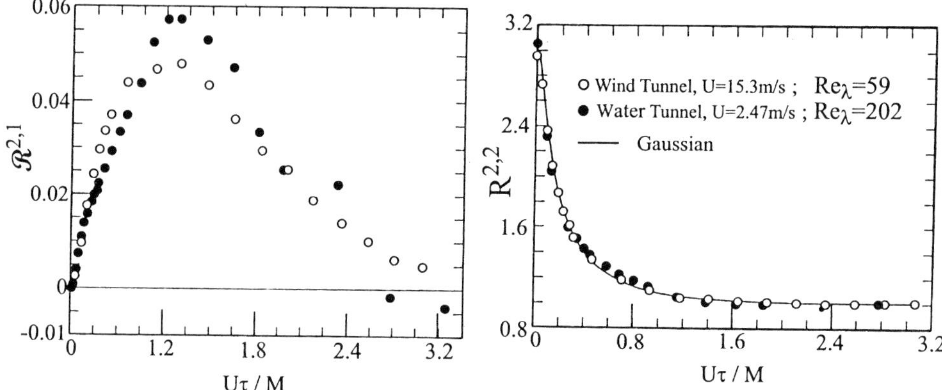

Figure 6.34. Third (left) and fourth (right) order time correlations for turbulent velocities. Adapted from Frenkiel et al. (1979). Here $\mathbf{R}^{m,n} = \frac{1}{2}\left[R^{m,n}(\tau) + R^{n,m}(\tau)\right]$. The correlation $\mathbf{R}^{m,n}$ was used instead of individual odd order correlations $R^{m,n}(\tau)$, since the latter exhibited *large dispersions from sample recording to sample recording* – a well known problem to everybody tried to evaluate odd order moments. Here M is the grid mesh and U is the mean velocity of the flow.

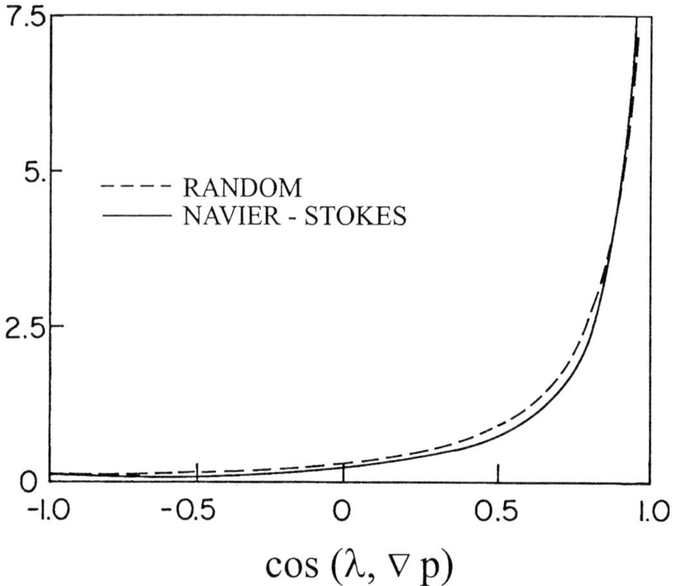

Figure 6.35. PDFs of the angle between the Lamb vector $\boldsymbol{\omega} \times \mathbf{u}$ and its potential part $\boldsymbol{\nabla}\alpha \equiv \boldsymbol{\nabla}p$ for a numerically simulated turbulence and a random (approximately Gaussian) velocity field with the same energy spectrum (Shtilman et al., 1993). The similarity is obvious.

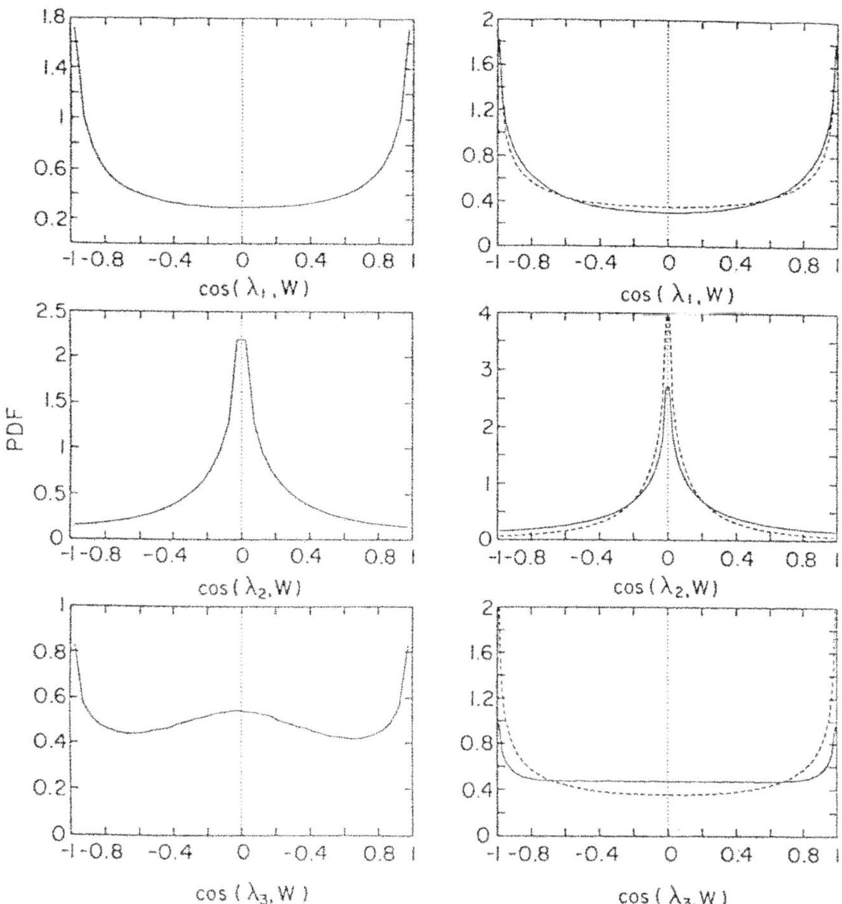

Figure 6.36. PDFs of the cosine of the angle between vortex stretching vector $W_i = \omega_i s_{ij}$, and the eigenvectors λ_i of the rate of strain tensor; left – grid turbulence, right - DNS and random Gaussian ($----$). $Re_\lambda \approx 75$. The behaviour for λ_2 seems to contradict the one shown in figure 6.11, since there is a tendency for alignment between ω and $\mathbf{W}$ and between ω and λ_2. However, closer inspection shows that the alignments shown in this figure and in figures 6.11 are associated with different regions in the flow.

relation between pressure and velocity fields. Much easier is to do this by looking at $\nabla^2 p$. Using the same method as in Shtilman et al. (1993), the PDF $\mathcal{P}(x)$, $x = \frac{\nabla^2 p}{\rho\langle\omega^2\rangle} = \frac{\omega^2 - 2s_{ij}s_{ij}}{2\langle\omega^2\rangle}$, is expressed in the following way (Spector 1996, private communication)

$$\mathcal{P}(x) = \{3^{1/2}5^{5/2}\}/(4\pi)\, x^2\, e^x\, [K_2(4x) - K_1(4x)], \qquad x < 0,$$

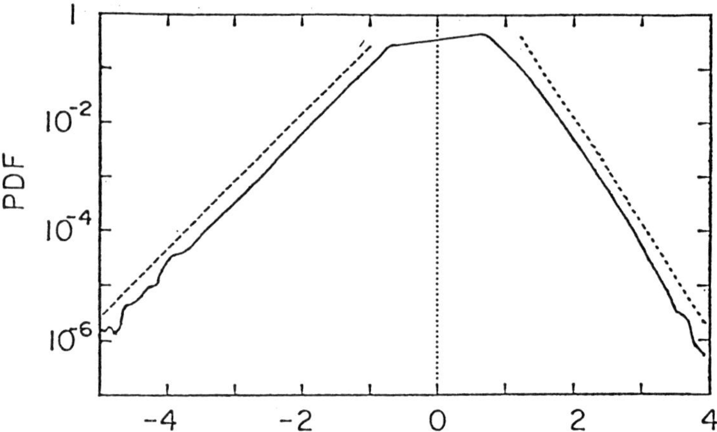

Figure 6.37.   PDF of the Laplacian of pressure $x = \dfrac{\nabla^2 p}{\rho\langle\omega^2\rangle} = \dfrac{\omega^2 - 2s_{ij}s_{ij}}{2\langle\omega^2\rangle}$. ———— - grid turbulence, $\mathrm{Re}_\lambda \approx 75$; - - - - - - - slope $+\,4.5$, — — — — - slope $-\,2.9$.

which for large $|x|$ has the asymptotics $\sim |x|^{1/2}\, e^{-3|x|}$, and

$$\mathcal{P}(x) = \{3^{1/2} 5^{5/2}\}/(4\pi)\; x^2\, e^{-|x|}[K_2(4x) + K_1(4x)], \qquad x > 0,$$

which for large x has the asymptotics $\sim x^{3/2}\, e^{-5x}$.

We see that the distribution of $\nabla^2 p$ for a Gaussian velocity field has exponential tails and is negatively skewed. This result is in agreement with the ones from laboratory and DNS experiments (figure 6.37). It shows that these effects are mostly of kinematical nature as many others like those given above usually are.

A related example is shown in figure 6.38. It is clear from this figure that the correlation function of the total acceleration, $\langle \mathbf{a}(\mathbf{x} + \mathbf{r})\mathbf{a}(\mathbf{x})\rangle$ is practically the same in real flows, as it is laboratory grid turbulence (Hill and Thoroddsen, 1997) and in DNS (Vedula and Yeung, 1999), and in artificial velocity fields under zero forth-cumulant assumption – Millionschikov hypothesis (Pinsky et al., 2000), and/or for the assumption of joint Gaussian velocities (Hill and Thoroddsen, 1997). This behaviour is of the same nature as the one shown in figure 6.34, right.

The examples shown in figures 6.35-6.38 together with other quasi-Gaussian manifeststions of turbulent flows should not be misinterpreted to conclude that turbulent flows are Gaussian. Indeed, we have seen that there are many non-Gaussian manifestations of turbulent flows, which are of special dynamical significance. As mentioned, turbulence – being essentially non-Gaussian – is such a rich phenomenon that it can 'afford' a number of manifestations which are Gaussian-like.

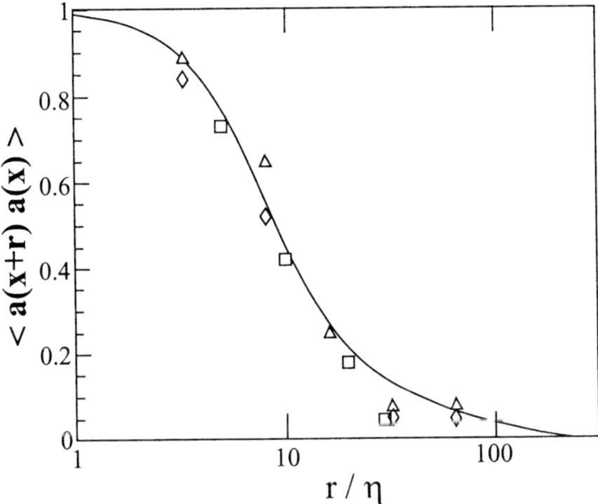

Figure 6.38.   The correlation function of the total acceleration $\langle \mathbf{a}(\mathbf{x} + \mathbf{r})\mathbf{a}(\mathbf{x}) \rangle$: $\diamond$ – grid turbulence (Hill and Thoroddsen, 1997); $\square$ – DNS (Vedula and Yeung, 1999); —— under zero forth-cumulant assumption – Millionschikov hypothesis (Pinsky et al., 2000); $\triangle$ – joint Gaussian velocities (Hill and Thoroddsen, 1997). The figure is from Pinsky et al. (2000).

6.8.3. IRREVERSIBILITY OF TURBULENCE

The equations (6.9, 6.10) and similar ones for $\langle s_{ij}s_{jk}s_{ki} \rangle$ can be seen as one of the manifestations of the statistical irreversibility of turbulent flows (Betchov, 1974; Novikov, 1974). The corresponding dynamical *instantaneous* (inviscid) equations are reversible. Hence, the term *statistical*. The (apparent) randomness of turbulent flows is important even at the kinematical level, e.g. the predominant tendency of stretching of material lines (see Chapter 4). There exist, at least, two different aspects of this problem. The first one is related to purely inertial behaviour governed by the Euler equations as mentioned above. Though these equations are reversible, it is (empirically) known that for Euler equations the enstrophy generation increases very rapidly with time – apparently without limit (see references in Tsinober, 1998b, 2000). This aspect is closely related to the (possible) formation of singularities in 3D Euler flows in finite or infinite time. The above example (equations [6. 9. 6.10] and similar ones for $\langle s_{ij}s_{jk}s_{ki} \rangle$) is closely related to this aspect. The second aspect is associated with the dissipative nature of turbulent flows. Viscosity provides a sink of energy, enstrophy, etc. moderating their unbounded growth in the inviscid case.

6.9. Summary

Turbulent flow in a box forced by a steady (or random) force exhibits large fluctuations in time of the overall quantities such as the energy input, dissipation, enstrophy, enstrophy production and others, of their long time averages. This process is similar to that observed experimentally in the 'French washing machine' at the level of energy dissipation and most probably occurrs in almost all turbulent flows.

One of the intrinsic processes of turbulence dynamics is the process of self-amplification of velocity derivatives. It involves both vorticity and strain, i.e. this process is characteristic of the whole field of velocity derivatives. It consists of two strongly (but nonlocally) interconnected processes: predominant vortex stretching/enstrophy production and predominant self-amplification of the rate of strain/production of total strain. While the former is essentially a process of interaction of vorticity and strain, the latter is more a self-amplification with indirect (but essential) aid of vorticity. The self-amplification of strain (hence enhanced dissipation) is a specific feature of the dynamics of three-dimensional turbulence. The essential dominance of the self-amplification of the velocity derivatives over the external forcing (or other factors driving the flow at the level of velocity field) occurs already at very moderate Reynolds numbers and becomes stronger with increasing Reynolds number. A similar phenomenon is observed in a turbulent shear flow in a channel over most of its cross section except in the proximity of the wall, and in the atmospheric surface layer (see chapter 8). This dominance occurs not only in the mean, but practically pointwise throughout the whole flow field, i.e. the self-amplification of velocity derivatives by turbulence is a process which is local in space. The property of self-amplification is a universal one not only qualitatively, it possesses a number of quantitative universal properties which are independent of the details of forcing. The self-amplification is a quasi-stationary process in the sense that the intergrals over the flow domain of the enstrophy (and strain) production and of its viscous destruction are approximately balanced at any moment in time. Hence the time derivative of the overall enstrophy (and strain) is much smaller than the overall enstrophy production and of its viscous destruction. It is not yet clear how to reconcile the self-amplification of velocity derivatives with the property of nonlocality of turbulent flows, especially the aspect of direct interaction between large and small scales.

Most of the enstrophy production, production of strain/dissipation and other nonlinear processes are associated with i – large strain, rather than with intense vorticity (large enstrophy), ii – alignment of vorticity with the largest eigenstrain (not with the intemediate one), iii – strong tilting of vorticity and finite curvature of vortex lines (not with Burgers-like objects with small curvature). The nonlinearities are an order of magnitude larger in the regions dominated by strain than in the enstrophy dominated regions. In this sense the enstrophy dominated regions are characterized by reduced nonlinearities. In other words the most intense nonlinear processes occur in the strain dominated regions. In particular, there is an approximate balance between the nonlinearities (e.g. vortex stretching) and the viscous terms (e.g. viscous destruction of vorticity) in the regions with concentrated vorticity, whereas in strain dominated regions the nonlinearities (e.g. the enstrophy production) are an order of magnitude larger than the viscous terms. All this also supports the view that regions of concentrated vorticity in turbulent flows are not as important as previously thought. The energy cascade (whatever this means) and its final result - dissipation are associated with the production of strain, i.e. with the quantity $-s_{ij}s_{jk}s_{ki}$, rather than with the enstrophy production. Vortex stretching suppresses the cascade (production of strain) and does not aid it, at least in a direct manner.

The primary effect of viscosity (apart from energy dissipation) is to balance the production of velocity derivatives, both strain and vorticity. The latter is also associated with vortex reconnection. The role of this process in turbulence remains not fully clear.

Turbulence is a nonlocal process. This nonlocality is manifested among other things in direct and bi-directional interaction/coupling of large and small scales. It is a generic internal property of turbulent flows and exists independently of the presence of mean shear or other external factors, but has different manifestations for different external factors. For example, in the presence of a mean shear the small scales become anisotropic, whereas if the small scales are artifically excited the overall dissipation and mixing rate of the turbulent flow increase substantially. The direct interaction/coupling of large and small scales is in full conformity and is the consequence of the generic property of Navier-Stokes equations, which are integro-differential due to the nonlocal relation between pressure and velocity fields. It is associated also with 'kinematics' due to the nonlocal relation between vorticity and strain. The material velocity derivative – the fluid particle acceleration is also related in a nonlocal manner to the velocity field, since the main contribution to the acceleration comes from the pressure gradient.

In view of the direct interaction/coupling of large and small scales it is not clear how meaningful (from the basic point of view) is the notion of an inertial range and related things as 'eddy viscosity' representations of the subgrid scales, etc. Eddy viscosity does not explain the enhanced transfer rates. It is just a purely *empirical* way of accounting for such rates.

From the statistical point turbulence is irreversible and essentially non-Gaussian, but possesses a number of quasi-Gaussian manifestations.

CHAPTER 7

STRUCTURE(S) OF TURBULENT FLOWS

> *Although symmetric causes must produce symmetric effects, nearly symmetric causes need not produce symmetric effects: a symmetry problem need have no stable symmetric solutions.* (Birkhoff, 1960).
>
> *I emphasized the concept of "broken symmetry", the ability of a large collection of simple objects to abandon its own symmetry as well as the symmetries of the forces governing it and to exhibit the "emergent property" of a new symmetry.* (Anderson, 1991).

7.1. Introduction

The nature and characterization of the structure(s) of turbulent flows are among the most controversial issues in turbulence research with extreme views on many aspects of the problem - in words of Richard Feynmann (1963, p. 41-12), *holding strong opinions either way.* For example, as mentioned in chapter 3, it is common in the vast literature on turbulence to consider the terms *statistical* and *structural* as incompatible or even contradictory:

...it became obvious that statistical averaging was in fact destroying the most interesting and important phenomena in turbulence – the formation, dynamics and persistence of vortex motion.

Its blindness to these structural facts is precisely the disability of the statistical idea. ...In place of theory without structure, the result to date has been structure without theory.

Following Lumley (1989), who provides many such examples, the references are suppressed to protect the guilty.

However, there are common points as well. For instance, it is mostly agreed that turbulence definitely possesses structure(s) (whatever this means) and that intermittency, which is addressed in the next section, is intimately related to *some* aspects of the structure(s) of turbulence.

It is argued in the section following the one on intermittency that it is a misconception to contrapose the *statistical* and the *structural* and that they represent different facets/aspects of the same problem, so that there is no gap between structure(s) and statistics[1]. Just like it seems impossible to separate the structure(s) from the so called 'random structureless background' or the *random processes from the nonrandom processes* (Dryden, 1948) due to strong interaction (and nonlocality), both between individual structures, and between structures and the 'background'. In other words there is no turbulence *without* structure, every part of the turbulent field just like the whole possess structure[2]. Structureless turbulence or any its part contradicts both the experimental evidence and the Navier-Stokes equations. It is noteworthy that the statement that turbulence has structure is in a sense trivial: *to say that turbulent flow is 'completely random' would define turbulence out of existence* (Tritton, 1988, p. 295) - after all turbulent flows seem to obey the Navier-Stokes equations.

7.2. Intermittency

> *At any instant the production of small scales is... occurring vigorously in some places and only weakly in the others* (Tritton, 1988).
>
> *Intermittency is a phenomenon where Nature spends little time, but acts vigorously* (Betchov, 1993).

The term intermittency is used in two distinct (but not independent) aspects of turbulent flows. The first one is the so-called external intermittency. It is associated with what is called here partially turbulent flows, specifically with the strongly irregular and convoluted structure and random movement of the 'boundary' between the turbulent and nonturbulent fluid. This kind of intermittency was studied first by Townsend (1948). We will address this matter in chapter 8.

The second aspect is the so-called small scale, internal or intrinsic intermittency. It is usually associated with the tendency to spatial and temporal localization of the 'fine' or small scale structure(s) of turbulent flows.

Our concern here is with the intermittency of this second kind. It is noteworthy that in a broad sense intermittency is a ubiquitous phenomenon occuring in a great variety of qualitatively different systems; see chapter 8 in Zeldovich et al. (1990) for a lively exposition of a wide number of different systems exhibiting intermittency, and also Vassilicos (2001). The main common features of all of them is (space/time) randomness and localization (both spatial and temporal) of their 'fine' structure. However, this is not

---

[1] Both issues are intimately related to the non-Gaussian nature of turbulence (and *some* of its quasi-Gaussian manifestations) and the necessity and the only objective means to handle the issues of turbulence structure(s) via statistics.

[2] There are proposals to *scan out* the structure(s). In fact there is no way to do so, since *structure* is everywhere. Even the so-called 'simple' structures (worms) are 'renormalized' by the background.

enough to define intermittency. For example, almost any nonlinear function or almost any nonlinear functional of a random Gaussian field is intermittent in the above sense, though random Gaussian fields by definition lack any intermittency.

7.2.1. WHAT IS SMALL SCALE INTERMITTENCY?

The phenomenon of small scale intermittency was discovered by Batchelor and Townsend (1949) in experiments with turbulent grid flows and in a wake past a circular cylinder[3]:

The basic observation which requires explanation is that activation of large wave-numbers is very unevenly distributed in space. These space variations in activation can be described as fluctuations in the spectrum at large wave-number... As the wave-number is increased the fluctuations seem to tend to an approximate on-off, or intermittent variation. Whatever the reason for the occurrence of these fluctuations, they appear to be intrinsic to the equilibrium range of wave-numbers. All the evidence is consistent with the inference that the fluctuations are small in the region of smallest wave-numbers of equilibrium range and become increasingly large at larger wave-numbers ... the mean separation of the visible activated regions is comparable with the integral scale of the turbuence, i.e. with the size of the energy-containing eddies. (pp. 252-3)

Batchelor and Townsend obtained some evidence that the deviation from Gaussianity is stronger as the Reynolds number is increased[4], which was confirmed by a number of subsequent experiments such as by Kuo and Corrsin (1971).[5]

An example of time records of the streamwise velocity component, and their derivatives obtained in a field experiment at $Re_\lambda = 10^4$ is shown in Plate 6. The increasingly intermittent behaviour of the signal with the derivative order is seen quite clearly. Also shown are records for the enstrophy ω^2, total strain $2s^2 \equiv 2s_{ij}s_{ij}$ and their surrogate $(\partial u_1/\partial x_1)^2$, and enstrophy production $\omega_i\omega_j s_{ij}$, $s_{ij}s_{jk}s_{ki}$ and their surrogate $(\partial u_1/\partial x_1)^3$. The experiment was performed in the atmospheric surface layer at a height $10m$

---

[3]The intermittent nature of the small scale structure of turbulent flows was foreseen by Taylor (1938b): *...the view frequently put forward by the author that the dissipation of energy is due chiefly to the formation of very small regions where vorticity is very high.* However, note that dissipation is high in regions where strain - not vorticity - is high (see section 6.2).

[4]However, they did not appreciate this effect and claimed that the flatness factors *seems to vary a little with Reynolds number*, though this factor changed from 5 to 7 for the third order derivative; see their figure 5 for the flatness factor of velocity derivatives of different orders and different Reynolds numbers.

[5]For an updated overview of the subsequent results see, e.g. Sreenivasan and Antonia (1997).

in approximately neutral (slightly unstable) conditions.

A qualitative summary is that small scale intermittency of turbulence is associated with its spotty (spatio-temporal) *structure* which among other things is manifested as a *particular* kind of non-Gaussian behaviour of turbulent flows. This deviation from Gaussianity, increases with both 1) increasing the Reynolds number and 2) decreasing the 'scale'. In other words, intermittency involves two (not independent) aspects of turbulent flows - their structure/geometry and statistics. These two aspects are reflected in attempts to 'define' intermittency. Two examples of such definitions are given below.

Structural/geometrical. A positively defined quantity μ (measure) is intermittent in space if for arbitrarily small μ_0, the fraction V_0/V of any volume V of fluid within which $\mu > \mu_0$ tends to zero as the Reynolds number $Re \to \infty$, i. e. $\mu(\mathbf{x}, t)$ becomes increasingly spiky, being concentrated almost entirely in this vanishingly small part V_0 of the volume V. This is a modified version of Moffatt's (1988) definition of intermittency of dissipation. In fact, all known dynamically important quantities including those, which are not positively defined (e.g. enstrophy generation), exhibit such a spiky behaviour. Note that the above definition refers to rather general property of turbulence structure - the 'increasingly spiky' structure of some variable can be realized in a great variety of ways: both the structure of these spiky regions and of the 'background' may be very different for the same V_0.

Statistical. A variable with zero mean will be called intermittent if it has a probability distribution such that extremely small and extremely large excursions are much more likely than in normally distributed variable [i.e. Gaussian]. *Therefore, the kurtosis* [flatness] *of intermittent variable with zero mean is large. Correspondingly, a non-negative variable is called intermittent if its variance is large compared to the square of its mean value* (Tennekes, 1973). This definition ignores Reynolds number dependence. For other definitions see Frisch (1995), Libby (1996).

It is important that intermittency implies non-Gaussianity, but not necessarily vice versa – practically any parameter can at most indicate the degree of intermittency of a flow *already* known to be intermittent (see next section).

7.2.2. MEASURES/MANIFESTATIONS OF INTERMITTENCY

Intermittency factor(s)

Loosely, an intermittency factor is defined as a fraction of volume (time) where the variable is 'active'. This is one of the most reliable *direct* measures of intermittency. The main deficiency is that intermittency factors depend on the choise of the threshold below which the variable is considered 'inactive' (Kuo and Corrsin, 1971; Kuznetsov et al., 1992; and references therein).

Flatness factor.
Flatness factor or kurtosis of a variable is defined as

$$F(a) = \frac{\langle a^4 \rangle}{\langle a^2 \rangle^2}$$

For a Gaussian field $F_G(a) = 3$, i.e. the flatness factor can be used as a measure of non-Gaussianity. The specific choice is justified by the fact that the inverse of flatness increases with and is roughly proportional to the fraction of volume/time where the variable is 'active'.

It has been established experimentally that the flatness increases with both the order of derivative (scale dependence) and the Reynolds number of the turbulence (Batchelor and Townsend, 1949; Kuo and Corrsin, 1971). Kuo and Corrsin interpreted this result (using also intermittency factor) as a decrease in the volume fraction occupied by fine-structure both as the Reynolds number is increased and as the structure becomes finer.

As mentioned, a statistical measure such as flatness may deviate strongly from a Gaussian value without any intermittency in the flow field. The simplest example is the Gaussian field itself, which by definition lacks any intermittency. However, any nonlinear function (or functional) of a variable, which is Gaussian, is non-Gaussian. For instance enstrophy, dissipation, pressure, etc. of a Gaussian velocity field possess *exponential* tails and their flatness is quite different from 3 (see chapter 6 and also figure 4 in Kennedy and Corrsin, 1961). For example, for a Gaussian velocity field $F_G(\omega^2) = \langle \omega^4 \rangle / \langle \omega^2 \rangle^2 = 5/3$ and $F_G(s^2) = \langle s^4 \rangle / \langle s^2 \rangle^2 = 7/5$. But this by no means indicate that, for a Gaussian velocity field, these quantities are intermittent. Moreover, the flatness of enstrophy is *larger* than that of total strain, $F_{\omega^2} - F_{s^2} = 4/15$. Does one have to conclude from the above result (as some authors did) that the enstrophy field is more intermittent than that of total strain in a *Gaussian* velocity field? Certainly not, since intermittency is, by definition, absent in any Gaussian field. This example shows that it is not sufficient to define intermittency as uneven distribution of enstrophy and dissipation in space, as is quite frequently done.

Similarly, the Reynolds stress $u_i u_j$ exhibits 'intermittency'. The main contribution to this intermittency comes from the fact that $u_i u_j$ is a product of two random variables both distributed close to Gaussian. For example, the PDF of the $u_1 u_2$ of the strongly intermittent signal obtained by Lu and Willmarth (1973) in a turbulent boundary layer is strongly non-Gaussian. However, the PDF $u_1 u_2$ is approximated with high precision by assuming both u_1 and u_2 to be Gaussian with a correlation coefficient between them adjusted from the experiment (-0.44) (see chapter 8)[6].

Passive objects (scalars like heat, vectors like magnetic field) in a random velocity field (real or artificially prescribed) are nonlinear functionals of the velocity field and forcing. Therefore, even when both the velocity field and forcing are Gaussian the field of a passive object is expected to be strongly non-Gaussian as usually (but not always) is the case (Majda and Kramer, 1999). Such kinematic intermittency is observed in a great number of theoretical and some experimental works (for a partial list of references see chapter 4). The term 'kinematic' is used here in the sense that there is no relation to the dynamics of fluid motion, which does not enter in the problems in question, and the velocity field is prescribed and often assumed to be Gaussian.

It is noteworthy that both criteria - the intermittency and the flattenss factors are of purely kinematic nature, i.e. they are not related - at least directly - to the dynamical aspects of turbulence. One would think that the situation is different with the so-called odd moments. However, this is not the case either.

Odd moments

Any odd moment of a Gaussian variable vanishes, for example skewness $S_G(a) \equiv \langle a^3 \rangle / \langle a^2 \rangle^{3/2} = 0$. Therefore, odd moments are very sensitive to deviations from Gaussianity, so that non-zero odd moments may be especially good indicators of intermittency. Build-up of odd moments is a result of both the (kinematic) evolution of a passive field in any random velocity field and the *dynamics* of turbulence itself. In the latter case, non-vanishing odd moments are the most important, dynamically significant manifestations of non-Gaussianity, i.e. they reflect *directly* the dynamic aspects of intermittency. The most prominent odd moments are the third order structure function for longitudinal velocity increments $S_3^{\|} = \langle \{ [\mathbf{u}(\mathbf{x}+\mathbf{r}) - \mathbf{u}(\mathbf{x})] \cdot (\mathbf{r}/r) \}^3 \rangle$ entering the 4/5 law, the enstrophy production $\langle \omega_i \omega_k s_{ik} \rangle$ and the third order moment of the strain tensor $\langle s_{ij} s_{jk} s_{ki} \rangle$ (see chapter 6). Note that all these and other odd moments vanish in a Gaussian velocity field. In contrast similar odd moments for passive objects, $\langle G_i G_k s_{ik} \rangle$, $\langle B_i B_k s_{ik} \rangle$ do not vanish (chapter 4). Hence, the passive objects in some respects are more intermittent, e.g. see figure 7.1.

The non-Gaussian nature of turbulent flows and the differences between the odd moments of passive fields and turbulence itself were discussed in chapter 6.

---

[6]The above examples may serve as a warning that multiplicative models enable us to produce intermittency for a purely nonintermittent field as is the Gaussian velocity field. See Zeldovich et al. (1990) on interesting observations on this and related matters.

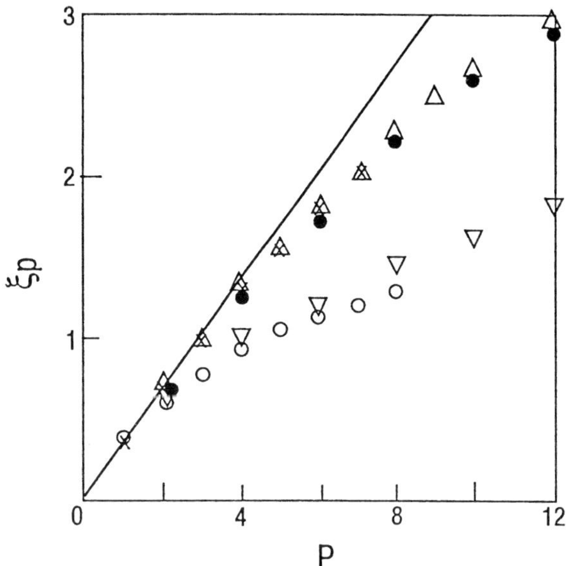

Figure 7.1. Exponents of structure functions for the longitudinal velocity component ($\triangle, \bullet, \times$) and temperature ($\triangledown, \circ$); $\triangle$ - Anselmet et al. (1984), $\triangledown$ – Antonia et al. (1984), $\circ$ – Ruiz-Chavaria et al. (1996), $\bullet$ – Vincent and Meneguzzi (1991); and exponents of structure functions for the transverse velocity component, $\times$ – Noullez et al. (1997). This figure is from Tsinober (1998b).

Scaling exponents and PDFs
It is commonly believed that among the manifestations of the small scale intermittency is the experimentally observed deviation of the scaling exponents for structure functions $S_p^{\parallel} = \langle \{ [\mathbf{u}(\mathbf{x} + \mathbf{r}) - \mathbf{u}(\mathbf{x})] \cdot (r/r) \}^p \rangle$ for $p > 3$ from the values implied by the Kolmogorov theory (i.e. anomalous scaling), which in turn is due to rare strong events. Namely,

$$S_p^{\parallel}(r) \propto r^{\zeta_p^{\parallel}}, \tag{7.1}$$

where $\zeta_p^{\parallel} = p/3 - \mu_p < p/3$ is a convex nonlinear function of p (see figure 7.1).

However, there are two major problems with scaling as follows.

First, there exists no one-to-one relation between simple statistical manifestations and the underlying structure(s) of turbulence[7]. A recent example is shown in figure 7.2.

Moreover qualitatively different phenomena can possess the same set of scaling exponents (see appendix C, section C.2), so that one needs more subtle statistical characterizations of turbulence structure(s) and intermittency. For example, until recently one of the common beliefs was that

---

[7]This issue is addressed in more detail in the next section.

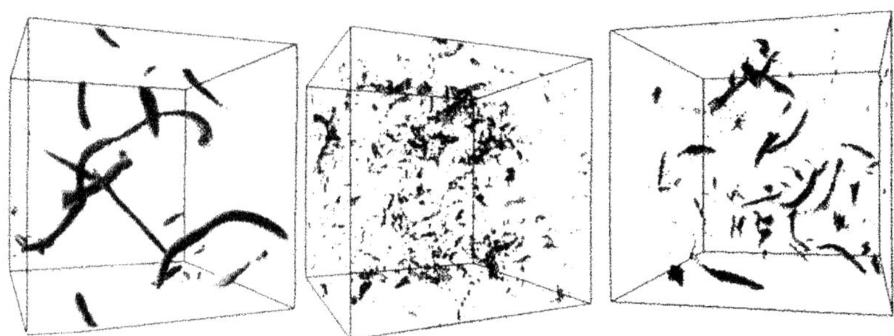

Figure 7.2. Iso-ω surfaces for three kinds of forcing of the RHS of DNS of NSE in a cubic box with periodic boundary conditions. Left - forcing in the lowest Fourier wave-numbers, middle – $\langle |f_k|^2 \rangle \sim k^{-3}$, and right – $\langle |f_k|^2 \rangle \sim k^{-5}$. In all cases the scaling properties are practically the same and are in agreement with other results (Sain, 1998).

the observed vortex filaments/worms are mainly responsible for the phenomenon of intermittency understood as anomalous scaling. However, it appears that this is not the case (see the evidence for that given in Tsinober, 1998a; and chapter 6). This was confirmed also by recent analysis by Roux et al. (1999) of the data from the experiments by Cadot et al. (1995), which *strongly suggests that the statistical contribution of vorticity filaments is not responsible for the intermittency phenomenon*, i.e. anomalous scaling. A similar result was obtained by Dernoncourt et al. (1998), (see also Chavanis and Sire, 2000; Min et al., 1996; and Sain et al., 1998)[8]. Likewise similar PDFs of *some* quantities can correspond to qualitatively different structure(s) and quantitatively different values of Reynolds number (Kraichnan and Kimura, 1994; Tsinober, 1998a,b). The emphasis is on *some* quantities like pressure or some other usually (but not necessarily) even order quantities in velocities or their derivatives, since the PDFs of *other* appropriately chosen quantities are sensitive to structure (see below). Another example is represented by numerous models attempted to reproduce the anomalous scaling (7.1) (a partial list of references is given in Sreenivasan and Antonia, 1997; and Tsinober, 1998b). These models followed the Kolmogorov (1962) refined similarity hypothesis (RSH) in which the mean dissipation $\langle \epsilon \rangle$ was replaced by 'local' dissipation ϵ_r averaged over a region of size r[9]. The scaling exponents obtained in all of these models are in good agreement with the experimental and numerical evidence, e.g.

---

[8]Jimenez and Wray (1998) hold an opposite view, that *the filaments are responsible for most of the intermittency effects of the higher moments of the velocity derivatives* (p.283).

[9]Kolmogorov proposed this hypothesis following the Landau objection to universality in the first Russian edition of *Fluid Mechanics* by Landau and Lifshitz (1944) about the role of large scale fluctuations of energy dissipation rate, i.e. nonuniversality of both the scaling exponents ζ_p and the prefactors C_p in (7.1).

these models exhibit the same scaling properties (and some other such as PDFs) as in real turbulence. It is noteworthy that many of these models are based on *qualitatively different* premises/assumptions and with few exceptions have no direct bearing on the Navier-Stokes equations[10]. The most common justification for the preoccupation with such models is that they (at least some of them) share the same basic symmetries (perhaps also some hidden symmetries), conservation laws and some other general properties, etc. as the Navier-Stokes equations. The general belief is that this - along with the diversity of such systems (there are many having nothing to do with fluid dynamics, e.g. granular systems, financial markets, brain activity) - is the reason for the above mentioned agreement. However, this is not really the case, e.g. in Kraichnan (1974) a counter example of a '*dynamical equation is exhibited which has the same essential invariances, symmetries, dimensionality and equilibrium statistical ensembles as the Navier-Stokes equations but which has radically different inertial-range behaviour*'! The majority of models exhibit *temporal* chaos only. Therefore, such and most other models hardly can be associated with the intermittency of *real* fluid turbulence, which involves essentially *spatial* chaos as well. Again, for the above reasons the agreement between such models and experiments (both laboratory and numerical) cannot be used for evaluation of the success of such models. There are proposals to use two sets of independent exponents $\zeta_p^{\parallel}$ and $\zeta_p^{\perp}$ (Chen et al., 1997), and there exist other 'universality' proposals involving 'much more' scaling exponents (see e.g. Frisch, 1995; Kurien and Sreenivasan, 2000; and references therein).

All the above is a clear indication that the question about the origins of intermittency (understood as anomalous scaling and/or in any other sense as above) in *real* turbulent flows remains open. Similarly open are the questions on universality of the intermittency manifestations (if such exist) in Navier-Stokes equations, though judging by the multitude of models of intermittency there is no universality whatsoever.

Phenomenology and models only will hardly be useful and convincing, since almost any dimensionally correct model, both right or wrong, will lead to correct scaling without appealing to NSE and/or elaborate physics. Scaling laws alone are not necessarily theories.[11] With all the importance of

---

[10]Therefore the success of such models can hardly be evaluated on the basis of how well they agree with experiments. For example, there exist many theories which produce the $k^{-5/3}$ energy spectrum for qualitatively and/or physically different reasons. A recent example is a *suggestion that the spectrum of fully developed turbulence is determined by the equilibrium statistics of the Euler equations and that a full description of turbulence requires only a perturbation, small in some appropriate metric, of a Gibbsian equilibrium,* (Chorin, 1996).

[11]For example, the Kolmogorov theory and many subsequent models used dissipation as a basic quantity, i.e. intimately related to strain. Several later theories are based

scaling, turbulence phenomena are infinitely richer than their manifestation in scaling and related things. Most of these manifestations are beyond the reach of phenomenology. Phenomenology is inherently unable to handle the structure of turbulence in general, and phase and geometrical relations in particular, to say nothing of dynamical features such as build up of *odd* moments, interaction of vorticity and strain resulting in positive net enstrophy generation/predominant vortex stretching. It seems that there is little promise for progress in understanding the basic physics of turbulence in going on asking questions about scaling and related matters only (Badii and Talkner, 2001; Feigenbaum, 1997; Tsinober, 1996b), without looking into the structure and, where possible, basic mechanisms which are *specific* to turbulent flows. In fact, the main question of principle which should have been asked long ago is: Why on earth should we perform so many elaborate measurements of various scaling exponents without looking into the possible concomittant physics and/or without asking why and how more precise knowledge of such exponents, even assuming their existence, can aid our understanding of turbulent flows?

Second, the very existence of scaling exponents, which is taken for granted, is a problem by itself. The existence of the scaling symmetry as any other symmetries of the Euler and *presumably*[12] of the Navier-Stokes equations at very large Reynolds numbers does not guarantee that various *statistical* characteristics should possess such symmetries too. Their restoring in the statistical sense is only a hypothesis (see chapter 5). The experimental support of the existence of scaling ranges from the existing evidence is rather weak, sometimes marginal. There are many difficulties in determining inertial ranges and the corresponding scaling exponents (see, for example, figures 6, 10 and 11 in Anselmet et al., 1984; and figure 8.6 in Frisch, 1995). In a recent work based on the experinetal data of Gagne (1987) and Malecot (1998), Arneodo et al. (1999) arrived to the conclusion that *because of the scale invariance breaking, the notion of inertial range is not well defined.*

A similar question arises in respect with multifractality which was designed to 'explain' (in fact it is another description of) the anomalous scaling (7.1), since there is no direct experimental evidence on the multifractal

---

on hierarchies of vorticity dominated structures. Most of both kinds of these theories agree with experimental results. However, while there is a basic reason, not only on dimensional grounds, for RSH $\Delta u_r^{\parallel} = \beta_1 (r\epsilon_r)^{1/3}$, since it can be seen as a 'local' version of the -4/5 Kolmogorov law $\langle (\Delta u^{\parallel})^3 \rangle = -4/5 \langle \epsilon \rangle r$, a similar claim (Chen et al., 1997), that $\Delta u^{\perp} = \beta_2 (r\Omega_r)^{1/3}$ $(\Omega = \nu\omega^2)$ remains just one more dimensionally - but not necessarily physically - correct relation (note that $\langle (\Delta u^{\perp})^3 \rangle \equiv 0$). Here, β_1, β_2 are stochastic variables independent of Re and r.

[12]Since it is not all clear why one can ignore the singular nature of the zero viscosity limit when dealing with the issue of scaling exponents.

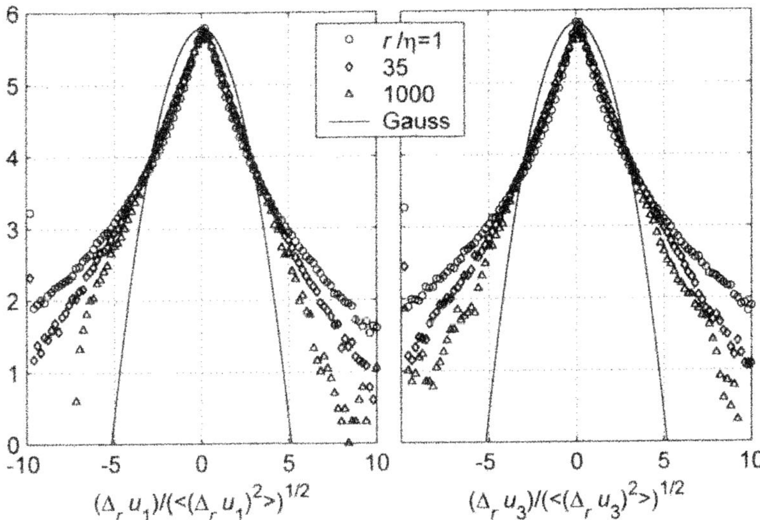

$(\Delta_r u_1)/(<(\Delta_r u_1)^2>)^{1/2}$

$(\Delta_r u_3)/(<(\Delta_r u_3)^2>)^{1/2}$

Figure 7.3.   PDFs of the increments of longitudinal (a) and transverse (b) velocity fluctuations obtained in a field experiment at $Re_\lambda = 10^4$ (Kholmyansky and Tsinober, 2000; Kholmyansky et al., 2001). The increasing deviation from a Gaussian behaviour with decreasing r is manifested quite clearly. The curve for $\eta \approx 1$ is essentially a PDFs of the longitudinal derivative. Note the asymmetry of the PDfs of the increments of longitudinal velocity fluctuations, which becomes largest at smallest separations.

structure of turbulent flows. So there is a possibilily that multifractality in turbulence is an artifact (see Frisch, 1995, p. 190).

The PDFs of an intermittent variable are quite useful, and they do not suffer from problems like scaling exponents. For example, using PDFs of velocity increments for different separations between two points it becomes clear that, the closer the two points, the more the PDFs deviate from a Gaussian distribution both in the centre and at the tails. An example is shown in figure 7.3.

However, PDFs also (both single point and two-point PDFs) contain rather limited information. Namely, such PDFs carry the information showing that that extremely small (the centre anomaly) and extremely large values (tails of the PDF) are much more likely than for a Gaussian variable. However, they contain no information on the structure of the underlying weak and strong events, nor on the structure of the background field. Hence, the same PDFs can have qualitatively different underlying structure(s) of the flow, i.e. 'how the flow looks'. For example, the qualitative difference in the behaviour and properties of regions dominated by strain and those with large enstrophy cannot be captured by such means and other conventional measures of intermittency. Also the PDFs, like scaling exponents, do not allow us to infer much about the underlying dynamics. This, however, is true of 'conventional' PDFs like those of velocity increments, but not of any PDFs such as those directly associated with geometrical flow properties (see below section 7.3).

Note that the largest deviation from Gaussianity occurs at small scales (smallest distances between two points; see figure 7.3). In this sense, the field of velocity derivatives, $\partial u_i/\partial x_k$, is more intermittent than the field of velocity, u_i, itself. One of the possible reasons for this is in the different nature of nonlinearity at the level of velocity field, i.e. in the Navier-Stokes equations and, for example, in the equation for vorticity (C.9). Namely, the nonlinearity in the Navier-Stokes equations, $(\mathbf{u} \cdot \boldsymbol{\nabla})\mathbf{u} = \boldsymbol{\omega} \times \mathbf{u} + \boldsymbol{\nabla}(u^2/2) = \boldsymbol{\nabla}(\alpha + u^2/2) + \boldsymbol{\nabla} \times \boldsymbol{\beta}$ contains a considerable potential part, $\boldsymbol{\nabla}(\alpha + u^2/2)$ (see chapter 6, sections 6.5–6.7). This potential part can be included in the pressure term, i.e. the solenoidal part of nonlinearity is reduced. There is no such reduction of nonlinearity on the level of vorticity. Hence the difference, since the nonlinearity can be seen as one of the general reasons for intermittency. We turn now to the discussion of more specific reasons.

7.2.3. ON POSSIBLE ORIGINS OF SMALL SCALE INTERMITTENCY

As one of manifestations of turbulence structure(s), intermittency has its origins in the structure of turbulence (see next section). Therefore we briefly address here the issue on possible origins of intermittency. There are roughly two kinds of origins of intermittency: kinematic and dynamic.

Direct interaction/coupling between large and small scales
As discussed in chapter 6, direct interaction/coupling between large and small scales is one of the elements of the nonlocality of turbulence. It is both of kinematic and dynamic nature. The first recognized manifestation of such interaction is that the small scales do not forget the anisotropy of the large ones. There is a variety of mechanisms producing and influencing the large scales: various external constraints like boundaries with different boundary conditions, including the periodic ones, initial conditions, forcing (as in DNS), mean shear/strain, centrifugal forces (rotation), buoyancy, magnetic field, external intermittency in partially turbulent flows, etc. Most of these factors usually act as organizing elements, favouring the formation of coherent structures of different kinds (quasi-two-dimensional, helical, hairpins, etc.). These, as a rule, large scale features depend on the particularities of a given flow that are not universal. Therefore the direct interaction between large and small scales leads to 'contamination' of small scales by the large ones, e.g. the edges of large scale structures are believed to be responsible for such 'contamination'. This contamination seems to

be unavoidable even in homogeneous and isotropic turbulence, since there are many ways to produce such a flow, i.e. many ways to produce the large scales. It is the difference in the mechanisms of large scale production which 'contaminates' the small scales. Hence, non-universality.

The direct interaction/coupling of large and small scales seems to be a generic property of all turbulent flows and one of the main reasons for small scale intermittency, non universality, and quite modest manifestations of scaling.

Near singularities

It is not known for sure whether Euler equations and/or Navier-Stokes equations at large Reynolds numbers develop a genuine singularity in finite time, though there is some evidence that, at least for Euler equations, this may be true. Whatever the real situation is, it seems a reasonable speculation that these 'near' singularities *trigger topological change and large dissipation events* (for Navier-Stokes equations); *their presence is felt at the dissipation scales and is perhaps the source of small scale intermittency* (Constantin, 1996). However, this does not help to understand the *inertial range* intermittency (if such exists) without invoking the *reacting back* of the dissipation range on the inertial range. As mentioned, such reaction back is possible due to the direct coupling between the large and small scales and other nonlocal effects. The experimentally observed phenomenon of strong drag reduction and change of structure of turbulent flows of dilute polymer solutions and other drag reducing additives is an example of such a 'reacting back' effect (see section 6.6 for other examples).

Near singular objects associated with non-integer values of the energy spectrum scaling exponents are thought to be closely related with some structure(s) and, consequently, with intermittency of turbulent flows (Vassilicos, 1996).

In any case, the 'near' singular objects may be among the origins of intermittency of a dynamical nature[13]. However, there is a problem with two-dimensional 'turbulence'. Namely, in this case everything is beautifully regular (Doering and Gibbon, 1995), but there is intermittency in the sense of the above definitions, with the exception of scaling exponents for velocity structure functions and corresponding quasi-Gaussian behaviour. However, non-Gaussianity is strong at the level of velocity derivatives of a second order (see chapter 8). Hence the possible formation of singularities in 3-D is not necessarily the underlying reason for intermittency in 3D turbulence.

---

[13]We mean singularities which appear at random in space and time and not in a strictly periodic (and fully coherent and mutually amplifying) fashion as in DNS with periodic boundary conditions.

Multiplicative noise, intermittency of passive objects in random media.

It has been known for about thirty years that passive scalars exhibit anomalous scaling behaviour and other strong manifestations of intermittency (see figure 7.1) even in pure Gaussian random velocity field (see Shraiman and Siggia, 1999, 2000; Sreenivasan and Antonia, 1997; Tsinober, 1998b; Warhaft, 2000; Zeldovich et al., 1988,1990; and references therein). Similar behaviour is exhibited by passive vectors (Kraichnan and Kimura, 1994; Rogachevski and Kleorin, 1997; Vergassola, 1996). These are dynamically linear systems, but they are of the kind which involve the so-called multiplicative 'noise', i.e. the coefficients in the equations that depend on the velocity field. Therefore, statistically they are 'nonlinear', since the field of passive objects depend nonlinearly on the coefficients of the equations; more precisely the field of passive objects is a nonlinear functional of the velocity field. Therefore, passive objects exhibit strong deviations from Gaussianity. In such systems, intermittency results either from external pumping (forcing term on RHS of the equations), or in systems without external forcing from instability (self-excitation) of a passive object in a random velocity field under certain conditions.

The velocity field does not 'know' about the passive objects. In this sense, problems involving passive objects are kinematic in respect with the velocity field in real fluid turbulence. They reflect the contribution of kinematic nature in real turbulent flows. It is noteworthy that the intermittency effects in such linear systems are stronger than in real fluid turbulence and exhibit anomalous scaling, which, generally, is nonuniversal (Falkovich, 1999; Shraiman and Siggia, 1999, 2000; Warhaft, 2000). In view of the recent progress in this field it was claimed that *investigation of the statistics of the passive scalar field advected by random flow is interesting for the insight it offers into the origin of intermittency and anomalous scaling of turbulent fluctuations* (Pumir et al., 1997; see also Majda and Kramer, 1999; Shraiman and Siggia, 1999, 2000). More precisely it offers an insight into the origin of intermittency and anomalous scaling of fluctuations in random media generally and independently of the nature of the random motion (Zeldovich et al., 1988), i.e. it gives some insight into the contributions of kinematic nature, but does not offer much regarding the specific dynamical aspects of strong turbulence in fluids. Moreover, anomalous diffusion (including scaling) of passive objects occurs in purely laminar flows in Eulerian sense (E-laminar flows) as a result of Lagrangian chaos (L-turbulent flows), i.e. in some cases intermittency of passive objects may have nothing to do with the random nature of fluid motion. For examples, see chapter 4.

Thus in real turbulent flows there are two contributions to the behaviour of passive objects, kinematic and dynamic. It seems hopeless to separate them in any sense. In a way, the problem of passive objects is more complicated than the dynamical one.

Summarizing, intermittency specifically in fluid turbulence is associated mostly with *some* aspects of its spatio-temporal structure, especially the spatial one. Hence, the close relation between the origin(s) and meaning of intermittency and structure of turbulence. Just like there is no general agreement on the origin and meaning of the former, there is no consensus regarding what are the origin(s) and what turbulence structure(s) really mean. What is definite is that turbulent flows have lots of structure(s). The term structure(s) is used here deliberately in order to emphasize the duality (or even multiplicity) of the meaning of the underlying problem. The first is about how turbulence 'looks'. The second implies the existence of some entities. Objective treatment of both requires use of some statistical methods. It is thought that these methods alone may be insufficient to cope with the problem, but so far no satisfactory solution was found.

7.3. What is(are) structure(s) of turbulent flows?

What we see is real. The problem is interpretation.

The difficulties of definition what the structure(s) of turbulence are (mean) are of the same nature as the question about what is turbulence itself. So before and in order to 'see' or 'measure' the structure(s) of turbulence one encounters the most difficult questions such as: what is (say, dynamically relevant) structure?, Structure of what? Which quantities possess structure in turbulence?, What is the relation between structure(s) and 'scales'? Can structure exist in 'structureless' (artificial) pure random Gaussian fields ? Which ones? All this - like many other issues - are intimately related to the skill/art to ask the right and correctly posed questions.

The meaning of structure(s) depends largely on what is meant by turbulence itself, and especially structure(s) of the particular field one is looking. For example, as discussed in chapter 4, the velocity field may have no structure, but the passive tracer may, simple laminar Eulerian velocity field (E-laminar) creates complicated Lagrangian field (L-turbulent). Purely Gaussian, i.e. 'structureless' velocity field creates structure in the field of passive objects. The structure(s) seen in the velocity field depend on the motion of the observer (see figure 7.4).

7.3.1. ON THE ORIGINS OF STRUCTURE(S) OF/IN TURBULENCE

This question - in some sense - is a 'philosophical' one. But its importance is in direct relation to even more important questions about the origin of turbulence itself.

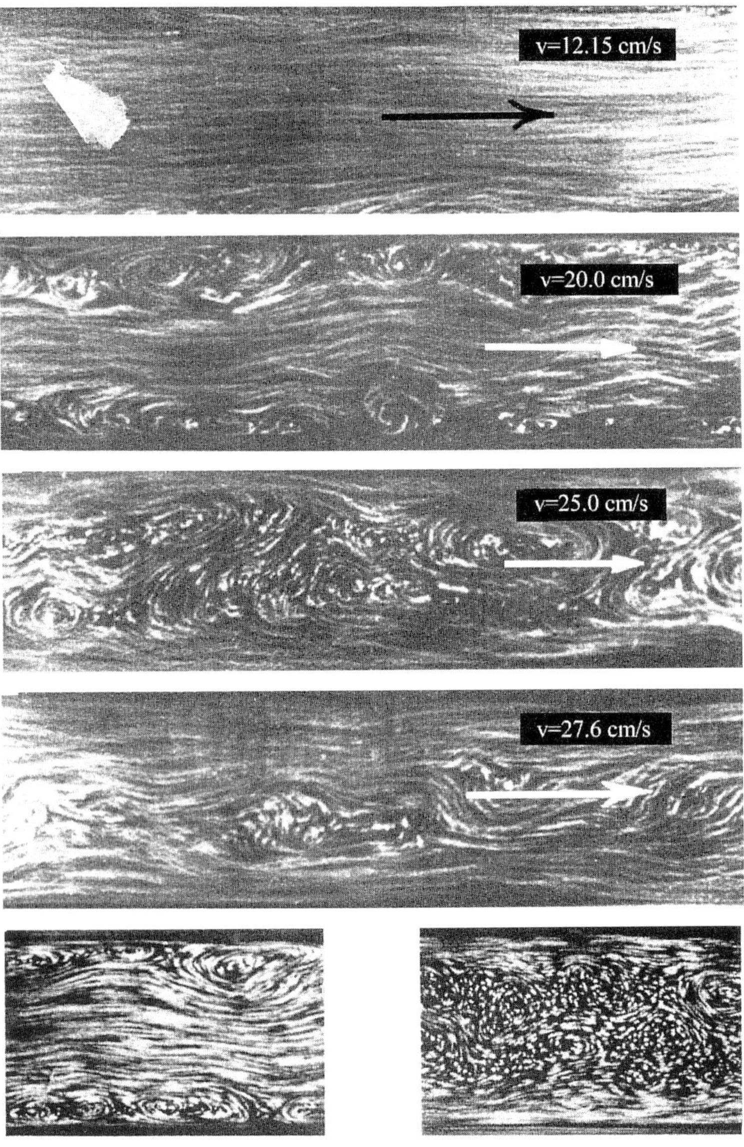

Figure 7.4. The four upper pictures, Tollmien (1931), correspond to the visualization of a turbulent water flow in an open 6cm wide channel photographed by a moving camera at different speeds. The mean velocity of the flow is 16.7cm/s. The two lower pictures are from Prandtl and Tietjens (1934). In the right picture, the camera moves with the speed equal to the velocity of water in the centre of the channel. In the left picture, the speed of the camera is small and close to the velocity of the water near the walls.

Instability

As mentioned in chapter 2, the most commonly accepted view on the origin of turbulence is flow instability. An additional factor is that instability is considered as one of the origins of structure(s) in/of turbulence. However, this latter view requires to assume that turbulence has a pretty long 'memory' of or, alternatively, that the 'purely' turblulent flow regime (i.e. at large enough Reynolds numbers) has instability mechanisms similar to those existing in the process of transition from laminar to turbulent flow state. Tritton (1988) defines turbulence as *a state of continuous instability*. The problem is that speaking about (in)stability requires one to define the state of flow (in)stability of which is considered, which is not a simple matter in the case of a turbulent flow.

Emergence

Another less known view holds that structure(s) emerge in large Reynolds number turbulence out of 'purely random structureless' background, e.g. via the so-called inverse cascades or negative eddy viscosity (see chapter 8). Among the spectacular examples, are the 'geophysical vortices' in the atmosphere, and ocean, as well as astrophysical objects. Another example is the emergence of coherent entities, such as vortex filaments/worms and other structure(s), out of an initially random Gaussian velocity field via the NSE dynamics[14]. An example of such structure(s) is shown in plate 5, for other examples see references in Tsinober (1998ab).

It 'just exists', or do flows become or are they 'just' turbulent?
To the flows observed in the long run after the influence of the initial conditions has died down there correspond certain solutions of the Navier- Stokes equations. These solutions constitute a certain manifold $\mathcal{M} = \mathcal{M}(\mu)$ (or $\mathcal{M} = \mathcal{M}(Re)$) in phase space invariant under phase flow (Hopf, 1948). Kolmogorov's scenario was based on the complexity of the dynamics along the atractor rather than its stability (Arnold, 1991; see also Keefe, 1990a; Keefe et al., 1992).
This view is a reflection of one of the modern beliefs that the structure(s) of turbulence - as we observe in *physical space* - is (are) the manifestation of the generic structural properties of mathematical objects in *phase space*, which are called (strange) attractors and which are invariant in some sense. In other words here the structure(s) assumed to be 'built in' the turbulence

---

[14]Recall P. W. Anderson (1971, 1995) who emphasizes *the concept of 'broken symmetry', the ability of a large collection of simple objects to abandon its own symmetry as well as the symmetries of the forces governing it and to exhibit the 'emergent property' of a new symmetry.* One of the difficulties in turbulence research is that no objects simple enough have been found so far such that a collection of these objects would *adequately* represent turbulent flows .

independently of its origin (hence the tendency to universality)[15]. It is noteworthy that the assumed strange-attractor existence makes sense for statistically stationary turbulent flows. However, for flows which are not such, e.g. decaying turbulent flows past a grid or a DNS simulated flow in box the attractor is trivial. Nevertheless, these flows possess many properties which are essentially the same as their statistically stationary counterparts provided that their Reynolds numbers are not too small ($Re_\lambda \sim 10^2$).

The above refers to the dynamical aspects of real turbulent flows. We mention again here also the

Emergence of structures in a passive objects in random media
in which the velocity field and the external forcing are prescribed. Whatever their nature - even Gaussian - structure is emerging in the field of passive objects (Zeldovich et al., 1988; Ott, 1999; and references therein). In other words structure(s) of passive objects emerges in structureless (artifical) random Gaussian velocity field.

7.3.2. HOW DOES THE STRUCTURE OF TURBULENCE 'LOOK'?

Until recently, very little was known about the nature of structure of turbulence and about the appearance of its structures (in physical space). The structure in question is the so-called fine structure and not the one which is promoted by various external factors and/or constraints like boundaries, mean shear, centrifugal forces (rotation), buoyancy, magnetic field, etc., which usually act as an organizing factors, favouring the formation of coherent structures of different kinds (quasi-two-dimensional, helical, hairpins, etc.). These structures, are, as a rule, large scale features which depend on the particularities of a given flow and thus are not universal. We will return to some of these mostly large scale structures including what is called 'coherent structures' or 'organized motion' in chapter 8.

Since the first DNS simulations by Siggia (1981), a number of computations have performed (see references in Tsinober, 1998 ab), which demonstrated clearly that even turbulence which is 'homogeneous' and 'isotropic' has structure(s), i.e. contains a variety of strongly localized events. The primary evidence is related to spatial localization of subregions with large enstrophy (i.e. intense vorticity) which are organized in long, thin tubes-filaments-worms. Such filaments were also directly observed in laboratory experiments (the ones mentioned in figure 5.1) employing the property of intense vorticity to be strongly correlated with regions of low pressure and using small air bubbles to visualize these regions (Douady et al., 1991;

---

[15]In the strange attractor theory, the experimental measurements are viewed as projections of these attractors onto low dimension that correspond to these measurements.

Villermaux et al., 1995; for more references see Tsinober, 1998a). This follows from the Poisson-like equation for pressure $2\nabla^2 p/\rho = \omega^2 - 2s_{ij}s_{ij}$. There is some evidence that in regions with moderate magnitude of vorticity it is organized in sheet-like structures. Much less is known about regions with large strain, $s_{ij}s_{ij}$, i.e. dissipation. They were tentatively identified as layered vortex sheets in Schwarz (1990), an observation that has not been confirmed by other observations or computations so far. Most common observations at Reynolds numbers accessible in DNS showed that isosurfaces of high strain are wrapped around the regions of strong enstrophy. However, in Tanaka and Kida (1993) and in recent computations by Boratav and Pelz (1997), the isosurfaces of large strain were observed as sheet-like objects with very sharp edges (razors/flakes). In fact such objects were observed already by Siggia (1981, figure 21). This does not mean that the vorticity field in these regions is simple and is necessarily sheet-like too. Some examples of the results mentioned are shown in figure 7.5.

The relatively simple appearance of the observed structures as shown above prompted a rather popular view that turbulence structure(s) is (are) simple in some sense and that essential aspects of turbulence structure and its dynamics may be adequately represented by a random distribution of simple (weakly interacting) objects, such as straight strained (Burgers-like) vortices (see chapter 6 and references in Tsinober, 1998a). In particular, it is commonly believed that most of the structure of turbulence is associated with and is due to various strongly localized intense events/structures, e.g. mostly regions of concentrated vorticity so that '*turbulent flow is dominated by vortex tubes of small cross-section and bounded eccentricity*' (Chorin 1994, p. 95) and that these events are mainly responsible for the phenomenon of intermittency (Belin et al., 1996; Frisch, 1995; Katul et al., 1994; Nelkin, 1995; Jimenez and Wray, 1998; and references therein). It is argued in Tsinober (1998a) that such views are inadequate (see chapter 6 for more details and latest references). It appears that - though important - these structures are not the most dynamically important ones and are the consequence of the dynamics of turbulence rather than its dominating factor. Namely, regions *other* than those involving concentrated vorticity such as: *i* – 'structureless' background, *ii* – regions of strong vorticity/strain (self) interaction and largest enstrophy and strain production dominated by large strain rather than large enstrophy, and *iii* – regions with negative enstrophy production are all dynamically significant (in some important respects more significant than those with concentrated vorticity), strongly non-Gaussian, and possess structure. Due to the strong nonlocality of turbulence in physical space all the regions are in continuous interaction and are strongly coupled. A similar statement can be made regarding the so called streamwise vortices observed in many turbulent flows (see Chapter

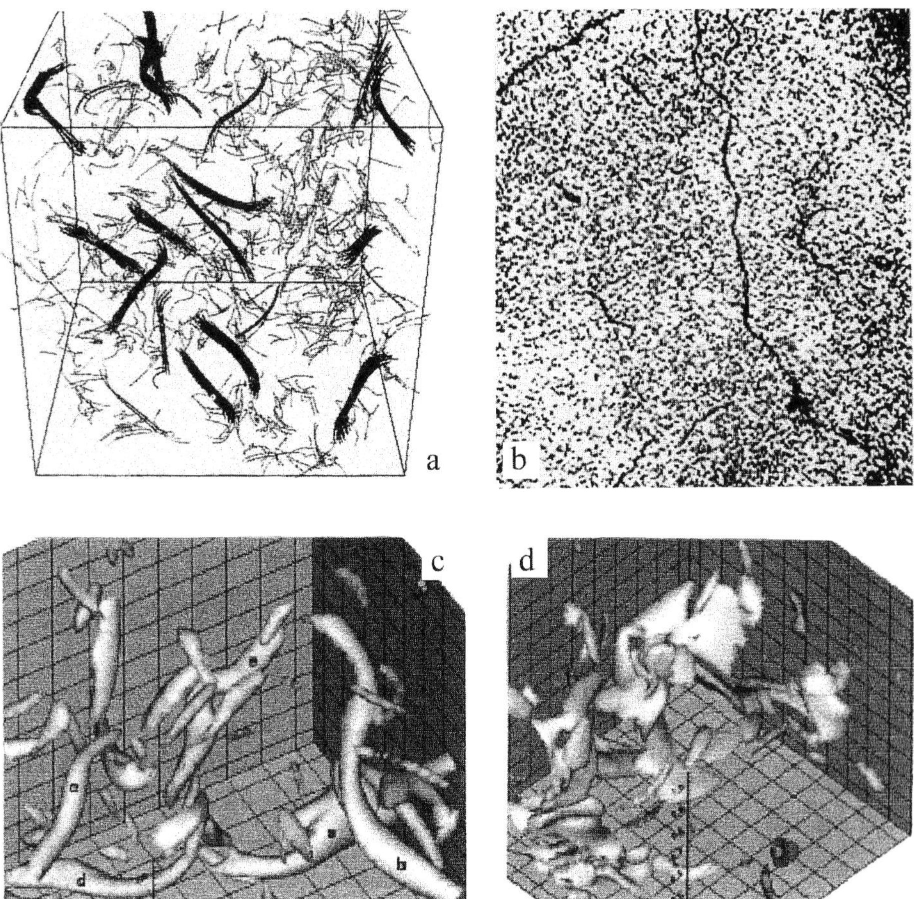

Figure 7.5. Vortex filaments in DNS (She et al., 1991; top left) and laboratory (Douady et al., 1991; top right). Isosurfaces of the second invariant of the velocity derivatives tensor $Q = \omega^2 - s_{ij}s_{ij}$ (bottom left) at 2 rms positive level, i.e. vorticity dominated regions, and isosurfaces of strain $s_{ij}s_{ij}$ (bottom right) at 2 rms level, i.e. strain dominated regions (Boratav and Pelz, 1997). The two bottom pictures were not included in Boratav and Pelz (1997), but are available at http://www.eng.uci.edu/~boratav/ and are used here by permission. The figure is from Tsinober (1998a).

8).

The above conclusions are the outcome of the use of quantitative manifestaions of turbulence structure, which just like intermittency are in the first place of statistical nature.

7.3.3. STRUCTURE VERSUS STATISTICS

On a qualitative level, it is widely recognized that fluid-dynamical turbulence (even 'homogeneous' and 'isotropic') has 'structure(s)', i.e. contains a variety of strongly localized events, which are believed to influence significantly the properties of turbulent flows. It is impossible to underestimate the observational information on the instantaneous structures of turbulent flows. Being extremely useful, the individual observations of such events/structures are inherently limited as compared to the statistical information, which requires us to employ the quantitative manifestations of turbulence structure. In order to proceed to the quantitative aspects of the problem it is not sufficient to look at pictures (however beautiful); one has to turn to numbers and quantitative relations such as in the above-mentioned anomalous scaling, which is one of many other more specific quantitative manifestations of turbulence structure.

The question about what structure(s) of turbulence mean(s) can be answered via a statement of impotence: speaking about 'structure(s)' in turbulence the implication is that there exist something 'structureless', e.g. Gaussian random field as a representative of full/complete disorder. Gaussian field is appropriate/natural to represent the absence of structure in the statistical sense. Hence all non-Gaussian manifestations of turbulent flows can be seen as some statistical signature of turbulence structure(s)[16]. However, simple probability criteria are insufficient, since *one can find in statistical data irrelevant structures with high probability* (Lumley, 1981). In other words the structure(s) should be relevant/significant in some sense. For example, it should be *dynamically* relevant for velocity field, and related quantities such as vorticity and strain. This does not mean that kinematical aspects of turbulence structure(s) are of no importance. For example, *anisotropy* is a typical *kinematic statistical* characteristic of turbulent flows which hardly can be applied to *individual* structures, e.g. a turbulent flow consisting mostly of 'anisotropic' individual structures can be statistically isotropic. Among the first statistical treatments of turbulence structure is, of course, the first paper by Komlogorov (1941a), the very title of which is *The local structure of turbulence in incompressible viscous fluid for very large Reynolds numbers.*

The advantage of such an approach is that it allows one to get insights into the *structure* of turbulence without the necessity of knowing much (if anything) about the actual appearance of it's *structures*. This is especially important in view of numerous problems/ambiguities in definitions of *individual* structures in turbulent flows, their identification and statistical characterization as well as their incorporation in 'theories'. The main reason is that there exist an intrinsic problem of both defining what the

---

[16]This does not imply that an exactly Gaussian field does not necessarily possess any spatial or temporal structures, see, e.g. fig. 3 in She et al. (1990) – any individual realization of a Gaussian field does have structures. However, an exactly Gaussian field does not possess *dynamically* relevant structure(s), it is dynamically impotent, see below.

relevant structures are (see Bonnet, 1996 for references and a review of existing techniques) which all are based on statistics anyhow, and of defining extracting/educing and characterizing the so-called coherent structures. For a number of reasons, it is very difficult, if not impossible, to quantify the information on the *instantaneous* structures of turbulent flows into dynamically relevant/significant form. The observed individual structures are not simple, neither are neither are neither weakly interacting between themselves or with the background. Indeed, *you can find structures, essentially arbitrary, which have equal probability to the ones we have latched onto over the years: bursts, streaks, etc....If structures are defined as those objects which can be extracted by conditional sampling criteria, then they are everywhere one looks in turbulence* (Keefe, 1990b). For instance, looking at a snapshot of the enstrophy levels of a *purely Gaussian* velocity field in She et al. (1990), one can see a number of filaments (the irrelevant ones) like those observed in real turbulent flows, i.e. pure Gaussian velocity field has some structure(s) too.

The next most difficult question is about the relevance/significance of some particular aspect of non-Gaussianity for a specific problem in question. It seems that here one enters the *subjective* realm: the criteria of significance (which is the matter of physics!) are decided by the researchers. However, the following examples show that objective choice of the structure sensitive statistics is dictated by general dynamical aspects of the problem[17].

For instance, the build up of *odd* moments is an important *specific* manifestation of structure of turbulence along with being the manifestation of its nonlinearity. The two most important examples are the third order velocity structure function $S_3(r) = \langle\{[\mathbf{u}(\mathbf{x}+\mathbf{r}) - \mathbf{u}(\mathbf{x})] \cdot \mathbf{r}/r\}^3\rangle$ and the mean enstrophy production $\langle\omega_i\omega_k s_{ik}\rangle$. The first one is associated with the $-4/5$ Kolmogorov law $S_3(r) = -4/5\langle\epsilon\rangle r$ (Kolmogorov, 1941b), which is the first strong indication of the presence of structure in the inertial range showing that both non-Gaussianity and the structure of turbulence are directly related to it's dissipative nature. It is remarkable that the title of this paper by Kolmogorov is *Dissipation of energy in the locally isotropic turbulence*. The $-4/5$ Kolmogorov law clearly overrules the claims that *'Kolmogorov's work on the fine-scale properties ignores any structure which may be present in the flow'* (Frisch 1995, p. 182) and that it is associated with near-Gaussian statistics, (Chertkov et al., 1999; Farge and Guyon, 1999; Katul et al., 1994; She et al., 1991). It is noteworthy that – as shown by Hill (1997) – the -4/5 Komlogorov law is more sensitive to the anisotropy,

---

[17]In the following, we will discuss the dynamical aspects of the problem. Various 'kinematic' issues, like the transport of passive objects (scalars, vectors, etc.), in which Gaussian or other *precribed* velocity fields are used rather successfully, are beyond the scope of this section. We mention only that structure(s) of the field of passive objects can be treated in a similar way as the one described in this section.

i.e. the third-order statistics (again odd moments), than the second-order statistics. Likewise the structure functions of higher odd orders $S_p^{\|}(r) = \langle(\Delta u_{\|})^p\rangle$ are essentially different from zero, see references in (Betchov, 1976; Sreenivasan and Antonia, 1997; Tsinober, 1998b).

The essentially positive value of the mean enstrophy generation $\langle\omega_i\omega_k s_{ik}\rangle$, discovered by Taylor (1938) is the first indication of the presense of structure in the small scales, where turbulence is particularly strongly non-Gaussian and intermittent (Kraichnan, 1967; Novikov, 1963; Sreenivasan and Antonia, 1997). The above two examples show that both the essential turbulence dynamics and its structure are associated with those aspects of it's non-Gaussianity exhibited in the build up of odd moments, which among other things means phase and geometrical coherency, i.e. structure (see section on the non-Gaussian aspects of turbulence in chapter 6). Hence, the importance of odd moments as indicators of intermittency. It is to be noted that the non-Gaussianity found experimentally both in large and small scales is exhibited not only in the nonzero odd moments, but also in strong deviations of even moments from their Gaussian values. Thus both the large and small scales differ essentially from Gaussian indicating that both possess structure.

We return to the question about what kinds of statistics are most appropriate to chacterize at least some aspects for turbulence structure. But first we must mention

7.3.4. EXAMPLES OF STATISTICS WEAKLY SENSITIVE TO STRUCTURE(S)

The first examples of this kind are energy spectra in which the phase (and geometric) information is lost. Hence their weak sensitivity to the structure of turbulence. This insensitivity, in particular, is exhibited in the scaling exponents when/if such exist. For example, the famous $-5/3$ exponent can be obtained for a great variety of *qualitatively* different real systems - not necessarily fluid dynamical - and theoretical models. A partial list of references contains the papers by Biferale et al. (1994), Cheklov and Yakhot (1995), Chorin (1994, 1996), Kiya and Ishi (1990), Lundgren (1982), Moffatt (1993), Nore et al. (1997), Pullin and Saffman (1997), Taguchi (1995), Tsinober (1998b), Vassilicos and Brasseur (1996), Zakharov et al. (1993). Of course, one can also construct a set of purely Gaussian velocity fields, i.e. lacking any dynamically relevant structure(s), with any desired length of the $-5/3$ 'inertial' range (Elliott and Majda, 1995). Vice versa the spectral slope can change, but the structure remains essentially the same *'yet retaining all the phase information'* (Armi and Flament, 1987; see figure 7.6). Moreover, not only *'the spectral slope alone is inadequate to differentiate*

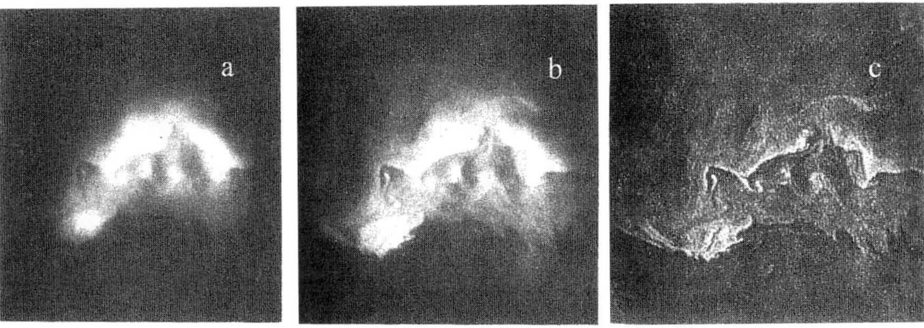

Figure 7.6. An example of spectral analysis of an infrared image of the ocean sea-surface temperature off the northern coast of California. The upper figure corresponds to the spectral density of slope k^{-4}, the middle - k^{-3}, and the lower - k^{-2}. An important point is that all the three correspond to the *same* original image and retain the phase information, *but place less emphasis on the observed spectrum.* Courtesy of Professor L. Armi; for more details see Armi and Flament (1987).

between theories' (Armi and Flament, 1987), *alone* it does not correspond to any particular structure(s) in turbulence or it's absence : there is no one-to-one relation between scaling exponents and structure(s) of turbulence. This is true not only of exponents related to Fourier decomposition with its ambiguity (Tennekes, 1976a), but of many other scaling exponents including those obtained in some wavelet space and in the physical space – a much overstressed aspect of turbulent flows.

Likewise, similar PDFs of *some* quantities can correspond to qualitatively different structure(s) and quantitatively different values of Reynolds number (Kraichnan and Kimura, 1994; Tsinober, 1998a,b). The emphasis is on *some* quantities like pressure or some other usually (but not necessarily) even order quantities in velocities or their derivatives, since the PDFs of *other* appropriately chosen quantities are sensitive to structure (see below).

7.3.5. STRUCTURE SENSITIVE STATISTICS

Use of odd order structure functions

This is an example how structure sensitive statistics can help in looking for the right reasons of measured spectra in the lower mesoscale range, Lindborg (1999). The procedure involves using the third order structure functions which are generally positive in the two dimensional case (contrary to the three-dimensional case). Calculations based on wind data from airplane flights, reported in the MOZAIC data set. It is argued that the k^{-3}-range is due to two-dimensional turbulence and can be interpreted as an enstrophy inertial range, while the $k^{-5/3}$-range is probably not due to two-dimensional turbulence and should not be interpreted as a two-dimensional energy inertial range. There is a competing hypothesis that the large scale $-5/3$ range is the spectrum of weakly non-linear internal gravity waves with a forward energy cascade (Van Zandt, 1982). A third claim is that the spectral slope in the enstrophy range is more shallow than -3 and is close to $-7/3$ (Tsinober, 1995a). This range and related anomalous diffusion is explained in terms of the phenomenon of spontaneous breaking of statistical isotropy (rotational and/or reflexional) symmetry - locally and/or globally.

Another example is the demonstration (mentioned in chapters 5 and 6) that the small scale structure of a homogeneous turbulent shear flow is essentially *anisotropic* at Reynolds number up to $Re_\lambda \approx 1000$ (Shen and Warhaft, 2000; see also Ferchichi and Tavoularis, 2000). In order to detect this anisotropy the authors measured the velocity structure functions of third and higher odd orders of both longitudinal and transverse velocity components and corresponding moments of velocity derivatives. In particular, they found a skewness of order 1 of the derivative of the longitudinal velocity in the direction of the mean gradient, which should be very small (or ideally vanish) for a locally isotropic flow. Similar results were obtained in DNS (see references in Shen and Warhaft, 2000; and Warhaft, 2000) and aslo Borue and Orszag, 1996; and Shebalin and Woodruff, 1997). We should recall that analogous 'misbehaviour' of large Reynolds-number turbulence regarding the skewness of temperature fluctuations in the atmospheric boundary layer is known since late sixties (Stewart, 1969; Gibson et al., 1970, 1977).

Geometrical statistics

This example shows how conditional sampling based on *geometrical* statistics can help to get insight into the nature of various regions of turbulent flow, e.g. those associated with strong/weak vorticity, strain, various alignments, and other aspects as described in the previous Chapter. The first general aspect is the qualitative difference in the behaviour and properties of regions with large enstrophy from strain dominated regions, which is also one of the manifestauions of intermittency. Another example is the PDF of the cosine of the angle between vorticity, $\boldsymbol{\omega}$, and the vortex stretching vector, $W_i \equiv \omega_k s_{ik}$, $\cos(\boldsymbol{\omega}, \mathbf{W})$. It is strictly symmetric for a Gaussian velocity field, whereas it is strongly positively skewed in real turbulent flows. It remains essentially positively skewed for any part of the turbulent field (see figure 6.13), e.g. in the 'weak background' (involving whatever definition based on enstrophy, strain, both and/or any other relevant quantity). Thus, contrary to common beliefs, the so called 'background' is not structureless, dynamically not inactive and essentially non-Gaussian, just like the whole flow field or any part of it. The structure of the apparently random 'background' seems to be rather complicated. The previous qualitative observations (mostly from DNS) about the '*little apparent structure in the low intensity component'* or the *'bulk of the volume'* with *'no particular visible structure'* should be interpreted as meaning that no *simple visible* structure has been observed so far in the bulk of the volume in the flow. It is a reflection of our inability to 'see' more intricate aspects of turbulence structure: intricacy and 'randomness' are not synonyms for absence of structure.

Pressure hessian

Recently, special attention has focused on the pressure hessian $\frac{\partial^2 p}{\partial x_i \partial x_j}$. Among the general reasons for such an interest is that the pressure hessian is intimately related to the nonlocality of turbulence in physical space (see chapter 6 and references in Tsinober, 1998a,b).

One of the quantities in the present context directly associated with the pressure hessian is the scalar invariant quantity $\omega_i \omega_k \frac{\partial^2 p}{\partial x_i \partial x_k}$. It is responsible for the nonlocal effects in the rate of change of enstrophy generation $\omega_i \omega_k s_{ik}$ (see equation [C.23] in the appendix C). What is special about this quantity, which is of even order in velocity, is that for a Gaussian velocity field $\left\langle \omega_i \omega_k \frac{\partial^2 p}{\partial x_i \partial x_k} \right\rangle_G \equiv 0$, whereas in a real flow it is essentially positive and $\left\langle \omega_i \omega_k \frac{\partial^2 p}{\partial x_i \partial x_k} \right\rangle \sim \frac{1}{3} \left\langle W^2 \right\rangle$, where $W_i \equiv \omega_k s_{ik}$ is the vortex stretching vector. Thus interaction between the pressure hessian and the vorticity is one of the essential features of turbulence structure associated with its nonlocality. It is noteworthy that a similar quantity involving strain is non-vanishing for a Gaussian velocity field, $\langle s_{ik} s_{kj} \frac{\partial^2 p}{\partial x_i \partial x_j} \rangle_G = -\frac{1}{20} \langle \omega^2 \rangle_G^2$.

7.4. Which quantities possess structure in turbulence and how to 'dig' them out?

We have seen that different quantities possess different structure(s) in the same flow, the velocity field may have no structure, but the passive tracer may well have one. Simple laminar Eulerian velocity field (E-laminar) creates complicated Lagrangian field (L-turbulent). A purely Gaussian, i.e. 'structureless' velocity field creates a structure in the field of passive objects. The structure(s) seen in the velocity field depend on the motion of the observer (see figure 7.4).

The most commonly used methods in looking at structure(s) are based on the so-called conditional sampling techniques, which employs some criteria to educe some structure(s). Among the simplest criteria is sampling based on equilevels of some function, e.g. enstrophy ω^2 (this is how the first evidence of concentrated vorticity/filaments/worms was obtained). More generally, this problem is related to pattern recognition and requires defining a conditional sampling scheme. This scheme is in turn based on what a particular investigator thinks are the most important physical processes, features, etc. This in turn opens a Pandora box of possibilities and contains an inherent element of subjectivity and arbitrariness, since the physics of turbulence is not well understood. In this sense, the circle is closed: in order to objectively define and educe some structure, one needs clear understanding of the physics of turbulence, which, it is in turn believed, can be achieved via study of turbulence structure(s).

The most popular method is to look for structure(s) using a criterion based on one parameter only, e.g. enstrophy ω^2. Though such an approach is useful and 'easy', it is inherently limited and reflects the simplest aspects of the problem. For example, even for characterization of *some* aspects of the *local* (i.e. in a sense 'point'-wise) structure of the flow field in the frame following a fluid particle requires at least *two* parameters.[18] Therefore attempts to *adequately* characterize finite scale structure(s) by one parameter only are unlikely to be successful, and one needs something like pattern recognition based on some conditional sampling scheme involving more than two parameters. A similar problem arises when attempting to characterize structure(s) of turbulent flows using *two*-point information but based on a *single* velocity component only, e.g. longitudinal structure functions $S_n^{\parallel}(\mathbf{r})$, since such an approach does not 'know' (almost) anything about the two other velocty components.

The 'not objective enough' nature of a variety of conditional sampling procedures resulted in a whole 'zoo' of 'structures' in different turbulent flows, which some people believe to be significant in some sense, but many do not. Among the reasons for such scepticism is some evidence that the attempts at representation of such a complicated phenomenon like turbulence as a collection of simple objects/structures only are unlikely to succeed (see chapters 5 and 6; and Tsinober, 1998a,b). As mentioned, until recently it was believed that concentrated vorticity/filaments is the dominating structure in turbulent flows in the sense that most of the structure of turbulence is associated with and is due to regions of concentrated vorticity. It appears that - though important - these structures are not the most dynamically important ones and are the consequence of the dynamics of turbulence rather than being its dominating factor.

Nevertheless, as shown above some 'objectivness' can be achieved using quantities appearing in the NSE and/or the equations which are exact consequences of NSE.

More information and references about various attempts to define, educe

---

[18] $Q = 1/4(\omega^2 - 2s_{ij}s_{ij})$ and $R = -1/3(s_{ij}s_{jk}s_{ki} + 3/4\omega_i\omega_j s_{ij})$. Here Q is the second – and R is the third invariant of the velocity gradient tensor $\partial u_i/\partial x_k$. The first is vanishing due to incompressibility (see Chacin and Cantwell, 2000; Ooi et al., 1998; and references therein). We return to this issue in chapters 8 and 9 in different contexts.

and characterize various 'coherent structures' can be found in Bonnet (1996) and Holmes et al. (1996, 1997).

7.4.1. STRUCTURE(S) VERSUS SCALES AND DECOMPOSITIONS

It is natural to ask how meaningful is it to speak about different scales in the context of 'structure(s)' and in what sense, especially when looking at the 'instantaneous' structure(s) of /in turbulence. The known structures indeed possess quite different scales. Vortex filaments/worms – have at least two essentially different scales, their length can be of the order of the integral scale, whereas their cross-section is of the order of Kolmogorov scale. Similarly, the ramp-cliff fronts in the passive scalar fields have a thickness much smaller than the two other scales. This fact is consistent with the observation by Batchelor and Townsend (1949), that *the mean separation of the visible activated regions is comparable with the integral scale of the turbuence, i.e. with the size of the energy-containing eddies.*

It is believed that appropriately chosen decompositions may represent structure(s) of turbulence, e.g. Holmes et al., 1996). Here again several notes are in order. First, this position depends strongly on what is meant by structure(s). Second, such a possibility is realistic when the flow is dominated by (usually large scale) structures, when many, or practically any reasonable decompositions will do anyhow. And third, structure(s) (and related issues such as geometry) emerging in the 'simplest' case of turbulent flows, in a box with periodic boundary conditions, is(are) are inaccessible via Fourier decomposition, the most natural one in this case.

7.5. Summary

Small scale intermittency of turbulence is associated with its spotty (spatio-temporal) *structure,* which among other things is manifested as a *particular* kind of non-Gaussian behaviour of turbulent flows. This deviation from Gaussianity, increases with both the i – Reynolds number and ii – as the 'scale' decreases. In other words intermittency involves two aspects of turbulent flows - their structure/geometry and statistics. Intermittency specifically in fluid turbulence is associated mostly with *some* aspects of its spatio-temporal structure. Hence, the close relation between the origin(s) and meaning of intermittency and the structure of turbulence.

There is no turbulence without structure(s). Every part (just as the whole) of the turbulent field – including the so-called 'structureless background' – possess structure. Structureless turbulence (or any of its part) contradicts both the experimental evidence and the Navier-Stokes equations. The qualitative observations on the *little apparent structure in the low intensity component* or the *bulk of the volume* with *no particular vis-*

ible structure should be interpreted as indicating that no *simple visible* structure has been observed so far in the bulk of the volume in the flow. It is a reflection of our inability to 'see' more intricate aspects of turbulence structure: intricacy and 'randomness' are not synonyms for absence of structure. This complexity of turbulence phenomena and its structure(s) makes it necessary and unavoidable to use statistical methods of description/characterization of turbulence phenomena and its structure(s). It is important to emphasize the distinction between statistics weakly dependent on structure(s) and structure sensitive statistics, which is intimately related to and underscores the non-Gaussian nature of turbulence. The latter allows one to obtain information on the structure of turbulence without knowing anything about its structures' actual appearance. Statistical description (not 'theories') is the only quantitative alternative to the visual qualitative methods. We remind that 'statistical' means not only the 'traditional' things like means/averages and other simple means, but all kinds of statistics including rather exquisite ones such as conditional statistics can be, depending on the nature of problems in question and the ability/skill of the researcher to formulate such questions.

The view that turbulence structure(s) is(are) simple in some sense and that turbulence can be represented as a collection of simple objects only seems to be a nice illusion which, unfortunately, has little to do with reality. It seems somewhat wishfully naive to expect that such a complicated phenomenon like turbulence can merely be described in terms of collections of only such 'simple' and weakly interacting objects.

CHAPTER 8

TURBULENCE UNDER VARIOUS INFLUENCES AND PHYSICAL CIRCUMSTANCES

Closer to real world turbulent flows

8.1. Introduction

As mentioned in chapter 1, there are many factors and influences which cause a real turbulent flow in nature and technology to deviate from the idealized homogeneous and isotropic state, sometimes strongly. In the latter case, turbulent flows may lose most of their resemblance to the three-dimensional homogeneous isotropic flow, but can be quite similar to the (quasi) two-dimensional one. Here we are inevitably back also to the origins of turbulence but with a different focus: the source sustaining the turbulence, such as mean shear, buoyancy, electromagnetic forces, shock waves. In other words there are different kinds of influences. The 'simplest' influences are 'one way'. They modify the turbulent flow in some way but are neither influenced by the turbulent flow nor sustain it, e.g. rotation under some conditions. Other kinds of influences are 'two way'. They are truly interactive in the sense that along with modifying the turbulent flow they are changed by the flow too, e.g. magnetic field, density stratification and other active scalars. Under certain circumstances, these influences contribute also to production and sustaining of turbulent flows, such as in the case of turbulent convection. Turbulent flows with mean shear (strain) belong to this latter category.

What follows is an overview of turbulent flows under various influences and physical circumstances, which include shear, buoyancy, rotation, (electro)magnetic field, compressibility and additives. For obvious reasons, the material of this chapter is limited by only the most important essential features, simple examples and qualitative aspects with references on comprehensive sources and some early and latest papers[1]. For the same reasons,

---

[1]For example, we do not include predominantly empirical material, such as the lively discussion whether the near wall turbulence has complete similarity, leading to the *log* law for the mean velocity, or it possesses incomplete simialrity leading to a *power* law for the mean velocity. It seems that with appropriate 'tuning' both - being dimensionally correct - do well, and that without deeper physical foundation than just similarity and dimensional analysis the 'controversy' cannot be resolved. For discussion and references see Österlund et al. (2000) and Wosnik et al. (2000).

no summary is given at the end of this chapter.

The two main common features of turbulent flows under various influences and physical circumstances are anisotropy and inhomogeneity[2]. Along with other consequences anisotropy results in nonzero off diagonal Reynolds stresses, $-\langle u_i u_j \rangle$, $i \neq j$, whereas inhomogeneity leads to nonzero gradients of the Reynolds stresses. Only in the latter case is there a two way coupling between the mean flow and the field of fluctuations, since the Reynolds averaged equations (C.43) for the mean flow contain the *gradients* of the Reynolds stresses[3]. Note that anisotropy can have a great variety of manifestations. For example, the flow in the proximity of turbulent channel flow cannot be considered neither approximately homogeneous nor isotropic in spite of the fact that in this region both the gradient of the mean velocity dU_1/dx_2 and the Reynolds stress are small, because the gradient of the Reynolds stress, $d\{-\langle u_i u_j \rangle\}/dx_2$, is not small.

The Reynolds stresses, $-\langle u_i u_j \rangle$, react back on the mean flow. This reaction results – among other things – in enhanced (turbulent) momentum transport, and consequently turbulent kinetic energy production $-\langle u_i u_j \rangle S_{ij}$. The simplest manifestation is much larger than in the laminar counterpart resistance, i.e. large gradients of mean velocity on the wall, and a flat velocity profile far from the boundaries in the turbulent channel flow.

The total turbulent energy balance for the whole flow domain is

$$\frac{d\mathcal{E}_T}{dt} = \mathcal{P} + \mathcal{W}_F - \mathcal{D} \qquad\qquad (C.49)$$

where $\mathcal{E}_T = \int e_T dV$, $e_T = \frac{1}{2}u^2$, is the total kinetic energy of turbulent fluctuations, $\mathcal{P} = \int -\langle u_i u_j \rangle S_{ij} dV$ - is the total rate of production/ destruction of energy turbulent fluctuations by the mean strain, S_{ij}, (mean velocity gradients), $\mathcal{W}_F = \int u_i F_i dV$, - is the total rate of production of energy of turbulent fluctuations by the external forces and $\mathcal{D} = 2\nu \int s_{ij} s_{ij} dV$, - is the total rate of dissipation (simply dissipation) of energy of turbulent fluctuations by viscosity. If the flow is statistically stationary and is a pure shear flow, i.e. $\mathcal{W}_F = 0$, the dissipation equals production $\mathcal{P} = \mathcal{D}$, i.e. $\mathcal{P} > 0$. If there is no mean flow to supply energy to the field of fluctuations, the statistically stationary state can be maintained by some external source (as, e.g. in thermal convection), and the energy balance equation takes the form $\mathcal{W}_F = \mathcal{D}$. It is noteworthy that in the presence of some energy supply *other* than the mean shear/strain the total rate of production of energy

---

[2]An inhomogeneous field is also anisotropic, but not vice versa.

[3]In case of a (hypothetical) homogeneous shear flow, the mean flow, which is just $\mathbf{U} = (Sx_2, 0, 0)$, does not 'know' about the turbulent fluctuations, since $d\langle u_1 u_2 \rangle/dx_2 = 0$. Therefore, without some additional sustaining mechanism turbulent fluctuations cannot be stationary and will decay. Moreover, it was shown by Harris et al. (1977) that such a flow as a whole (i.e. mean plus fluctuations) is impossible.

turbulent fluctuations by the mean strain, $\mathcal{P}$, does not have to be positive even in the case of statistically stationary turbulent flow, since in this case the balance is $\mathcal{W}_F + \mathcal{P} - \mathcal{D} = 0$. This leads to the possibility that the field of turbulent fluctuations 'feeds' the mean flow, as in examples described in section 8.5.

8.2. Shear flows

We limit the following discussion to 'simple' turbulent shear flows, which have a (quasi) one dimensional mean $\mathbf{U} = (U\mathbf{i}, V, 0)$, $V \ll U$, with slow streamwise and no spanwise variations, $\partial\langle\cdots\rangle/x_2 \ll \partial\langle\cdots\rangle/x_1$, $\partial\langle\cdots\rangle/x_3 = 0$, and axisymmetric analogues, where $\langle\cdots\rangle$ means some mean of any quantity. These are turbulent flows in channels (pipes) and boundary layers (on weakly curved bodies), which are called wall bounded turbulent flows, and turbulent flows in jets, plumes, wakes and mixing layers, which are called boundary-free (or simply free) turbulent shear flows. Moreover, for qualitative purposes it is convenient and sufficient to assume that $\mathbf{U} = (U\mathbf{i}, 0, 0)$, and $\partial\langle\cdots\rangle/x_1 = \partial\langle\cdots\rangle/x_3 = 0$ as in turbulent channel flows with x_1, x_2, x_3 for the streamwise, wall-normal and spanwise coordinates in which all statistical properties depend on the coordinate, x_2, normal to the channel boundary only. Such a flow possesses a mean vorticity $\boldsymbol{\Omega} = (0, 0, dU/dx_2)$ having only a spanwise component, and a mean rate of strain, S_{ij}, with nonzero components $S_{12} = S_{21} = \frac{1}{2}dU/dx_2$, and with the eigenvectors along the axes inclined at $\pi/4$ and $3\pi/4$ to the streamwise direction, x_1, and corresponding eigenvalues $\Lambda_1^S = \frac{1}{2}dU/dx_2$, $\Lambda_2^S = -\frac{1}{2}dU/dx_2$ and $\Lambda_3^S = 0$.

The turbulent fluctuations are exposed to the persistent action of this mean strain, which leads to anisotropy of the Reynolds stress tensor of turbulent fluctuations $-\langle u_1 u_2 \rangle$ in such a way that the eigenframe of the instantaneous Reynolds stress tensor $-u_1 u_2$ tends to be aligned with the eigenframe of S_{ij}, see left column (especially bottom) in figure 8.20. This alignment occurs in such a way that the term $-u_i u_j S_{ij}$ responsible for turbulent energy production (see equation [C.47]) is positively skewed, so that its mean is (usually) positive and so is the total rate of production/destruction of energy turbulent fluctuations by the mean strain, $\mathcal{P} = \int -\langle u_i u_j \rangle S_{ij} dV$. In fact, the above alignment is nothing but the tendency for alignment between vector u_i (more precisely its projection on the plane x, y) and eigenvector of S_{ij} corresponding to its negative eigenvalue, which is seen from the relation $-u_i u_j S_{ij} = u^2 \Lambda_i^S \cos^2(\mathbf{u}, \boldsymbol{\lambda}_i^S)$. It is noteworthy that this kind of behaviour is of more general nature[4].

---

[4]For example, the 'source' term, $-\Delta_i^u \Delta_j^u s_{ij}$, in the RHS of (C.59), appendix C, responsible for the production of the energy of error appears to be a positively skewed quantity (Tsinober and Ortenberg, 2001). That is, just like the mean, $-\langle u_i u_j \rangle S_{ij}$, is positive in turbulent shear flows, the integral of the production of the energy of error,

Similarly, vorticity (more precisely its projection on the plane x, y) tends to align in this region with the stretching eigenvector of the mean strain S_{ij}, which is inclined at the angle $\pi/4$ to the streamwise direction (Moin and Kim, 1985). This tendency is also of more general nature[5]. In wall bounded flows, both alignments are influenced by the wall. Very close to the wall, the eigenframe of $-\langle u_1 u_2 \rangle$ tends to coincide with the x_1, x_2, x_3, so that the Reynolds stress $-\langle u_1 u_2 \rangle$ vanishes in the immediate proximity of the wall. Similarly vorticity, 'vortices' and other associated 'structures' tend to be closely aligned with the streamwise direction x_1, forming a pattern of 'streamwise vortices' with low speed 'steaks' in between. Among other things the importance of the streaks is in the bursts of intensity of the fluctuating motion, in which most of turbulent kinetic energy is produced (see figure 8.1 and below). These bursts arise on the background of the streaks in the low speed regions with inflection points in the profile of the streamwise velocity component.

The meaning/definition(s) of the quasi-streamwise vortices close to the wall and structures in the outer region such as horseshoe or hairpin vortices with two- and/or one leg, asymmetric staggered vortices and many others, vary considerably among authors in the turbulent shear flows community (see Panton, 1997; and references therein ; also Chacin and Cantwell, 2000). A recent conception of hairpin vortex packet is promoted by Adrian et al. (2000). It is certain that all turbulent shear flows produce streamwise, ω_1, (and also wall normal, ω_2) vorticity, in patterns possessing some structure with a number of robust features, which cannot be identified as purely random[6].

The quasi-streamwise vortices are believed to be the dynamically important feature of turbulent flows that is mostly responsible for the turbulent momentum transport, i.e. the Reynolds stress $-\langle u_1 u_2 \rangle$, and consequently turbulent kinetic energy production. However, this interpretation is not necessarily the correct one as is shown in the example of the fully developed turbulent flow such as the flow in a plane channel considered in section 6.6.

$P_{\Delta^u} = -\int \Delta_i^u \Delta_j^u s_{ij} dV$, over the flow domain at any time moment is positive. There is a strong tendency of alignment between the error vector, Δ^u, and the eigenvector of the rate of strain tensor of the instantaneous velocity field, s_{ij}, corresponding to its negative eigenvalue. Concomittantly Δ^u tends to be normal to the two other eigenvectors of s_{ij}.

[5]Namely, vorticity tends to align with the streting eigenvector of the *large*-scale 'part' of the rate of strain tensor (Kevlahan and Hunt, 1997; Porter et al., 1998). This does not contradict the tendency of vorticity to align with the eigenvector corresponding to the intermediate eigenvalue of the instantaneous rate of strain (and its fluctuative part).

[6]Creation of streamwise vorticity is an inherent property of many flows possessing backgound spanwise vorticity with a primary instability that is two-dimensional. The secondary instability leads to the formation of streamwise 'vortices' (Brown, 1970; Pierrehumbert and Widnall,1982; Phillips et al., 1996; and references therein).

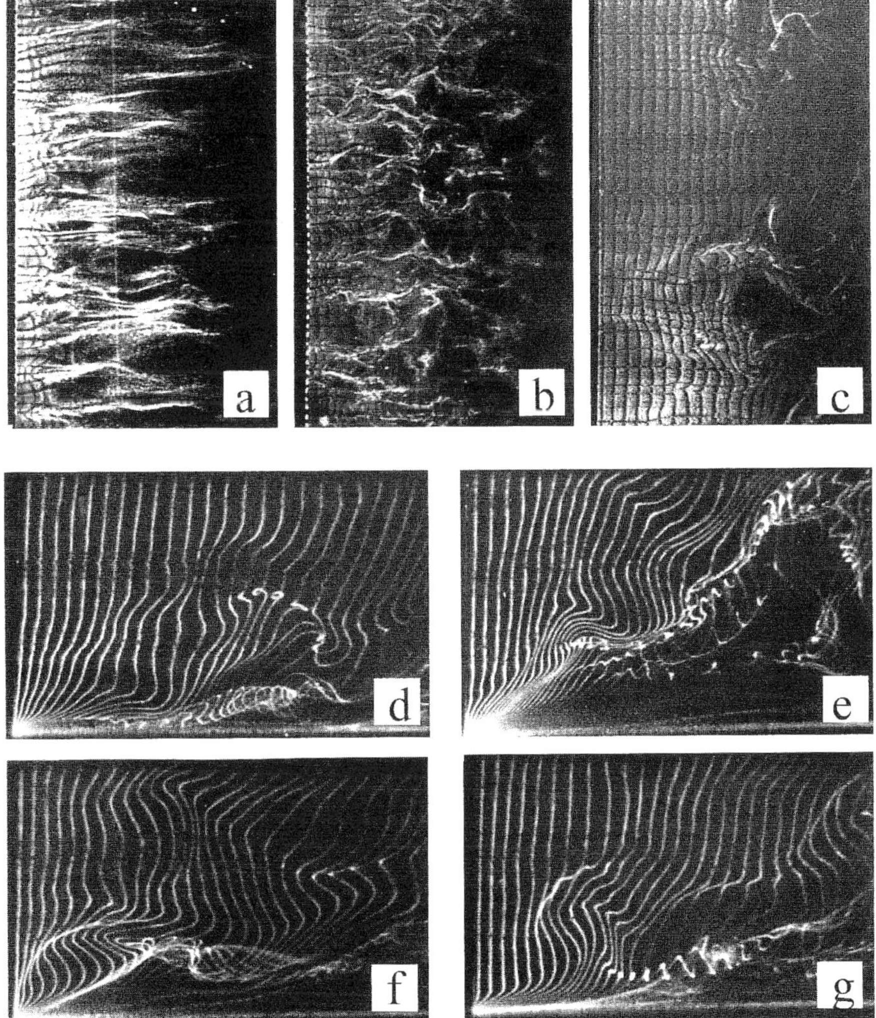

Figure 8.1. Turbulent boundary layer as visualized by hydrogen bubbles. Top view: a) $y^+ = 2.7$, b) $y^+ = 38$, c) $y^+ = 407$ (Kline et al., 1967). Side view: d), e) - selected frames showing formation and breaking of a streamwise vortex motion, f) - same for a transverse vortex motion, g) - showing a 'wavy mode' of bursting (Kim et al., 1971).

In such a flow

$$\langle u_1 u_2 \rangle \equiv \int_0^{x_2} \langle \omega_2 u_3 - \omega_3 u_2 \rangle dx_2, \tag{8.1}$$

i.e. the Reynolds stress is associated *directly* with the wall normal and spanwise vorticity components ω_2 and ω_3, but not with the streamwise vorticity component ω_1. In fact, more important is the x_1-dependence, since a flow having also a ω_1-component, but lacking the x_1-dependence is impotent in the sense that, in such a flow, there is only one way coupling between the flow in the cross-stream plane, x_2, x_3 and the streamwise flow. Namely, the flow in the cross-stream plane influences the streamwise flow, but the streamwise flow does not affect the flow in the cross-stream plane. Consequently there is no source of energy to sustain the flow in the cross stream plane. Moreover, a pure two-dimensional 'turbulent' channel flow[7], possessing no streamwise vorticity at all (it has only spanwise vorticity, ω_3) *is* capable to produce considerable Reynolds stresses (see figure 8.16). In any case the relation (8.1) or the one from which it follows , $d\langle u_1 u_2 \rangle / dx_2 = \langle (\boldsymbol{\omega} \times \mathbf{u})_1 \rangle$, (6.5), show the importance of vorticity in maintaining the Reynolds stresses in turbulent flows (Tennekes and Lumley, 1972). We recall that in turbulent shear flows $\langle (\boldsymbol{\omega} \times \mathbf{u}) \rangle \neq 0$, whereas it is vanishing in homogeneous turbulent flows.

The origin of the (coherent) structures in turbulent shear flows is usually associated with some kind (not well defined) instability of imaginary flows with turbulence but without coherent structures (Lumley and Yaglom, 2000). This kind of instability is sometimes considered to be the driving instability of the underlying mean flow in the fully developed turbulence (Roshko, 1993). A constructive example of such an approach is given by (Nikitin and Chernyshenko, 1997). They looked at the instability of the mean flow in the near wall region resulting from the action of a ' body force' $\partial Q / \partial x_2, Q = \langle u_3 u_3 \rangle - \langle u_2 u_2 \rangle$. The resulting spacing between the fastest growing modes is the same as experimentally observed between the low speed streaks. It is noteworthy that this agreement is achieved by using an *empirical* expression for Q by approximating the data from DNS. Another view is that coherent structures result from the preferential amplification of a particular class of perturbations (Farrell and Ioannou, 1998; Marasli et al., 1991). However, this approach is based on the *linear* stability theory, which means that turbulent flow should have quite long 'memory' and ability for selective amplification in the sea of broadband excitation (including direct excitation of small scales) occurring in naturally arising turbulent flows. Alternatively this may mean that at least in some flows (as mixing layers, wakes) the large scale structure(s) result(s) from a large scale instability not related directly to the turbulent nature of the flows under consideration.

The origin of coherent structures in the wall turbulent shear flows is also associated with the phenomenon of bursting. This name was given by Kline et al. (1967) to the sequence of events happening to the near wall structures:

---

[7]It is noteworthy that the interaction of fluctuations and the mean flow is essentially three-dimensional, i.e. it involves u_3 fluctuations. Therefore, it was believed that pure two-dimensional flows are incapable to develop appreciable Reynolds stressess (Tennekes and Lumley, 1972, p. 41).

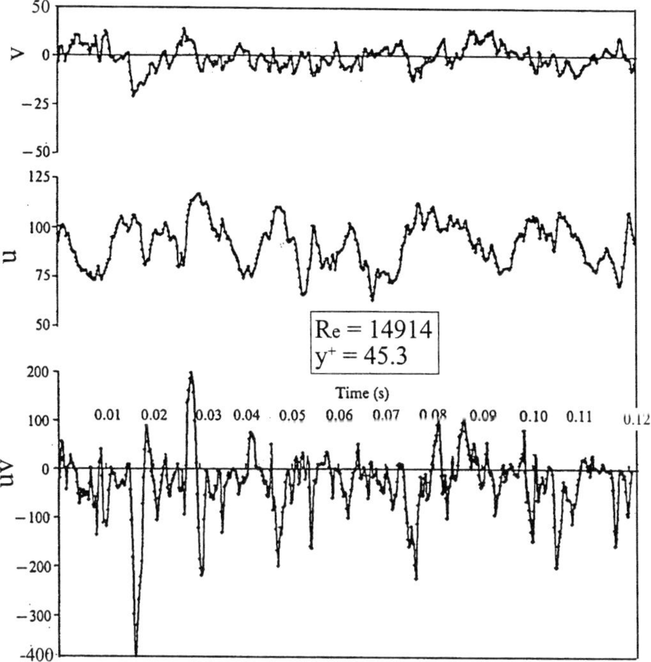

Figure 8.2. Time records of the streamwise, u, and normal, v, components of velocity fluctuations, and the Reynolds stress, uv, the latter exhibiting an intermittent behaviour (Wei and Willmarth, 1989).

lift up, oscillation and break up (or down; see figure 8.1). Other people found an 'ejection sweep cycle' similar to bursting; see Cantwell (1990), McComb (1990) and Holmes et. al (1996) for an overview of what is called coherent structures and associated phenomena in turbulent shear flows. The first evidence was obtained from flow visualizations. These were interpreted as some sort of secondary instability producing a burst of instantaneous Reynolds stress, $u_1 u_2$, mainly responsible for the turbulence production and maintance in the wall bounded flows. The bursty behaviour of $u_1 u_2$ was observed directly in measurements, for example by Lu and Willmarth (1973; see figure 8.2) and many others.

It is the right place to be reminded (see section 7.2.2) that the intermittent behaviour of $u_1 u_2$ can be accounted solely by the multiplicative nature of the Reynolds stress (as a product) assuming both u_1 and u_2 to be Gaussian with the correlation coefficient between them adjusted from the experiment (-0.44). In such a way the PDF of $u_1 u_2$ is approximated with high precision (Lu and Willmarth, 1973). This means that dealing with such signals one has to be able to separate such 'false' intermittency from the one inherent to the flow field. Relatively simple low dimensional

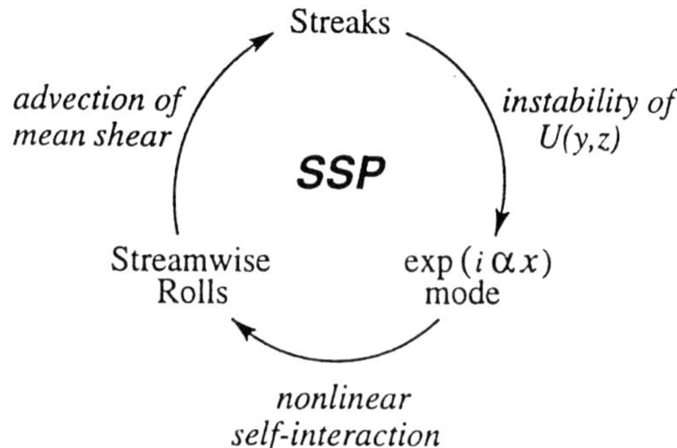

Figure 8.3.   The self-sustaining process, courtesy of Professor F. Waleffe.

systems also exhibit features like the bursting phenomenon (Holmes et al. 1996, 1997; and Knobloch and Moehlis, 2000). There seems to be little doubt that in real turbulence the bursting phenomenon is a result of its dynamics, but the above examples (see also Tsui and Dhruva, 1999), show that such phenomena may arise not necessarily for the 'right' reasons. The bursting process involves most of the scales from those represented by the velocity field to those related to the velocity derivatives. In terms of time scales, the duration of the bursts is pretty short; it is a fast process. Therefore, it is likely that the bursting process is associated with strain dominated regions, just like in the case of nonsheared turbuent flows the most intense nonlinear activity is associated with the regions dominated by strain (see chapter 6 and the discussion below at the end of this section).

The near wall structure of wall bounded turbulent flow is closely related to the process of production and self-sustaining of turbulence in such flows. Attempting to get an insight into the details on how energy is fed in the field of fluctuations in turbulent shear flows such a self-sustaining mechanism was proposed by Waleffe (1990; see figure 8.3) the essence of which is a nonlinear mechanism consisting of creation, destruction and regeneration of streaks.

The proposal by Waleffe (1990) was followed by a convincing confirmation of such mechanisms in direct numerical simulations, though not without a variety of disagreements (Jimenez and Pinelli, 1999; Panton, 1997; and Waleffe and Kim 1998). Waleffe (1990) also proposed that the streak spacing of ~ 100 wall units, i.e. $x_2^+ = x_2 u^+/\nu$, $u^+ = (\nu dU/dx_2|_{x_2=0})^{1/2}$, should be considered as a critical Reynolds number for transition from laminar 1D flow to a 3D finite amplitude state in shear flows. The 100^+ spacing would then correspond to the smallest Reynolds number at which a flow can be maintained in a state different (not necessarily turbulent) from unidirectional laminar flow. This idea is based on the computations by Jimenez and Moin (1991) for different Reynolds numbers, the main result of which is that turbulent flow cannot be maintained in boxes which are narrower than 100 wall units in the spanwise direction. The self-sustaining processes are known to exist in other flows, e.g. in the near wake of a bluff body (see Huerre and Rossi, 1998).

The three-dimensional self-sustaining process (SSP) is reminiscent of enstrophy and strain self-production, discussed in section 6.3. The difference is that in the SSP energy is fed into the system directly, whereas, in case of self-amplification of the field of velocity derivatives, they are produced entirely by the fluctuative field itself, once created and supported by the velocity field. In case of turbulent shear flows, there are many terms contributing to production of enstrophy and strain (see equations [C.51, C.53] in the appendix C). Well known order of magnitude estimates (Tennekes and Lumley, 1972) show that at large Reynolds numbers production of enstrophy, $\frac{1}{2}\langle\omega^2\rangle$, is mainly associated with the term $\langle\omega_i\omega_k s_{ik}\rangle$, i.e. with the self-amplification of the field of vorticity/strain fluctuations. According to these estimates contributions to the enstrophy production associated with the mean velocity gradient, $\langle u_k\omega_i\rangle\partial\Omega_i/\partial x_k$, $\langle\omega_i\omega_k\rangle S_{ik}$, $\Omega_k\langle\omega_i s_{ik}\rangle$, i.e. due to presence of mean vorticity Ω_i and strain S_{ij}, are small compared to $\langle\omega_i\omega_k s_{ik}\rangle$. Similar estimates remain valid for the production of the total mean squared strain $s^2 \equiv \langle s_{ij}s_{ij}\rangle$ (see equation [C.53]). Namely, its production is mainly due to the term $-\langle s_{ij}s_{ki}s_{ki}\rangle$, whereas contributions to the strain production associated with the mean velocity gradient, $-\langle u_k s_{ij}\rangle\partial S_{ij}/\partial x_k$, $\langle s_{ij}s_{jk}\rangle S_{kj}$, are small compared to $-\langle s_{ij}s_{ki}s_{ki}\rangle$.

The field experiments by Kholmyansky et al. (2001a) at $Re_\lambda = 10^4$ showed that this is really the case. The largest terms among the mentioned above, $\langle u_k\omega_i\rangle\partial\Omega_i/\partial x_k$ and $-\langle u_k s_{ij}\rangle\partial S_{ij}/\partial x_k$ are two orders of magnitude smaller than $\langle\omega_i\omega_k s_{ik}\rangle$ and $-\langle s_{ij}s_{ki}s_{ki}\rangle$.

It appears that the dominance of $\langle\omega_i\omega_k s_{ik}\rangle$ and $-\langle s_{ij}s_{ki}s_{ki}\rangle$ and 'smallness' of the RDT-like terms may occur already at rather moderate Reynolds numbers. Such an example is given by Sandham and Tsinober (2000) for a turbulent channel flow at the overall $Re = 3300$, based on the half channel width and the mean velocity at the centreline (see figure 8.4).

The main result, shown in figure 8.4, is that the terms $\langle\omega_i\omega_k s_{ik}\rangle$ and $-\langle s_{ij}s_{ki}s_{ki}\rangle$ are indeed the dominant ones, except in the proximity of the wall, $x_2^+ < 20$, $x_2^+ = x_2 u^!/\nu$, $u^! = (\nu dU/dx_2|_{x_2=0})^{1/2}$. In the region close to the wall the terms $\langle\omega_i\omega_k s_{ik}\rangle$ and $-\langle s_{ij}s_{ki}s_{ki}\rangle$ remain of the same order as some of the RDT-like terms. This result is consistent with the one obtained by Kim (1989) in his analysis of pressure fluctuations in simulated channel flow. Contrary to the common belief that the RDT-like contribution to pressure is the dominant component, Kim found that the pure nonlinear pressure is comparable near the wall and is larger away from the wall than the RDT-like contribution.

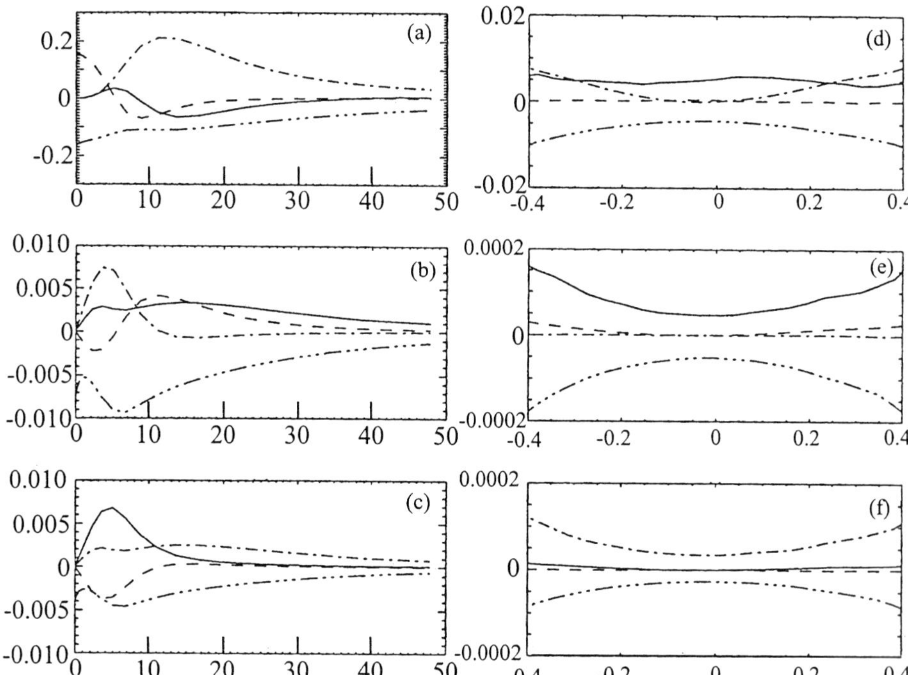

Figure 8.4.   Budgets of from top to bottom kinetic energy, enstrophy and strain magnitude showing near-wall behaviour on the left and channel central region on the right. Solid, dashed, chain dot, and chain triple dot stylelines refer respectively to terms 1-4 in the equations (C.48'), (C.51') and (C.53'). The approximate sign means that other terms in the equations (C.48), (C.51) and (C.53) in the appendix C turn out to be small in the channel flow. From Sandham and Tsinober (2000).

We mention also another result of importance for section 8.9. Namely, there is a strong correlation between vorticity and strain in the proximity of the wall, $x_2^+ < 20$, so that $\omega^2 = 2s^2$ instantaneously at $x_2^+ \leq 10$. Far away from the wall, they are decorrelated, as in homogeneous turbulence. In this latter case (see Chapter 6), most of enstrophy production, production of strain/dissipation and other nonlinear processes are associated with large strain, rather than with intense vorticity (large enstrophy). This was observed also by Kholmyansky et al. (2001a) at $Re_\lambda = 10^4$ in the field experiments, and by Sandham and Tsinober (2000) the turbulent channel flow.

$$\frac{D_U\langle e_T\rangle}{Dt} \approx -\frac{\partial}{\partial x_j}\left\{\langle u_j e_T\rangle - 2\nu\langle u_i s_{ij}\rangle\right\} - \langle u_i u_j\rangle S_{ij} - 2\nu\langle s^2\rangle. \qquad (C.48')$$

$$\frac{1}{2}\frac{D_U\langle\omega^2\rangle}{Dt} \approx \langle\omega_i\omega_j s_{ij}\rangle + \langle\omega_i\omega_j\rangle S_{ij} + \langle\omega_i s_{ij}\rangle\Omega_j + \langle\nu\omega_i\nabla^2\omega_i\rangle. \qquad (C.51')$$

$$\frac{D_U \frac{1}{2}\langle s_{ij}s_{ij}\rangle}{Dt} \approx -2\langle s_{ij}s_{ik}\rangle S_{kj} - \frac{1}{2}\langle\omega_i s_{ij}\rangle\Omega_j - \langle s_{ij}s_{jk}s_{ki}\rangle - \frac{1}{4}\langle\omega_i\omega_j s_{ij}\rangle + \nu s_{ij}\nabla^2 s_{ij}.$$
$$(C.53')$$

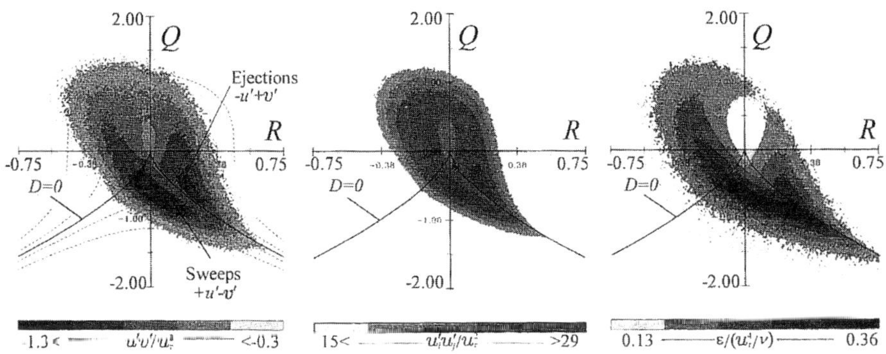

Figure 8.5. Time averaged Reynolds stress (left), turbulent kinetic energy generating events (centre) and dissipative events in the $R - Q$ plane. From Chacin and Cantwell (2000).

A similar phenomenon was observed in a in recent analysis of flow in a turbulent boundary layer by Chacin et al. (1996), Chong et al. (1998) and Chacin and Cantwell (2000). These authors used directly the eigenvalues and invariants of the velocity gradient tensor $\partial u_i/\partial x_j$. In particular, they looked at various flow properties in association with the invariant map of the second invariant, $Q = \frac{1}{4}\omega^2 - \frac{1}{2}s_{ik}s_{ik}$ (vertical axis) versus the third invariant, $R = -\frac{1}{3}s_{ik}s_{km}s_{mi} - \frac{1}{4}\omega_i\omega_k s_{ik}$ (horizontal axis) of the velocity gradient tensor. One of the most interesting (from our point of view) findings is that the main contribution to the shear stress, turbulent energy production, and dissipation comes from the regions with $Q < 0$ with larger contribution from the lower right quadrant, i.e. $Q < 0$ and $R > 0$, not only dominated by strain, but also by production of strain, $-s_{ik}s_{km}s_{mi}$, see figure 5, also figures 8, 11 and 14 in Chacin and Cantwell (2000).

It should be emphasized that these regions are mainly not the ones corresponding to vortices (hairpins or whatever), which are located mostly in the regions with $Q > 0$, or a bit more precisely in regions with $D > 0$, where $D = \frac{27}{4}Q^3 + R^2$ is the discriminant of $\partial u_i/\partial x_j$. That is the regions of major nonlinear activity are really associated with large strain (mainly corresponding to what Chacin and Cantwell call 'blank' spaces) rather than with regions of concentrated vorticity with lower dissipation (see Tsinober, 2000 for similar results in quasi-isotropic turbulence). In other words, it seems that concentrated vorticity is not that important also in turbulent shear flows and that structure(s) associated with turbulence (not only its energy) production are mainly due to the large strain rather than large vorticity[8]. Structure(s) associated with the latter seem to be the consequence of the turbulent dynamics rather than its dominating factor[9]. A final remark is that these results do not contradict (6.5), (8.1) stating the importance of vorticity in maintaining the Reynolds stress. First, these are relations for the mean quantities, and second, there is no turbulent flow without vorticity. However, important details of the relations between Reynolds stress, vorticity, strain and their production remain not clear enough.

In closing this section we mention that, in wall-bounded turbulent flows, there are two factors influencing turbulence. The first factor is the mean shear, whereas the second factor is the boundary. Hence in order to 'isolate' the second factor, one can look at the shear-free turbulent flows near different boundaries, solid and free, permeable and not (see Aronson et al., 1997; and references therein).

8.3. Partly turbulent flows – entrainment

> *... the intermittent character of the disturbance. The disturbance would suddenly come on through a certain length of the tube and pass away and then come on again, giving the appearance of flashes, and these flashes would often commence at one point in the pipe... This condition of flashing was quite as marked when the water in the tank was very steady, as when somewhat disturbed. (Reynolds, 1883).*
>
> *One of the properties of the region of rotational turbulent flow is that the exchange of fluid between this region and the surrounding space can occur only in one direction. The fluid can enter this region from the region of potential flow, but can never leave it (Landau and Lifshits, 1959).*
>
> *...entrainment - the erosion by turbulence of the underlying non-turbulent fluid... (Phillips, 1966).*

Partly turbulent flows have already been mentioned in chapters 1 and

---

[8]This fact is, in way, not so unexpected because it is the strain that is responsible for local deformation of fluid elements (and, consequently, dissipation in Newtonian and non-Newtonian fluids) , whereas vorticity acts locally only as a rigid rotation. It is, therefore, likely that that the turbulent strain is the cause of quite a facsinating phenomenon of bioluminescence (known, at least, since 15th century; Harvey, 1952) in turbulent wakes of ships and dolphins in warm seas (Vasil'kov et al., 1992; Herring, 1998). Rohr et al. (1997) found that *turbulent flow is necessary to provide significant bioluminscence stimulation.* They suggested to use bioluminscence as a flow diagnostic. This same idea was communicated to the author by Y. Couder and S. Douady in 1995 in Paris.

[9]The interpretation of the results by Chacin and Cantwell (2000) given here is not in full agreement with their conclusions, especially regarding the role of vortices and concentrated vorticity.

2. The main special features of these flows are the coexistence of regions with laminar and turbulent states of flow and continuous transition of laminar flow into turbulent via the entrainment process through the boundary between the two. In fact, most of turbulent flows are partly turbulent: boundary layers, all free shear turbulent flows (jets, plumes, wakes, mixing layers), penetrative convection in the atmosphere and in the ocean, gravity currents, avalanches, clear air turbulence, and many others. Transitional flows consisting as a rule of turbulent regions growing in a laminar environment are also partly turbulent flows.

The so called external intermittency is associated with the coexistence of laminar and turbulent flow regions – an observer located in the proximity of (either side of) the 'mean' boundary between these regions observes intermittently laminar and turbulent flow in the form of a signal similar to that as, say ω^2, in plate 6, see also figure 21.4 in Tritton (1988), clearly demonstrating the external intermittency of in the wake past a circular cylinder. Here we again encounter the question about what turbulent is. When we look at a flow like the one in figure 1.3 or figure 151 in van Dyke (1982), we clearly see what is turbulent and what is laminar. But the question is how one can say whether a *small* part of flow is turbulent. In other words if turbulence is to be identified by statistical means, then what is the meaning of 'turbulent' locally? This involves taking decisions about what is turbulent using some (conditional) criterion (see discussion and references in Kuznetsov et al., 1992).

This distinction starts with Reynolds (1883) in the form of such qualitative description of transitional phenomena as flashes of turbulence in the pipe, i.e. he makes a clear distinction between laminar and turbulent flow regions (quasi-)locally without invoking any statistical characteristics. The first physically qualitative distinction between turbulent and nonturbulent regions made by Corrsin (1943) and Corrsin and Kistler (1954, 1955), is that turbulent regions are rotational, whereas the nonturbulent ones are (practically) potential, thus employing one of the main differences between turbulent flow and its random irrotational counterpart on the 'other' side of the interface separating them. It is difficult to implement such a distinction, since it requires information on vorticity which until recently was not accessible, and no experiments are known to employ adequately vorticity in studying the properties of the laminar-turbulent interface and the entrainment across it[10]. A recent analysis by Bisset et al. (1998, 2001) of the data from direct numerical simulation of a temporally developing plane wake by Moser et al. (1998) confirmed the Corrsin's approach (figure 8.6). One can see an abrupt change of vorticity across the laminar-turbulent interface,

---

[10]An attempt was made by Foss and Klewicki (1984) to measure and use for this purpose the spanwise vorticity component in a plane shear layer.

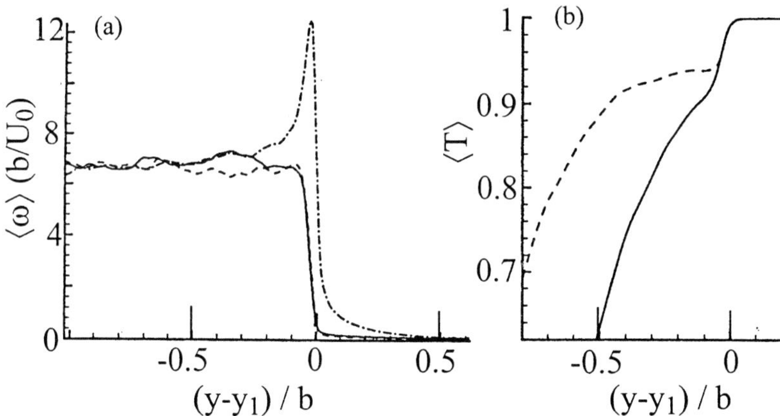

Figure 8.6. Variation of vorticity magnitude (a) and temperature (b) across of laminar-turbulent interface of a temporally developing mixing layer (Bisset et al., 1998). Different curves correspond to different ways of interface detection.

and a much less sharp change in temperature.

Corrsin discovered that *i* - the boundary between the two regions is essentially a thin interface which he called the 'viscous superlayer', in which viscosity plays a dominant role, and *ii* - the 'effectiveness' of the entrainment process is strongly enhanced by the large scale undulations of the interface due to large scale motions which result in engulfment of irrotational fluid into the turbulent flow.

The main mechanism by which nonturbulent fluid becomes turbulent as it crosses the interface is believed to involve viscous diffusion of vorticity across the surface. As this process is associated with small scales it is thought to be the reason why the interface appears sharp compared to the scale of the whole flow.

However, at large Reynolds numbers, the entrainment rate and the propagation velocity of the interface relative to the fluid are known to be independent of viscosity (see for example Ricou and Spalding, 1961; Hinze, 1975; Townsend, 1976; Tritton, 1988; and Bisset et al., 2001 for more information and references). Therefore *the slow process of diffusion into the ambient fluid must be accelerated by interaction with velocity fields of eddies of all sizes, from viscous eddies to the energy-containing eddies so that the overall rate of entrainment is set by large-scale parameters of the flow* (Townsend, 1976). That is *although the spreading is brought about by small eddies* [viscosity] *its rate is governed by the larger eddies. The total area of the interface, over which the spreading is occurring at any instant, is determined by these larger eddies* (Tritton, 1988). This is analogous to independence of dissipation of viscosity in turbulent flows at large Reynolds numbers. In other words, small scales do the 'work', but the amount of work is fixed by the large scales in such a way that the outcome is independent of viscosity. This shows that independence of some parameter of viscosity at large Reynolds numbers does not mean that viscosity is unimportant. It means *only* that the (cumulative) effect of viscosity is Reynolds number independent.

It is important that becoming rotational is only a necessary condition of becoming turbulent. Once the irrotational fluid acquired some vorticity via viscous diffusion this vorticity is amplified by the process of predominant vortex stretching due to the random nature of the motion in the proximity of the interface, though there is no direct evidence that this really happens. This points to another possibility. Due to the random nature of the flow at both sides of the interface a small amount of 'seeding' of vorticity will bring into action the mechanism of the predominant vortex stretching in the 'irrotational' part of the flow. There is no evidence that this really happens either. Some indication comes from the equation (6.10) showing that an initially Gaussian and *potential* velocity field with small seeding of vorticity will produce - at least for a short time - an essentially positive enstrophy production (as well as production of strain) though strictly this is true for homogeneous turbulence. This mechanism can be effective only in the proximity of the interface where large strain should exist (Bisset et al., 1998, 2001), because the fluctuations attenuate exponentially with the distance from the interface in the irrotational part of the flow[11]. Since in real flows it is likely that some small amount of vorticity is always present in the 'potential' part of the flow this second possibility looks quite realistic. In any case, it seems that the small scale nibbling of the nonturbulent flow region by the turbulence at the interface consists of both diffusion of vorticity by viscosity and vorticity amplification via predominant vortex stretching in the proximity of the interface. These processes are enhanced due the strongly convoluted shape of the interface caused by motions on a wide range of scales which includes what is sometimes called large scale engulfment. It is necessary to be reminded that concomitant to the process of vorticity amplification is always the process of amplification of strain.

A final note is that entrainment is an essentially three-dimensional process - vorticity and strain cannot be amplified in two-dimensional flows.

Apart practical aspects, the process of entrainment is of special interest from the basic point, as an irreversible process of transition from laminar to turbulent state of motion. However, apart from the above speculations, little is known about the details of the underlying physics of entrainment (see Bisset et al., 2001; and references therein).

[11]See Landau and Lifshits (1959, p.129) for a simple explanation of this attenuation.

8.4. Variable density

There are several effects causing variations of fluid density. These are due to variations of temperature, presence of additives (salt, moisture, small particles) and compressibility. All of them can have profound influence on the properties of turbulent flows and lead to a number of new effects. For example, in many cases (but not always) fluids with variable density can support waves, so that regions with turbulence can radiate such waves and can interact with waves radiated from other regions (see figure 8.10). Turbulence can be produced locally by the breaking of such waves, which also may transfer momentum and energy to the mean turbulent flow. Another important effect is anisotropy appearing, for example in the presense of gravity. Finally, additional mechanisms of vorticity production and dissipation arise in some flows with variable fluid density.

8.4.1. CONVECTION

The term (free) *convection* refers usually to buoyancy induced flows produced by inhomogeneity of fluid density in the presence of gravity. Thus, convection can be thermogravitational if the density variations are caused by changes of temperature, thermohaline (double-diffusive) if the density inhomogeneity is due to both variations in temperature and salt content (or salts with *different* diffusivities)[12], and compositional if the density variations are caused mainly by changes of composition such as those occurring in the presence of cristallization (ice, salt, magma) or resulting from dissolution/melting. Convection is a most typical example of a situation when turbulence is produced and driven by a mechanism totally different from mean shear. The simplest situation is when heavier fluid is overlying lighter fluid, a state referred to as unstable stratification. Closely related are turbulent motions resulting from the so-called Rayleigh-Taylor instability (Dalziel et al., 1999) and Richtmyer-Meshkov instability (Prasad et. al., 2000).

The term *convection* is used to emphasize the fact that the only reason for the motion is the spatial inhomogeneity of fluid properties, which in turn is changed by the fluid motion and which therefore contrast with passive objects, which are not felt by turbulent motions.

Turbulent convection sets in at large enough imposed gradients of, say,

---

[12]In case, of several species contributing to fluid density, convection arises also when the fluid density is homogeneous and even statically stably stratified, but the content of individual species is not homogeneous. The flow motion then is induced by the differences in diffusion properties of the species and subsequent instability leading to a release of the potential energy of the unstably (top-heavy) distributed component(s) (see Turner, 1985).

temperature or heat flux, reflected in large values of appropriately defined Rayleigh numbers, $Ra = g\alpha\Delta T d^3/\nu\kappa$, – a nondimensional imposed temperature difference ΔT, where g is the acceleration due to gravity, d is some external characteristic length (depth of the heated fluid layer), and α, ν and κ are the fluid properties, respectively coefficient of thermal expansion, kinematic viscosity and thermal diffusivity. The flow properties of both turbulent convection and its onset[13] depend on a further parameter - the Prandtl number, $Pr = \nu/\kappa$. Systematic description and review (with plenty of references) of turbulent convection is found in Tritton (1988), Siggia (1994), Zimin and Frik (1988); see also Grossman and Lohse (2000) and Niemela et al. (2000). Here as usually in this chapter we deal mostly with simple examples and qualitative aspects.

Several examples of convection are shown in figure 8.7.

Just like in turbulent shear flows it was only recently recognized that there also exist in *turbulent* convection large-scale objects called thermal plumes, which are similar to those observed in transitional stages like those shown in figure 8.6c. These structures are believed to be important in the heat transfer and interaction with the boundary layers and in some cases with the induced mean flow called 'turbulent wind' (see section 8.5). This recognition was mostly used in order to cope with the problem of the dependence of the Nusselt (and also Reynolds) number, $Nu = Hd/(\rho C_p \kappa \Delta T)$, the nondimensional heat flux, H, on the Rayleigh number, Ra. This dependence is usually *assumed* as a power law, $Nu \sim Ra^\beta$, with the exponent β and a prefactor both depending on the Prandtl number, Pr. In this respect, turbulent convection is not different from some other fields in turbulence research: scaling behaviour to a large extent has monopolized the attention and resulted in several different theoretical explanations of the observed behaviour. These are described in Siggia (1994), Grossman and Lohse (2000), Niemela et al. (2000) and references therein. The only point we want to repeat here is that scaling laws by themselves are not sufficient for testing theories, since there is no one-to-one relation between scaling exponents and physical processes, which among other things is reflected in qualitatively different theories leading to the *same* scalings. As an example we mention the 2/7 law ($Nu \sim Ra^{2/7}$) corresponding to the so called 'hard turbulence' at very large Ra. In reality this may be equally not a 2/7 exponent at all: as noted by Grossmann and Lohse (2000) the expression $Nu = 0.27Ra^{1/4} + 0.038Ra^{1/3}$ (with the prefactors obtained from *experiment*) mimics the 2/7 power-law over ten orders of magnitude

---

[13]The transitional regimes in thermal convection may exhibit an entire sequence of states of flows. For example, convection in a horizontal layer heated from below exhibits at least eight transitions judging by the behaviour of the dependence $Nu(Ra)$ (Zimin and Frik, 1988; Siggia, 1994; and references therein).

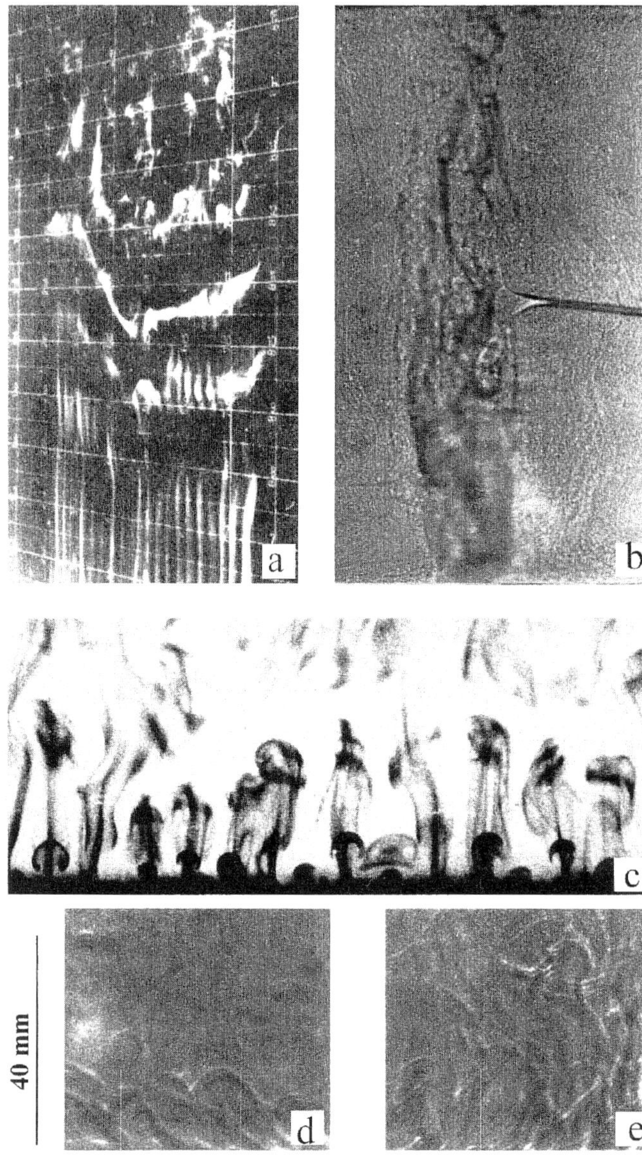

Figure 8.7. Examples of convective motions. a) Smoke visualization of transition in free convection boundary layer on a vertical heated plate (Čolak-Antić, 1964). b) Violent (double-diffusive) convective motion induced by injection of sugar solution into a salt solution of approximately the same density (Turner and Chen, 1974). c) Thermals rising in water from a heated horizontal surface (Sparrow et al., 1970). d) thermals tilted by the 'thermal wind' along the bottom, and e) the same flow as in d) with the thermals prevented from tilting by vertical inserts (Ciliberto et al., 1996).

in Ra in the range $10^5 < Ra < 10^{14}$ (for $Pr \sim 1$), see their figure 4[14]. Similarly, Niemella et al. (2000) in their cryogenic helium gas experiments obtained a relation $Nu = 0.124Ra^{0.309}$ in the range $10^6 < Ra < 10^{17}$ and Pr between 0.7 and 12, which is described equally well by another relation $Ra = 0.0587(Ra^{3/2}lnRa^{3/2})^{1/5}$. A third example refers to the so called ultimate regime predicted by Kraichnan (1962a), in which $Nu \sim Ra^{1/2}$. This regime is characterized by the dominant contribution of turbulence (as compared to that by conduction) to the heat transport in the boundary layer on the heated (cooled) wall and is a dimensional necessity following from the assumption that the heat flux is independent of molecular properties, i.e. viscosity and thermal conductivity. The independence of heat flux of molecular properties is expected to occur at very large Rayleigh numbers and indeed was observed in cryogenic experiments in cells with smooth and rough walls (Roche et al., 2001) at $Ra > 10^{12}$ and $Pr = 1.45 \div 4.9$ (see also figure 3 in Siggia, 1994). This regime was also observed in numerical simulations by Vincent and Yuen (2000) in the range $10^{10} < Ra < 10^{12}$ at $Pr = 1$, but for *two-dimensional* flow. In addition they observed the conventional relation $Nu \sim Ra^{1/3}$ in the range $10^8 < Ra < 10^{10}$. Does this mean that the three-dimensional nature of turbulence is unimportant to the overall heat transfer in turbulent convection? Since turbulent convection is essentially an interaction of hydrodynamic and thermal fields the next question is what properties of the hydrodynamic field are important in this interaction. Whatever the answer, it seems fundamentally important to study the properties of real *three-dimensional* convective turbulence.

The second important development in turbulent convection was the recognition of the existence and importance of a 'turbulent wind' discovered by Krishnamurti and Howard (1981). This is a mean turbulent shear flow which also stirs the fluid in the bulk of the convective flow. The turbulent wind is fed by the thermal plumes via the so called negative eddy viscosity (see section 8.5).

8.4.2. STABLE STRATIFICATION

Unstable stratification, as described above, is the cause itself of the turbulent flow. In case of stable stratification, i.e. lighter fluid overlying heavy fluid, the situation is drastically different. In the presence of stable stratification a fluid particle displaced from its equilibrium state experiences a restoring force due to buoyancy. This leads to two major effects. The first one is that stably stratified fluids can support waves called internal waves. Internal waves at scales ranging from parts of meters to several kilometers

---

[14]High precision measurements by Xu et al. (2000) clearly indicate that there does not seem to exist any single exponent describing the $Nu(Ra)$ relation.

are observed in various natural environments[15]. The second effect is that stable stratification limits the vertical motions thereby tending to suppress turbulence (with an additional sink of kinetic energy by turning it into potential energy of the system in the process of mixing) and leading to a pancake-like anisotropy of turbulence. This is in contrast with cigar-like anisotropy in the case of unstable stratification. A more important difference is that stable stratification itself, cannot be the cause of a (turbulent) motion, so that an additional factor such as mean shear, is necessary to support turbulent flows in the presence of stable stratification. However, *due to* stable stratification, turbulence can be *produced* in a special way via breaking of the internal waves (see figure 8.10 *b* and *c*)[16]. The so-called CAT (clear air turbulence; Pao and Goldburg, 1969) is believed to be associated mainly with the breaking of internal waves in the atmosphere. Similarly breaking waves both internal and surface may be important in production of turbulence in the ocean and air-sea interaction as well (see references in McIntyre,1993; Melville, 1993; and Staquet and Sommeria, 1996). For review and references on turbulence in the presence of stable stratification see Hopfinger (1987), Riley and Lelong (2000) and Thorpe (1987), and also Smyth and Moum (2000a,b) for some latest references and very useful expositions of the issues of scales and anisotropy is stably stratified mixing layers.

Examples of fluid phenomena in the presence of stable stratification are shown in figures 8.8 and 8.9.

Coexistence and especially interaction of turbulence and waves in stably stratified fluids makes it difficult to distinguish between the two. The wave field is usually associated with that part of flow that propagates, whereas turbulence is identified with the nonpropagating part of the motion. However, it is not clear up to now how to separate *random* gravity-wave motion (which do not produce vertical transport) and genuine turbulence (which do) in a stably stratified fluid (Stewart, 1959). This issue becomes more serious in cases of strongly nonlinear internal waves, and interaction between turbulence and waves, and, of course, when the breaking of latter produces turbulence, which in turn can radiate internal waves.

When the stratification is strong enough, turbulence becomes strongly anisotropic and sometimes is identified as quasi-two-dimensional due to strong suppression of the vertical velocity component. However, the similarity to two-dimensional flow, generally, ends at the level of velocity field.

---

[15]Internal waves play an important role in a variety of flows in technological, geophysical and astrophysical contexts, see, e.g. Staquet and Sommeria (1996). Internal waves in the atmosphere are thought to be used by some birds surfing on them while crossing the Atlantic, Mollo-Christensen (1980, private communication).

[16]Breaking of internal waves leads to irreversible mixing. In that it is different from wave breaking at an interface between two immiscible fluids such as surface water waves.

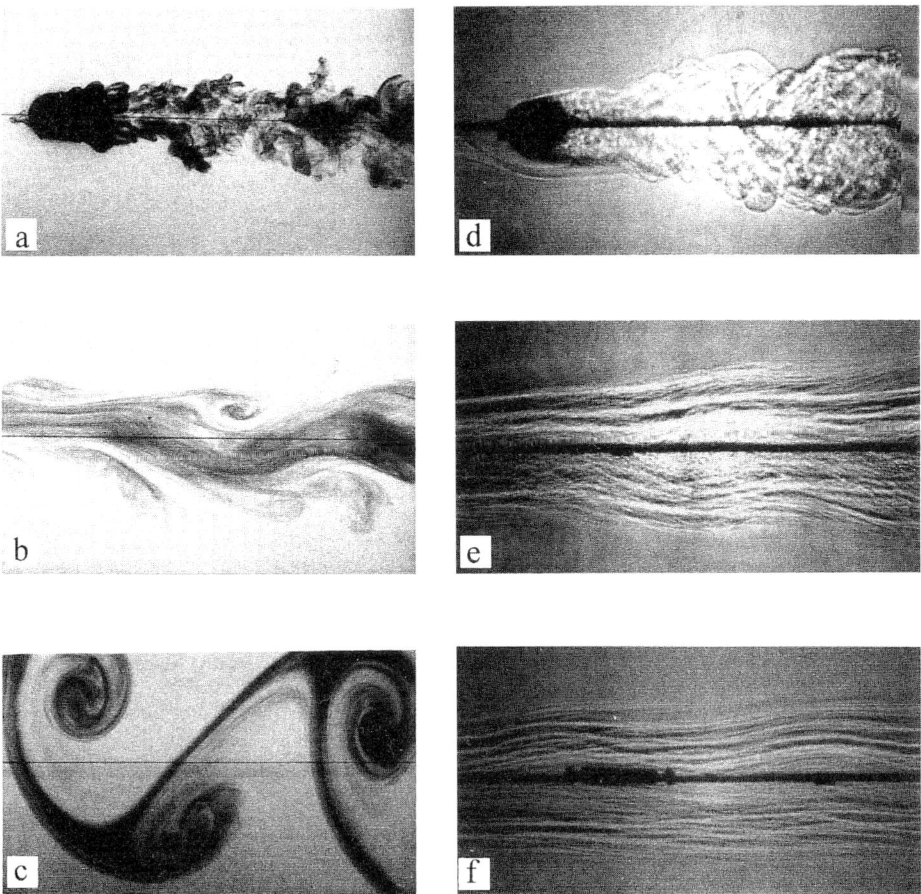

Figure 8.8. Turbulent wake past a sphere in stably stratified (by salt) flow. Left - top view, Hopfinger (1997), right - side view Pao (1969). The distance from the sphere increases from top to bottom. It is seen that, farther from the sphere, the turbulence is suppressed (collapses) leaving striations in the vertical density structure due to incomplete mixing after multiple overturnings, and degenerates into a (quasi) horizontal large scale flow with vortices resembling those observed in a wake past a cylinder in a fluid with constant density. No internal wave radiation is discernible in these images, but see figure 4 in Spedding et al. (1996) in which the internal wave component is clearly seen, as it is 'separated' from the vortical one.

The vertical velocity gradient (i.e. horizontal vorticity) may be quite large, as observed by Fincham et al. (1996) and Spedding et al. (1996) in the form *of a complex 3D network of structures in which layers of eddies cannot evolve independently of one another.* Herring and Metais (1989), in their numerical experiments, observed formation of small scales in the vertical and other attributes of three-dimesionality at pretty strong stratification. Among others the reasons for such essentially three-dimensional nature are

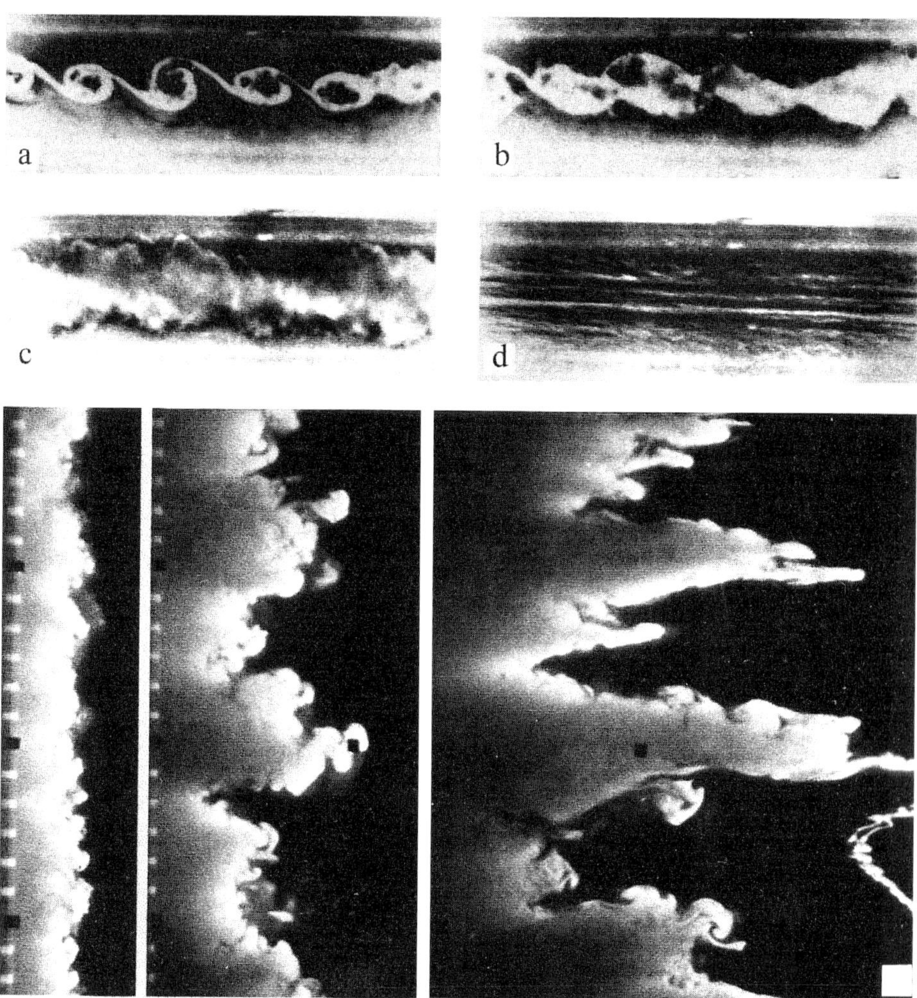

Figure 8.9.　　Turbulence suppression and collapse in a stably stratified fluid. Top four figures - collapse of turbulence developed from Kelvin-Helmholz-type instability (Thorpe, 1971), time increases from left to right and from top to bottom (a→d). The three bottom figures show the time evolution (from left to right) of turbulence produced by a vertical (on the left of each frame) oscillating grid (Browand et. al., 1987). One can see formation of turbulent intrusions due to limitation on the vertical scale of turbulent motions by the stratification and subsequent collapse of turbulence.

believed to be associated with instabilities specific for strongly stratified flows. A recent example is the so-called 'zigzag' instability (Billant and Chomaz, 2000). This instability is suspected as a generation mechanism responsible for the observed multiple-layer phenomena in laboratory, field and numerical experiments in strongly stratified flows and other 'unusual' phenomena such as 'pancake' eddies. In other words, strongly stratified fluids along with the 'intrinsic' stability caused by the restoring force, possess some specific instabilities leading to essentially three-dimensional structure of such flows especially in small scales. At even stronger stratification active turbulence collapses into a state of nearly horizontal 'fossil' motion (figures 8.8 and 8.9).

Many complications and additional effects arise in the presence of boundaries (rigid or free) and other influences such as rotation.

8.4.3. COMPRESSIBLE FLOWS

Compressibility influences turbulence in several ways. For example, there are several mechanisms influencing the dissipation of turbulent energy. It appears that the so-called dilatational dissipation associated with compressibility is usually unimportant (see footnote after equation (C.19) in appendix C).

In shear flows, the dissipation is mainly reduced due to reduced level of turbulence production and not due to dilatational effects (Sarkar, 1995).

Compressibility can lead to an increase in dissipation as well. This happens in the presence shock waves[17]. After such a wave vorticity increases, and generally turbuence is amplified (Andreopulos et al., 2000), and most probably so does strain, so that the dissipation increases. As the Mach number of turbulence $M_t = u'/V_s$ increases (u' - is the turbulence intensity and V_s - is the speed of sound), eddy shocklets - shock waves associated with the turbulent motion - arise in the flow. They provide an additional way for the dissipation of energy possibly important in astrophysical contexts.

Turbulence acts as an amplifier of almost any disturbance that it is subjected to, so no wonder that turbulence is a source of noise (see figure 8.5f). Finally, turbuence can be manipuated by acoustic excitation.

For review and references see Friedrich and Bertolotti (1997), Lele (1994, 1997), Andreopulos et al. (2000) and a special issue on aeroacoustics of *Theoretical and Computational Fluid Dynamics*, **6**, Nos. 5-6, October 1994.

8.5. Rotation

Rotating fluids have much in common with stably stratified ones. They also can supprort waves called inertial waves[18] (see figure 8.5 d, e), since the Coriolis force also acts as a restoring force, though somewhat differently than that of buoyancy[19]. Turbulence in the presence of rotation becomes

---

[17]But may also occur in subsonic flows, (Briassulis et al., 2001).

[18]The term is derived from Bjerknes et al. (1933).

[19]In the case of buoyancy, a fluid particle slightly displaced from its equilibrium state moves along a straight line, whereas in a rotating fluid it moves in a circle, since the Coriolis force acts in the direction normal to the direction of velocity.

anisotropic and tends to acquire in some sense a quasi-two-dimensional structure in the plane normal to the axis of rotation. Strong rotation in most cases leads to turbulence suppression and collapse. There are differences too. For example, strong rotation also reduces the gradients along the axis of rotation, so that turbulence acquires cigar-like anisotropy (in contrast with the pancake-like anisotropy of stably stratified flows) and becomes close to two-dimensional at the level of velocity derivatives. The second difference involves generation of intense vortices in the presence of local forcing, for example at the boundaries (see figure 8.17b).

The presence of rigid boundaries, shear and stratification leads to additional effects, resulting both in stabilizing and destabilizing influences. We mention here three examples. The first one is turbulent flow in a rotating pipe. Most observations show the stabilizing influence of rotation leading among other things to considerable drag reduction (see Orlandi, 1997; and references therein). On the other hand experiments by Nagib et al. (1971) clearly showed that under rotation the pipe flow becomes *less* stable, so that the critical Reynolds number drops from 2500 in the absence of rotation to 900 in the presence of rotation. The second example comes from the experiments of Ibbetson and Tritton (1975). Contrary to other observations, they observed *faster* decay of turbulence past a towed grid in the presence of rotation. One of the possible explanations is that the rotation was not strong enough, so that the additional dissipation in the so called Eckmann layers on the walls normal to the axis of rotation won the competition with the effect of turbulence two-dimensionalization by rotation. The third example is provided by computations of Dritshel et al. (1999). They show clearly that structure of rotating and stratified flows is intrinsically three-dimensional on small-to-intermediate scales reflecting the competition between the pancake-like anisotropy of stratified flows with the cigar-like anisotropy of rotating flows.

For review and references on turbulent rotating flows see Cambon (1994), Hopfinger (1989), Tritton (1985), and Godeferd and Lollini (1999).

The examples shown in figure 8.10 do not exhaust the list of waves existing in fluids. For example, in addition to sound waves in compressible fluids, there are also shock waves. Apart from inertial waves in rotating fluids there exist Rosby waves arising in particular geometrical arrangements. A variety of waves exist also in flows of electrically conductive fluids in the presence of magnetic field (MHD flows).

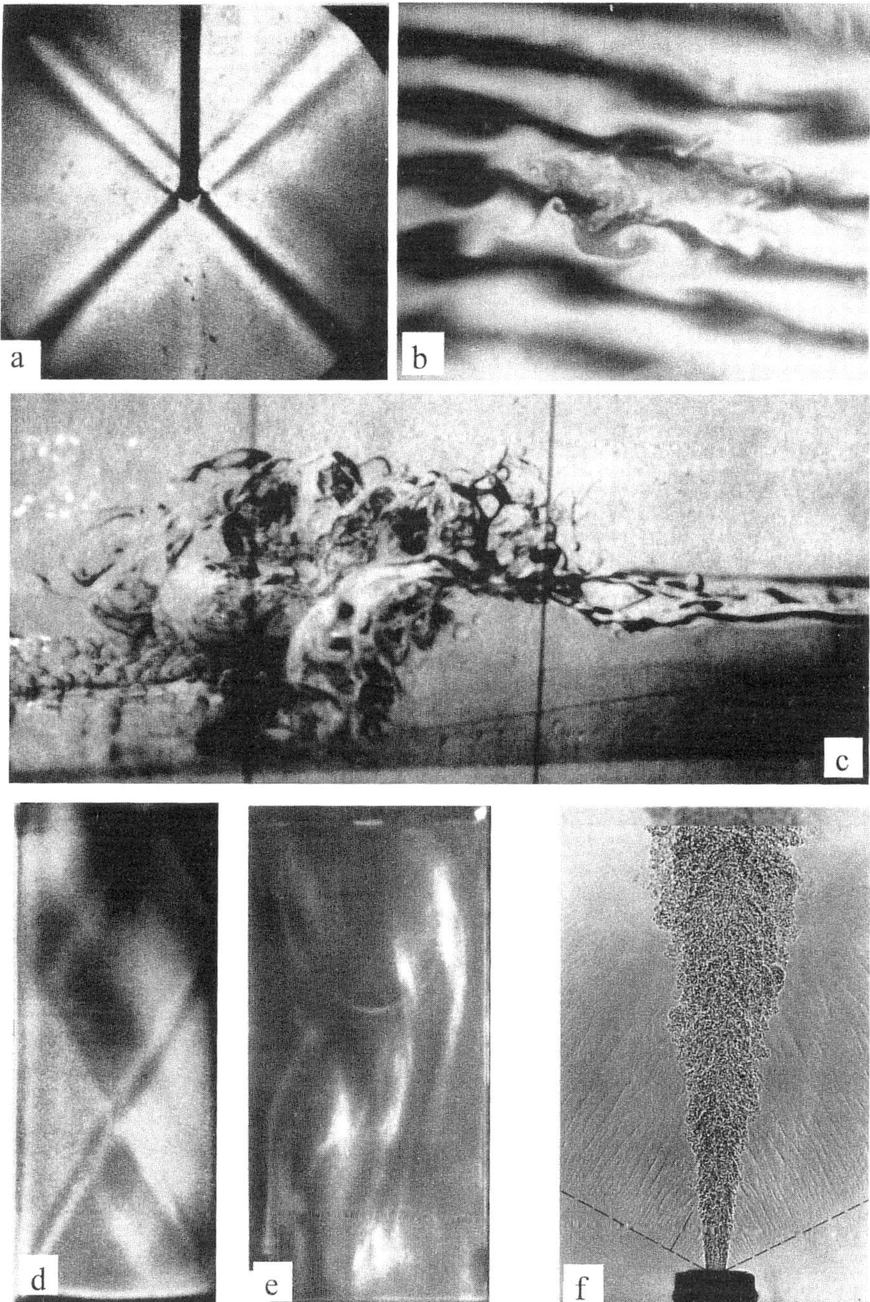

Figure 8.10. Turbulence and waves. *a*) - internal waves propagating from an oscillating body in a stratified fluid (Mowbray and Rarity, 1967). A similar pattern (*d*) is seen in a rotating fluid (inertial waves) which under certain conditions degenerates into state of disorder (*e*) due to the phenomenon of resonant collapse (McEwan, 1970). *b*) shows an internal wave breaking in a laboratory tank filled by a linearly stratified fluid, courtesy of Dr. J. Sommeria. A similar phenomenon is shown in *c*) for the case when the density variation is confined in a relatively thin layer (Pao, 1969) which is analogous to the breaking waves on a water surface. *f*) shows a supersonic jet and sound waves radiated by turbulence in the jet (Tam, 1972).

8.5.1. HELICITY

Rotating fluids lack reflexional symmetry. As a consequence, the quantity called helicty[20], see section 2.2.1 in appendix C, is nonvanishing. Large helicity may lead to reduction of nonlinearity and consequently reduction of drag and dissipation. A recent example is the numerical study of turbulent flow in a rotating pipe by Orlandi (1997). He obsreved a clear positive correlation between increase in helicity and decrease in dissipation. More subtle issues include spontaneous breaking of reflexional symmetry and production of helicity 'out of nothing', and the role of helicity when its mean vanishes. For review and references on this issue and in general, on helicity in turbulent flows see Droegemeier et al. (1993), Khlomyansky et al. (2000b), Kida and Takakoka (1994), and Moffatt and Tsinober (1992).

As noted in chapter 6, section, 6.6, nonzero mean helicity $\langle \mathbf{u} \cdot \boldsymbol{\omega} \rangle$ is an indication of stronger coupling between large and small scales favouring creation of large scale structures out of small scale turbuence - a process which is frequently called 'inverse energy cascade'. A similar phenomenon is observed in the so called turbulent dynamo when small scale turbulent flows of an electrically conducting fluid are able to generate large scale magnetic fields (Childress and Gilbert, 1995; and references therein). Such phenomena are akin to a class of fluid flows described in the next section.

8.6. Negative eddy viscosity phenomena

The term *negative eddy viscosity* is used to denote flow situations in which the turbulent transport of momentum occurs against the mean velocity gradient, i.e. from regions with low momentum to regions with high momentum, so that the Reynolds stress, $-\langle u_1 u_2 \rangle$, and the mean velocity gradient, dU/dx_2, are of *opposite sign*[21]. In other words, the role of the Reynolds stress as one of the agents of coupling the fluctuations with the mean flow is not necessarily one-directional as in pure turbulent shear flows it may act also in the 'opposite' direction. Concomitantly, kinetic energy moves in the 'opposite' direction too - from fluctutaions to the mean fow. It should be stressed that the above does not imply anything more than what is said, e.g. no eddy viscosity in the narrow or any other such thing.

---

[20]The term 'helicity' in the fluid dynamic context was introduced by Betchov (1961) for the quantity $\epsilon_{ijk}\langle u_i(x)\frac{\partial u_j}{\partial x_k}\rangle \equiv \langle \mathbf{u} \cdot \boldsymbol{\omega} \rangle = 6E(0)$, where $E(r)$ is the third scalar function defining the second order correlation of homogeneous and 'semi-isotropic' turbulence, i.e. invariant under rotations but not reflexionally invariant: $\langle u_i(\mathbf{x})u_j(\mathbf{x} + \mathbf{r}) \rangle = C(r)\delta_{ij} + S(r)r_i r_j + E(r)\epsilon_{ijk}r_k$.

[21]For the sake of simplicity, it is assumed here that the mean flow is one dimensional, $\mathbf{U} = U\mathbf{i}$, and that the only nonvanishing gradient of the mean velocity is in the direction x_2.

Such a situation is possible in the presence of some energy supply other than the mean strain. Then the total rate of production/destruction of energy turbulent fluctuations by the mean strain, $\mathcal{P} = -\langle u_1 u_2 \rangle dU/dx_2$, does not have to be positive even in the case of statistically stationary turbulent flows, since in such cases the balance is $\mathcal{W}_F + \mathcal{P} - \mathcal{D} = 0$. Such situations are observed both in laboratory experiments and in the large scale flows of geo- and astrophysics. Selected examples and references are given below. Additional examples and references can be found in Tsinober (1990b) and Paret and Tabeling (1998).

8.6.1. LABORATORY EXPERIMENTS

The popular example is the jet in which the turbulent energy production, $-\langle u_1 u_2 \rangle dU_1/dx_2 < 0$, is negative in the region between the locations in the flow cross section where $dU_1/dx_2 = 0$ and $-\langle u_1 u_2 \rangle = 0$. This region is rather narrow and the mean velocity gradient there is rather small. It was therefore argued that, though the mean velocity may gain energy locally, when integrated over the entire cross section, the effect, however, will be a decrease in the total kinetic energy of the mean motion. Another example is the boundary layer at a convex wall. The effect of curvature on the outer part of the layer leads, in some cases, to a reversal of the turbulent stress.

In this connection, of special importance are experiments in highly coherently forced mixing layer (see Weisbrot and Wygnanski, 1988 and references therein). In these experiments, a whole region in the flow field was found to have turbulent shear opposing the mean velocity gradient over the entire flow cross section (see figure 8.11). It is noteworthy that, unlike other flows, this phenomenon, is especially strong close to the middle of the mixing layer cross section,' where the mean velocity gradient is maximal. Several simple kinematic explanations of the process were proposed in terms of orientation changes (tilting) of elliptically shaped vortices similar to the explanation given by Starr (1968) in a geophysical context (see also Busse, 1983). Though it is tempting to explain the above phenomenon in simple kinematic (two-dimensional) terms, or more generally in terms of (also dynamical) properties of two-dimensional turbulence (see section 8.7), it should be emphasized that the above phenomenon seems to be essentially three-dimensional. A clear indication of this is found in measurements by Oster and Wygnanski (1982). It follows from their results that, in the zone of opposing turbulent shear, the ratio $w/u \sim 0.4 \div 0.5$ $(v/u \approx 2)$, while for a regular (non-forced) mixing layers, it is $0.9 \div 1.0$ $(v/u \approx 1)$; here u, v, w - are the intensities of turbulent velocities correspondingly in the streamwise, lateral and spanwise directions. In other words, the forced mixing layer becomes *more anisotropic*, but still remains far from being

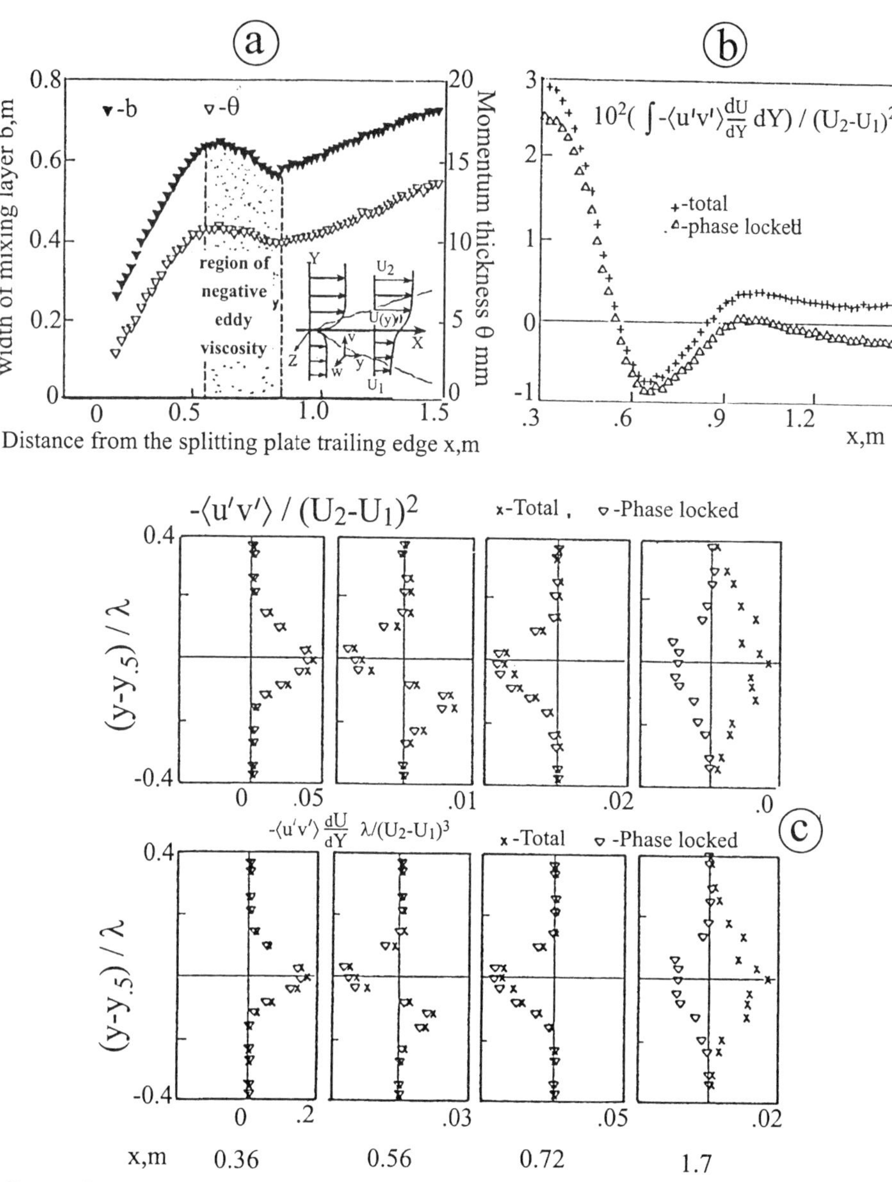

Figure 8.11. The negative eddy viscosity phenomenon in a forced mixing layer. a) - width and momentum thickness. b) - integrated turbulent energy production. Note the region $0.55 < x < 0.84m$ where the mixing layer becomes narrower and the Reynolds stress and the turbulent energy production reverse their signs. c) - profiles of Reynolds stresses and turbulent energy production at several streamwise locations. Adapted from Weisbrot and Wygnanski (1988). The figure is from Tsinober (1990b).

even quasi-two-dimensional.

Krishnamurti and Howard (1981) discovered the so-called 'turbulent wind', i.e. a mean flow in experiments on turbulent convection in a horizontal layer of fluid heated from below and cooled from above. Namely, at high enough Rayleigh number, there exist a nonzero horizontally averaged velocity along the bottom in one direction and along the top in the opposite direction. The direction of this mean flow is random, as it would be in a symmetry breaking bifurcation. This, however, happens only when the flow is essentially turbulent and the direction of the momentum transport by the Reynolds stress is up and not down the gradient of this mean velocity, i.e. the Reynolds stress contributes to the maintenance of the mean flow. The main feature is that almost all plumes rising from the heated bottom were tilted away from the vertical in such a way as to aid the momentum transport to the mean flow. Concomitantly, the energy of the fluctuations is transferred to the mean flow. Later Krishnamurti and Howard (1983) performed a similar experiment in an annular cell in which they measured both the mean velocity and the Reynolds stress, which appeared to be of the sign opposite to that of the gradient of the mean velocity in most of the vertical positions from the heated bottom in conformity with the negative eddy viscosity. It should be emphasized that the opposite sign of the Reynolds stress to that of the mean velocity gradient is only a manifestation of the negative eddy viscosity phenomenon, but not an explanation of its existence[22].

The 'turbulent wind' was observed subsequently in other experiments (for references see Grossman and Lohse, 2000; and Siggia, 1994). For example, the tilted plumes observed in experiments by Ciliberto et al. (1996) are shown in figure 8.7d, whereas the mean horizontal velocity profiles obtained by Xin et al (1996) are shown in figure 8.12.

Systems in which small scale flows are driven by electromagnetic forces in a particular way may develop a mean flow essentially in the same way (see references in Paret and Tabeling, 1998; and Tsinober, 1990b,c).

Wei and Willmarth (1992) also observed negative Reynolds stresses and negative production of turbulent energy in the near-wall region in a turbulent channel flow with drag-reducing polymer injection (see section 8.9).

Finally, we mention a related phenomenon in stably stratified turbulent flows - the so-called PCG, persistent countergradient fluxes. The essence of PCG is the countergradient transport of momentum and active scalar. It is observed at large scales when stratification is strong, but in small scales it is present with weak stratification as well (see Gerz and Schumann, 1996; Komori and Nagata, 1996; and references therein).

---

[22]Note that in the absence of a driving source, such as pressure gradient, the simplest form of the Reynolds equation is $\nu \, d^2U/dx_2^2 = d\langle u_1 u_2 \rangle/dx_2$, i.e. $\nu \, dU/dx = \langle u_1 u_2 \rangle + const$. This does not allow us to claim that the sign of $-\langle u_1 u_2 \rangle$ is opposite to that of dU/dx, since it is not a simple matter to find the *const*.

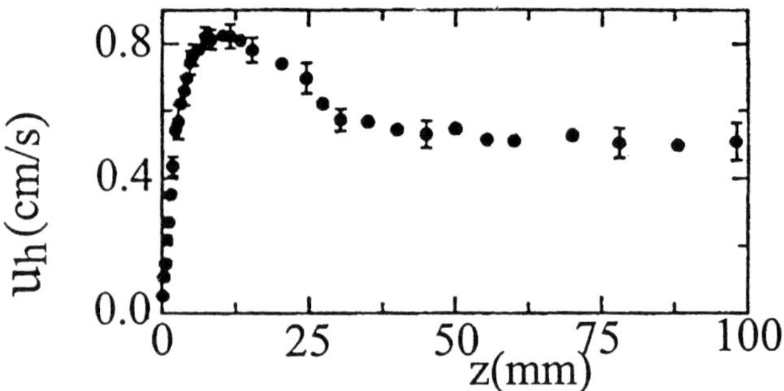

Figure 8.12. Vertical profile of the mean horizontal velocity, u_h, in a turbulent convection experiment (Xin et al, 1996); z is the distance from the heated bottom. The intensity of turbulent fluctuations was pretty high and was about $0.2u_h$.

8.6.2. EXAMPLES FROM GEOPHYSICS

There exists considerable evidence that a number of processes taking place in large scale flows in geo- and astrophysics are associated with the phenomenon of the negative eddy viscosity. For example, the generation and maintenance of mean flows by large scale fluctuations, such as jet flows in the atmosphere and in the ocean, as well as zonal circulation in the atmospheres of some planets of the solar system and of the sun (see Busse, 1983; Monin, 1987; Monin and Yaglom, 1996; Starr, 1968; and references therein).

There are two main kinds of fluctuative motions feeding such mean flows. The first kind are the convective motions (similar to those producing the 'turbulent wind' in laboratory experiments), such as produced by the supply of solar energy to the Earth atmosphere. The second kind are the so-called Rossby waves (see Tritton, 1988 for a simple explanation of what they are). Propagation and breaking of these waves, are thought to be responsible for the generation of the large scale mean motions mentioned above (Monin, 1987; McIntire 1993)[23]. Along with turbulence production the wave propagation and breaking lead to transferring of momentum and energy to the mean flows. At least some of these phenomena are analogous to the so-called acoustic wind in which a mean flow is produced by acoustic waves (Lighthill, 1978). There is an essential difference too, since the acoustic streaming is a (quasi-) linear phenomenon, whereas the wave breaking, mostly responsible for the wave contribution to the negative eddy viscosity phenomenon, is essentially nonlinear. This makes (almost) impossible to make a clear distinction between turbulence and waves - a difficulty noted already by Stewart in 1959 in the context of the issue of distinguishing between internal waves and turbulence in the atmospheric flows (see section

[23]This idea dates back to the proposal by Rossby in 1947.

8.3.2).

8.6.3. POSSIBLE EXPLANATIONS

An explanation (at least qualitative) of the 'anomalous' negative eddy viscosity phenomena is usually given via properties and by analogy with two-dimensional turbulence (see section 8.7), which, under certain conditions, exhibit negative eddy viscosity and other 'anomalous' properties. The first attempt of this kind was made by Lorenz (1953). He considered a two-dimensional (turbulent) flow consisting of mean and fluctuative components, both unsteady, and the interaction between the two. In case when the scales of the smaller motions of the mean flow and of the largest ones in the fluctuations overlap, the energy of the fluctuations can be transferred to the mean flow, i.e. the eddy viscosity becomes negative. For other two-dimensional examples, see Paret and Tabeling (1998) and Tsinober (1990b). The analogy, however, is qualitative only, and geophysical and other flows with negative eddy viscosity (and energy production reversal) at best can be considered as coexistence of quasi-two-dimensional structures (mostly in large scales) with more three-dimensional smaller scales. What seems to be certain is the fact that all negative eddy viscosity flow configurations are considerably anisotropic (more than 'normal' flows) due to some external influences (geometrical constraints, rotation, density stratification, magnetic field, etc.). There exist a number of theoretical models of three-dimensional flows in simple geometry (see references in Tsinober, 1990b). The common feature of these models is that the background small scale motions should be in some sense *anisotropic* to be able to develop a large scale instability. At present, it is not clear whether any of the existing approaches enable researchers to explain observations about negative eddy viscosity phenomena. In fact, no theoretical framework is available to describe the kinematical and dynamical features of these flows. *The more fundamental problem, to explain from first principles why there is a mean flow... is beyond reach... The effect is intrinsic and not understood* (Siggia, 1994; but see Malkus, 1996).

8.7. Magnetohydrodynamic flows

An electrically conductive fluid flowing in the presence of an electromagnetic field experiences a (ponderomotive) body Lorenz force per unit mass $\mathbf{j} \times \mathbf{B}$, where $\mathbf{B}$ is the magnetic field induction (or simply magnetic field) and $\mathbf{j}$ is the electric current density induced by an externally applied electrical field and/or by the electrical field induced by the fluid motion in the presence of the magnetic field. The magnetic field is in turn changed by the electrical currents induced by the fluid flow and governed by an equation

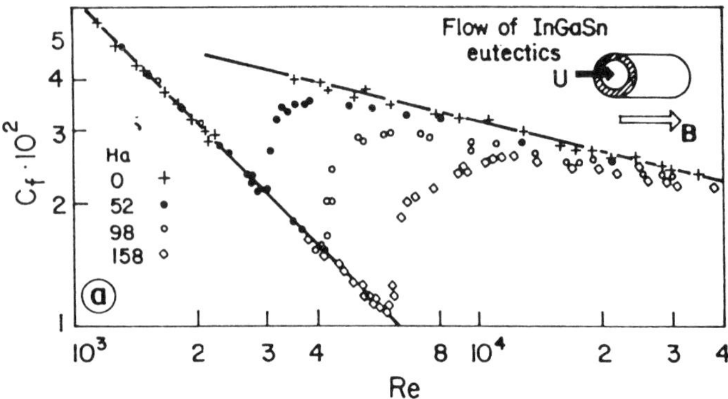

Figure 8.13.   Drag reduction in a circular pipe by a longitudinal magnetic field, Krasil-nikov et al. (1971). Here $Ha = BL(\sigma/\rho\nu)^{1/2}$ is the nondimensional parameter proportional to the intensity of the magnetic field B. The figure is from Tsinober (1990c).

such as (C.36) without a forcing term. Concomitant to the interaction between the fluid flow and the electromagnetic field is the Joule dissipation, j^2/σ, σ is the fluid electrical conductivity, and corresponding losses. Nevertheless, the total losses, (e.g. total drag and dissipation) can be smaller than in a flow without electromagnetic field by virtue of the dramatic changes in stability properties and/or structure of the fluid flow (figure 8.13).

There is a vast variety of possible configurations of MHD-flows even in the simplest geometries. We restrict ourselves in the following to examples in the situation when the changes of the magnetic field induced by the fluid flow are negligible (i.e. the case of small magnetic Reynolds number, $Re_m = \sigma UL \ll 1)^{24}$ and when this magnetic field is homogeneous. For review and references, see Moreau (1990) and Tsinober (1990c).

Our main purpose here is to illustrate some typical effects of turbulent flows arising in the presence of a magnetic field.

Apart from the above-mentioned, one of the prominent effects of magnetic field on a turbulent flow of electrically conducting fluid is again anisotropy. Namely, in the presence of an externally imposed magnetic field, gradients of various flow properties in the direction of the magnetic field are reduced. Hence, there is a possibility of a quasi-two-dimensional flow when the magnetic field is strong enough provided that boundaries (and boundary conditions) favour such a flow. Indeed, such flows have been observed in a number of cases (see references in Tsinober, 1990c; and Porthérat et al., 2000). Like in some other quasi-two-dimensional flows, the MHD quasi-two-dimensional flows exhibit reduced dissipation, by analogy with purely

---

[24]The opposite situation, i.e. $Re_m = \sigma UL \gg 1$, is typical in astrophysical contexts. In systems containing large amounts of liquid metal, such as fast reactors Re_m, can be of the order 10^2. It is many orders of magnitude larger in astophysical objects.

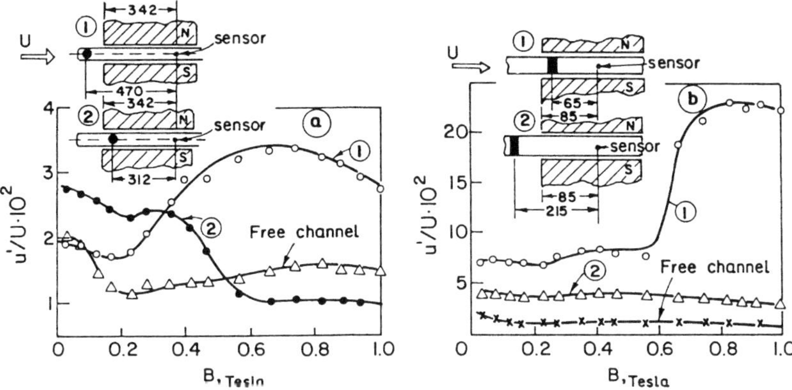

Figure 8.14. The effect of a magnetic field on the perturbations in the wake behind a cylinder. a) cylinder axis perpendicular to the magnetic field, b) cylinder axis parallel to the magnetic field (Kit et al., 1970).

two-dimensional flows which lack strain (i.e. dissipation) production. An example of such an effect is shown in figure 8.14. It indicates that in cases when the geometry of boundaries and the boundary conditions favour quasi-two-dimensional flows in the plane perpendicular to the magnetic field, i.e. when the axis of the cylinder is parallel to the magnetic field, the turbulence intensity in the wake of the cylinder is three times *larger* than in the absence of magnetic field. This happens because a quasi-two-dimensional turbulent flow is unable to dissipate the energy of the disturbances created in the proximity of the cylinder in the way in which a three-dimensional flow in the absence of magnetic field.

Another effect of quasi-two-dimensionalization of some MHD-flows is the anisotropic transport. Namely, in the presence of magnetic field, one observes suppression of the transport of a passive scalar (e.g. indium dissolved in mercury) in the direction of the magnetic field and its enhancement in the plane perpendicular to the magnetic field (see figure 11b in Tsinober, 1990c). A related recent example is displayed in figure 8.15, which shows some results of convection of sodium-potassium alloy driven by a horizontal temperature gradient in the presence of a horizontal magnetic field. It is seen that a moderately strong magnetic field causes an increase in heat transport between the two vertical walls. The reduction of turbulent energy content in the small scales and its increase in the large scales is consistent with the tendency to quasi-two-dimensionalization. In larger magnetic fields, the flow is suppressed by the so-called Hartmann effect on the walls perpendicular to the magnetic field. In the absence of such a breaking effect, the enhancing influence of the magnetic field on the transport properties of the turbulent flow in the plane perpendicular to the magnetic field would

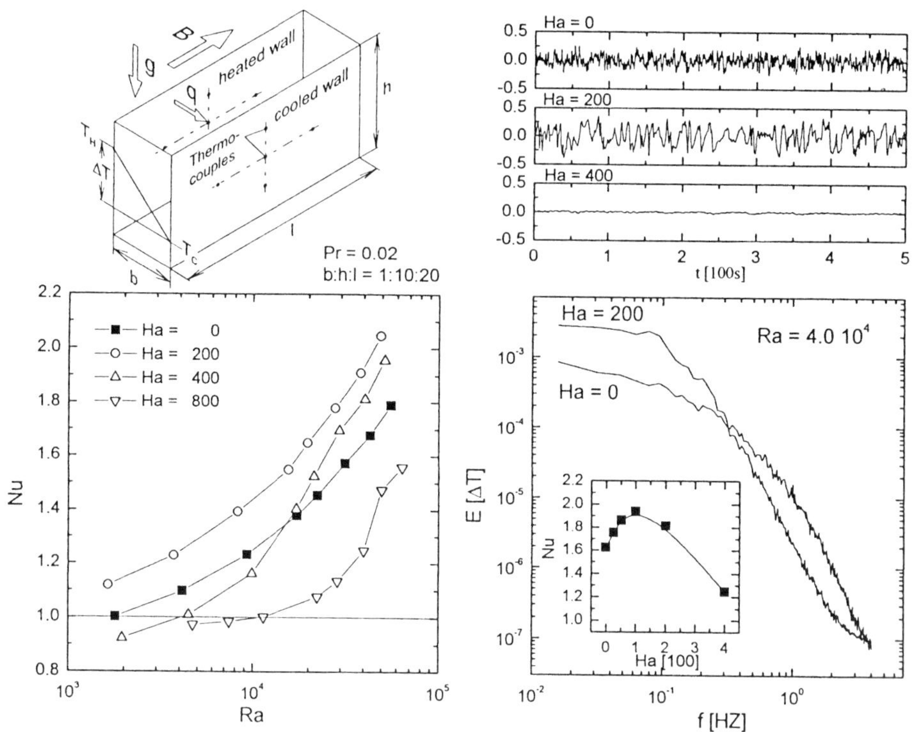

Figure 8.15. The effect of a horizontal magnetic field on convection of sodium-potassium alloy driven by a horizontal temperature gradient (Burr et al., 2000). Hartmann number, *Ha*, is defined as in figure 8.13.

be stronger.

A final note is that there exist many turbulent MHD-flows in which the influence of magnetic field is totally different from that shown in the above examples and that depends on geometry and properties (electrical conductivity) of the flow boundaries, the nature of magnetic field (AC, DC, inhomogeneous) and other factors.

8.8. Two-dimensional turbulence

> *This trend towards well-defined large-scale structures can make it questionable if the 2-D flow should be described as 'turbulent' and it casts some doubts on the concept of inertial range and the relevance of energy spectra... Random initial vorticity distribution quickly assumes a stringlike pattern, which persists as the flow simplifies into a few 'cyclones' or 'finite area vortex regions'. (Fornberg, 1977).*

Two-dimensional turbulence (whatever this means) can be seen as one of the extreme anisotropic states of fluid flow. Though not strictly realizable (except in direct numerical simulations), two-dimensional turbulence is of interest for several reasons. First, it is believed that properties of pure two-dimensional turbulence *may* be useful in treating turbulence in a number of quasi-two-dimensional systems. Such flows include large scale geophysical flows (Lindborg, 1999); flowing soap films (Rivera *et al.*, 1998); some flows in rotating or/and stratified systems (Riley and Lelong, 2000), and some magnetohydrodynamic flows (Kraichnan and Montgomery, 1980), Porthérat et al., 2000). Second, pure two-dimensional turbulence is accessible, at least in part, by methods of statistical physics, contrary to the case of three-dimensional tubulence (Kraichnan and Montgomery, 1980; Onsager, 1949; Pomeau, 1995; Sommeria, 2001; and refcrences therein). Third, in two dimensional turbulence, there exist a process which is to some extent analogous to vortex stretching in three-dimensional turbulent flows. Namely, this is the process of predominant stretching of the vorticity gradient $\boldsymbol{\zeta} = \boldsymbol{\nabla}\omega$ ($= \partial\omega/\partial x_i$) or equivalently $\boldsymbol{\xi} = rot\boldsymbol{\omega}$ (see equation [C.22] in appendix C; Herring et al., 1974; Novikov, 1997; Weiss, 1991). For example, in numerical simulations of decaying two-dimensional turbulence the mean palinstrophy production $\langle\xi_i\xi_k s_{ik}\rangle$ becomes pretty large before starting to decay (see figure 8.16). The nonzero $\langle\xi_i\xi_k s_{ik}\rangle$ is a clear indication of non-Gaussian nature of two-dimensional turbulence, though at 'lower levels' (velocity, velocity increments for points at not too small distances), it is close to Gaussian (see Boffeta et al., 2000; Paret and Tabeling, 1998; and references therein). It is noteworthy that the predominant stretching of vorticity gradients and positiveness of palinstrophy production is a genuinely nonlinear (inviscid) process and not the consequence of the (approximate) balance between palinstrophy production and its viscous destruction.

Nevertheless, one of the main arguments against qualifying two dimensional chaotic flows as turbulence is the absence of the vortex stretching process and generally of the process of self-amplification of the field of velocity derivatives. This qualitative difference leads to reduced ability of such flows to dissipate energy and a number of other important differences (see end of this section and (8.8.2). For example, in order to numerically simulate

two-dimensional turbulence, it is common to do this by starting with *random* initial conditions for decaying turbulence and adding a *random* forcing in the right hand side of the Navier-Stokes equation (and some dissipation at large scales too) when looking at a statistically stationary situation, just like in case of Burgers equation. In three-dimensional turbulence the results are not sensitive (at least qualitatively) to whether the forcing is random or deterministic or even time independent. This does not seem to be the

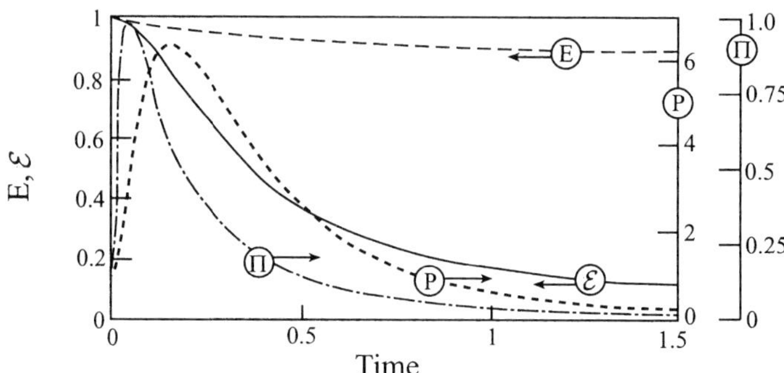

Figure 8.16.   Time behaviour of total energy, $E = \frac{1}{2}\int u^2 dA$; enstrophy, $\mathcal{E} = \frac{1}{2}\int \omega^2 dA$; palinstrophy, $P = \frac{1}{2}\int \xi^2 dA$; and palinstrophy production, $\Pi = \int \xi_i\xi_k s_{ik} dA$, in 2-D decaying turbulence from an initially random state with $Re_\lambda = 1300$. Note the slow decay of energy and the increase of palinstophy to six times its initial value before starting to decay. The palinstrophy production is normalized on $P\mathcal{E}^{1/2}$ at $t = 0$. The two-dimensional skewness $\Pi P^{-1}\mathcal{E}^{-1/2}$ reaches an approximately constant value after a relatively short transient time as in Herring et al. (1994). Courtesy of Dr. Y. Kimura.

case in two-dimensional turbulent flows. For instance, a two dimensional flow in a plane channel driven by a constant pressure gradient or with fixed mean flow rate (Jimenez, 1990; Lomholt,1996) exhibits a very different kind of chaotic behaviour than that which is observed in simulations with random forcing. Namely, the two dimensional channel flow develops mostly large-scale and quite slow-in-time and organized in space variations, which hardly anybody will recognize as turbulent. This is a reflection of the fact that, chaotic properties of two-dimensional flows seem to be qualitatively different from those of three-dimensional turbulence.

The latter example is interesting also in the context of the similarities and differences between pure two-dimensional and quasi-two-dimensional turbulent flows.

8.8.1. PURE TWO-DIMENSIONAL VERSUS QUASI-TWO-DIMENSIONAL

The common view was, until recently that the essential aspects of quasi-two-dimensional turbulent flows (Q2D) can well be described by pure two-dimensional ones (P2D), both globally and locally, i.e. that ε in $Q2D = P2D + \varepsilon$ is small in some sense.

One of the popular beliefs is that locally this is true of Q2D regions with concentrated vorticity (vortex filaments). However, it appears that locally quasi-two-dimensional regions corresponding to large $\cos(\boldsymbol{\omega}, \boldsymbol{\lambda}_2)$ (see section 6.5.2.), to which belong the regions of concentrated vorticity, are qualitatively different from purely two-dimensional ones, in that they possess essentially nonvanishing enstrophy generation $\omega_i \omega_j s_{ij}$ and intermediate eigenvalue Λ_2 of the rate of strain tensor, which are identically zero in P2D flows. Moreover, in these regions both $\omega_i \omega_j s_{ij}$ and Λ_2 are *larger* than in the whole field and in this sense ε is not small in $Q2D = P2D + \varepsilon$.

Another example relates to the behaviour of the wall-bounded turbulent shear flows in the wall proximity, namely, *that flow in the viscous sublayer is almost two-dimensional and two-component over fairly long periods of time* (Fischer et al., 2001; and references therein). However, this seems to be true (at best) of the level of the velocity field. The field of velocity derivatives and related quantities remains far from being quasi-two-dimensional even in the closest proximity of the wall (Tsinober et al., 1995).

In the case of globally Q2D turbulent flows, the matters seem to be even controversial. For example, the experimental (and recent numerical) results obtained for turbulent MHD flows in channels with large aspect ratio in the presence of an azimuthal magnetic field showed that, in such flows, at $Re \leq 10^4$ (which are Q2D) on the one hand, the drag is indistinguishable from its laminar value, and on the other hand, the level of turbulence may be substantially higher than that in the same flow without magnetic field (see e.g. the review in Tsinober,1990c). However, the examination of the results of Jimenez (1990) concerning the DNS of NSE of a plane Poiseulle turbulent flow (which is P2D) at $Re \sim 10^4$ shows that its drag is about twice as large as the purely laminar value and is only twice as small as its value for the 3D turbulent flow. in other words, the P2D plane Poiseulle turbulent flow is not that low dissipative. Moreover, the Reynolds stresses in this flow are not small either (as had been expected) and contribute about a half to the total stress (see figure 8.17).

These results were confirmed using an essentially different code by Lomholt (1996)[25].

The problem of the relation(s) between Q2D and P2D turbulent flows is complicated further by the multiplicity of Q2D states: there exist several Q2D flows such as flows in rotating frames, flows with stable density stratification, MHD-flows and some others, which along with being kinematically/ geometrically similar are in many respects dynamically very different (see figure 8.18). An example of the difference in the dynamical nature is exhibited in the above-mentioned zigzag instability in strongly stably stratified flows (Billant and Chomaz, 2000). Another example is related to the MHD

---

[25]It is noteworthy that since the common belief was that *any* two-dimensional turbulent flow should be low dissipative along with the results on really low dissipative nature of Q2D flows in MHD channels with spanwise magnetic field, the author of this book tended to think that the results by Jimenez (1990) were erroneous. It was, however, a calculation of the same flow undertaken by Sune Lomholt (1996) using an essentially different code that confirmed unequivocally the results of Jimenez (1990).

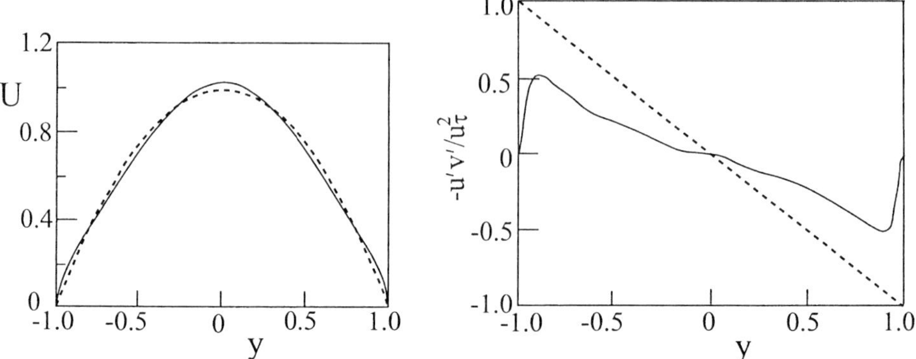

Figure 8.17. A two-dimensional 'turbulent' flow in a channel. Left - mean velocity, the continuous line corresponds to the actual flow, the dashed line is the Poiseulle parabolic velocity profile; note that, along with similarity of the two velocity profiles, there is large velocity gradient at the wall. Right - the Reynolds stress (continuous line) and the total stress (dashed line). The figure is from Lomholt (1996). These results are practically the same as those obtained by Jimenez (1990).

convection flow shown in figure 8.15. It appears that the P2D analogue to this flow studied numerically by Burr (1999, private communication) exhibits totally different behaviour (see also Dolzhansky, 1999).

There is little doubt about the qualitative difference between Q2D states produced by physically different processes, e.g. the ones in MHD are of dissipative nature (Joule dissipation), whereas those with rotation are not. Strong anisotropy is a necessary condition only for Q2D and/or low dissipative behaviour, e.g. shear turbulent flows with strong shear are both strongly anisotropic and strongly dissipative. Similarly, strong correlations along some direction, i.e. Q2D behaviour, do not exclude the possibility of vorticity stretching in this direction.

8.8.2. SOME ADDITIONAL DIFFERENCES BETWEEN TWO-DIMENSIONAL AND THREE-DIMENSIONAL TURBULENCE

Apart from the above-mentioned differences between three-dimensional and two-dimensional turbulence, there are some additional ones which deserve special mentioning.

There is a positive net production of s^2 in the 3-D case, whereas it is conserved (inviscidly) in the 2-D case. Note that it is not a pointwise Lagrangian invariant, as are vorticity and enstrophy in two-dimensional flows.

In 3-D turbulence vorticity, ω_i, and strain, s_{ij} are equal partners, both are self-amplified, 'live' on the scales of the same order, and are related by a conjugation symmetry relation (Ohkitani, 1994). In 2D, it is $\boldsymbol{\xi} = rot\boldsymbol{\omega}$,

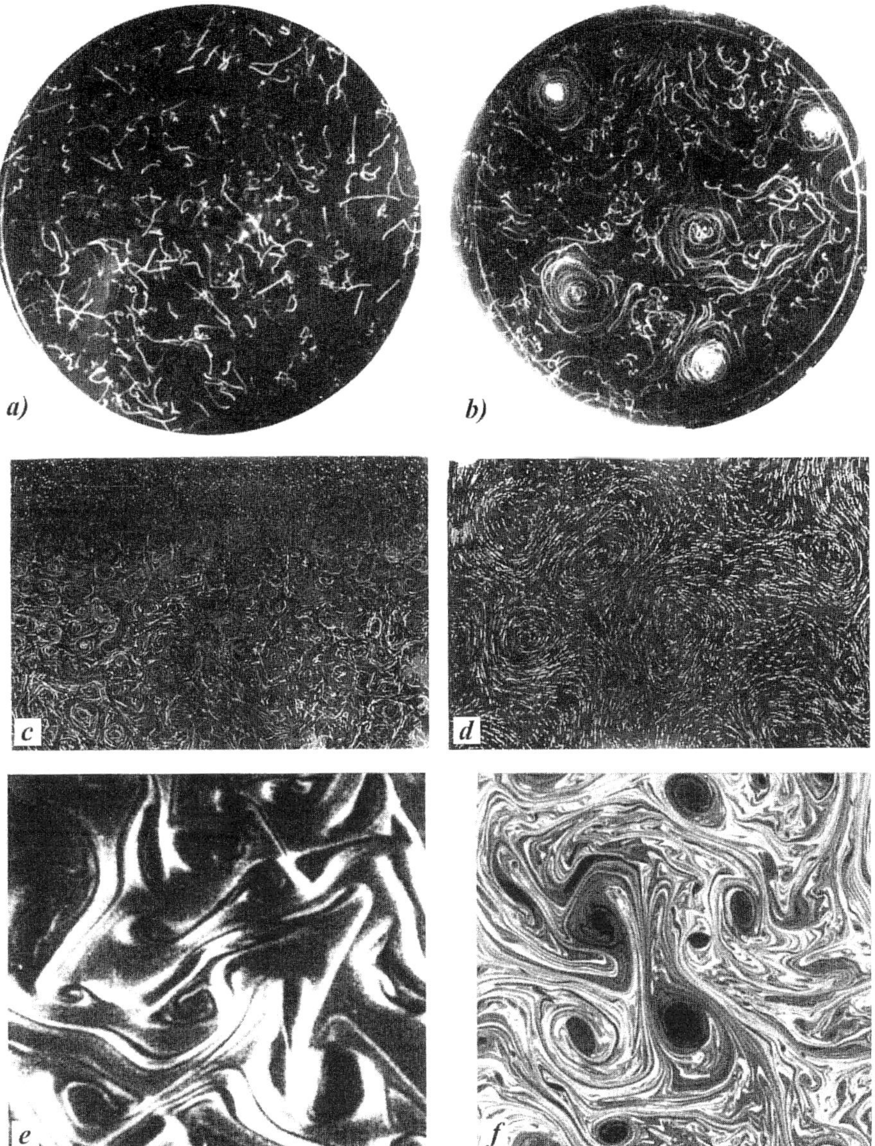

Figure 8.18. Two-dimensional and quasi-two-dimensional turbulence. a) three-dimensional and b) quasi-two-dimensional turbulence induced by excitation at the bottom, in b) the tank is rotating (McEwan, 1976); c) and d) quasi-two-dimensional turbulence (as seen from the top) past a grid towed in a stably stratified fluid (Maxworthy et al., 1987). e) quasi-two-dimensional turbulence in a soap film (Rivera et al., 1998); f) numerically simulated two-dimensional turbulence, courtesy of Dr. B.L. Hua.

or equivalently vorticity gradient, $\boldsymbol{\zeta} = \boldsymbol{\nabla}\omega \; (= \partial\omega/\partial x_i)$ that is amplified via interaction with strain, so that the partners are not equal anymore:

the strain is not amplified and the characteristic scales of ξ (and/or ζ) are much smaller than those of s_{ij}, which can be interpreted as a kind of scale separation. It is this scale separation and the absence of amplification of strain that makes the 2D problem more regular than in 3-D, i.e. in 2-D the nonlinearity is 'less nonlinear'. Indeed, it is known that the solution to the 2-D Navier-Stokes equations with smooth intial and boundary conditions at any Re is smooth for all times, i.e does not have any singularities (Doering and Gibbon, 1995). This is, of course, true of vorticity gradients. As mentioned, the reason for this is that the behaviour of strain (which plays a crucial role in amplification of vorticity gradients) is different in 2-D case than in 3-D case. Namely, in the 2-D case the strain is an inviscid invariant, whereas in the 3-D case it is not due to the predominant its production due to term $s_{ij}s_{jk}s_{ki}$ in the equation (C.18). This seems to be also the reason for some similarity in the behaviour of vorticity and passive scalar in two-dimensional turbulent flows along with some differences (Lapeyre et al., 2000 and references therein). The main reasons for such differences are that, just like in three-dimensional case, the whole flow field is defined by vorticity with apropriate boundary conditions on velocity, and that the equation for a passive scalar is linear, whereas vorticity dynamics is governed by a nonlinear equation. Along with the absence of the self-amplification of velocity derivatives this nonlinearity may lead to formation of large scale structure(s) ('vortices') out of small scale one(s) - a process which is usually called inverse cascade of energy (Bracco et al., 2000; Paret and Tabeling, 1998; Sommeria, 2001; and references therein). In this respect, the behaviour of passive scalar is qualitatively different from that of vorticity; instead of 'vortices' sharp fronts are formed in the field of a passive scalar (e.g. Celani, 2001).

8.9. Additives

Turbulent flows can be strongly modified by additives in even extremely small concentrations. The most spectacular changes occur with only few parts per million of flexible polymers added to the solvent (see references in Cadot et al., 1998; Gyr and Bewersdodff, 1995; McComb, 1990; and Sreenivasan and White, 2000). These changes are exhibited in a number of flow parameters both large scale and small scale, though the direct action of the dissolved polymers is obviously in the small scales. The large scale manifestations are represented in the first place by strong reduction of drag (up to 80%) in turbulent shear flows[26]. An example for smooth and rough

---

[26]There is evidence that the concentration of even 0.5 wt ppm of polyethileneoxide in water can reduce the drag up to 40% (McComb, 1990). There are polymers of extremely high molecular weight which lead to the same effect with only 0.05 wt ppm (Bewersdorff et al., 1993; Gyr and Bewersdorff (1995). Along with turbulent momentum, transport of

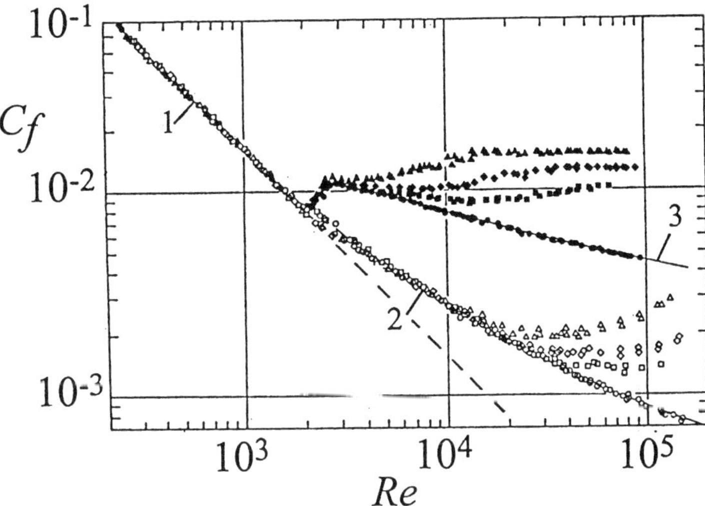

Figure 8.19. Friction factor for maximum drag reduction in smooth and rough pipes. Solid points refer to solvent, hollow points to polymer solutions yielding maximum drag reduction. (1) - laminar, Poiseuillle's law , (2) - turbulent for smooth pipes, (3) - 'maximum' drag asymptote; Values of k/R - relative roughness: $\triangle$- 14.6, $\diamond$ - 22.8, $\square$ - 35, circle - smooth (Virk, 1971). The dashed line is a continuation of (1) and it was added to indicate the difference between the minimal achievable drag with that of purely laminar flow. Note that the drag reducing effect is considerably diminished at large Reynolds numbers when the wall roughness is large enough. This effect was brought to an extreme by Cadot et al. (1998) in experiments with the facility schematically shown in figure 5.1: in the presence of baffles at the top and bottom of the rotating disks the polymer has no effect on the drag at all.

pipes is shown in figure 8.19. Apart from drag reduction, the figure displays the phenomenon of maximum drag reduction or the maximum drag reduction asymptote (MDR). The essence of this phenomenon is that one cannot achieve drag reduction beyond MDR either by increasing the concentration of polymer or switching to another polymer. In other words, there is a saturation of drag reduction effect, which is limiting the drag reduction leaving the drag considerably larger than its purely laminar counterpart.

Another phenomenon is the threshold effect. Namely, the drag in flows of polymer solution follows the normal behaviour of the solvent until the deviation starts at some Reynolds number beyond which the drag reduction occurs. Sometimes this is thought to be connected with a threshold in the wall shear stress.

Along with global there are effects on turbulence structure. The first effect is directly related to the drag reduction – it is a strong decrease of other quantities (heat, mass) is inhibited as well.

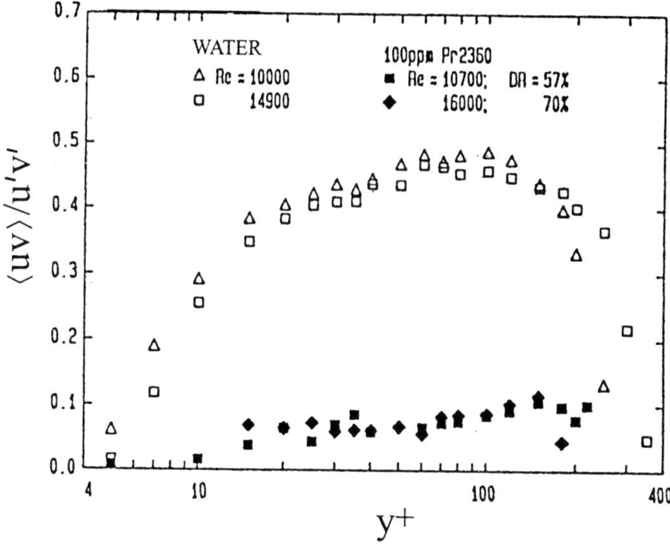

Figure 8.20.   An example of suppression of the Reynolds stresses in a turbulent channel flow (Gampert and Yong, 1990; see also Warholic et al., 1999).

the Reynolds stresses (and turbulence production) as can be seen from an example given in figure 8.20.

However, the strong suppression of the Reynolds stresses occurs *without* substantial reduction of the energy of turbulent fluctuations. The suppression of the Reynolds stresses is due to decorrelation of the streamwise ($u \equiv u_1$) and wall-normal ($v \equiv u_2$) components of the velocity fluctuations (see figure 8.21). The turbulent energy in flows of drag reducing solutions may be somewhat smaller than in normal flows, but also may be increased (see references in Tsinober, 1990b). This is not in contradiction with substantial reduction of turbulent energy production, since concomitantly the dissipation is also strongly reduced as well. An important effect is the increased anisotropy: the wall normal velocity fluctuations are considerably suppressed (McComb, 1990; Tsinober, 1990b; Tong et al., 1990; see also figure 8.21). Anisotropy was also observed in grid generated turbulence (Hibberd and Dohmann, 1988; Doorn et al., 1999; see also the references mentioned above). However, this latter anisotropy seems to be mostly associated with the effects in the process of turbulence production on and in the closest proximity of the grid. Some authors, the latest example being in Doorn et al. (1999), observe a reduced rate of decay the grid generated turbulence[27] others do not (e.g. Hibberd and Dohmann, 1988).

---

[27]One of the possible reasons for the slower decay is the initial anisotropy. For example, a cigar-like turbulence was created past a honeycomb installed after a conventional grid (Hidenaru et al., 1988). The streamwise velocity component was considerably larger than

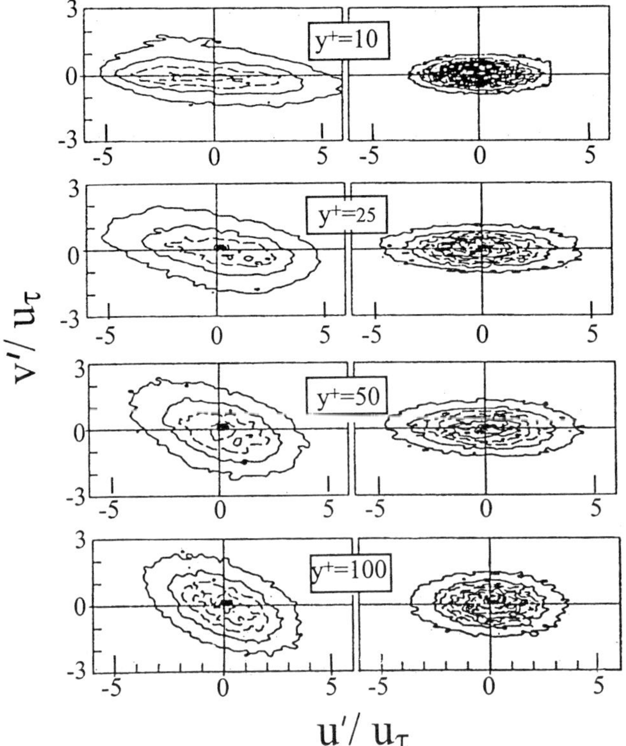

Figure 8.21.   Joint PDFs of the streamwise and wall-normal velocity fluctuations in a turbulent channel flow. Left column – water; right column – polymer solution (Gampert and Yong, 1990).

Visual observations by Cadot et al. (1998) show that polymers have an effect on the structure of the flow both when there is drag reduction and when such a reduction does not occur in the case of inertial forcing by baffles (see figures 8.19 and 5.1). In both cases, the low pressure filaments (vortices) are smaller in numbers and larger in scale. Cadot et al. (1998) also report changes on larger scales. This latter effect was observed in various forms in previous studies. An example is shown in figure 8.22 (see also figure 14.1 in McComb, 1990). Spectral and correlation measurements show that the energy content of turbulent flows is shifted to larger scales; the small scales carrying much less energy than without polymers.

Thus, there are clear indications that the polymer drag reduction is not associated with suppression of turbulence, but with qualitative changes of some of its structure and production. In other words, there exist turbulent flows with strongly reduced drag and consequently dissipation. This implies

the two other components. The rate of decay of this turbulence was observed to be substantially slower than that for quasi-isotropic turbulence.

that in such turbulent flows the nonlinearities should be strongly reduced. Suppression of the Reynolds stresses is an effect of this kind. One can expect that such a reduction of nonlinear processes should occur also at the level of velocity derivatives, i.e. the process of (self-) production of velocity derivatives (vorticity and strain) should be suppressed in flows of drag reducing polymers. A direct indication that this is the case was obtained by Gyr and Tsinober (1996). They compared the quantity $-\langle(\partial u/\partial x)^3\rangle$ in turbulent flows of polymer and surfactant solutions and of water in a pipe of square cross section. We are reminded that in isotropic turbulent flows $\langle\omega_i\omega_j s_{ij}\rangle = -\frac{4}{3}\langle s_{ij}s_{ik}s_{ki}\rangle = -\frac{35}{4}\langle(\partial u_1/\partial x_1)^3\rangle$, so that the latter can be used as a 'surrogate' of the enstrophy and strain production. The main result is that $-\langle(\partial u/\partial x)^3\rangle$ is an order of magnitude smaller in turbulent flows of polymer and surfactant solutions both in the flow bulk and in the near-wall region, thus indicating that enstrophy and strain production in drag reducing flows is strongly inhibited. As expected, a similar behaviour is observed for the 'surrogate' of the dissipation $\langle(\partial u/\partial x)^2\rangle$ in the near wall region, but not in the bulk. This is in conformity with the view that the major contribution to drag reduction process comes from the near wall region as was clearly shown in the experiments by Cadot et al. (1998).

There exist several attempts to explain the phenomenon in flows of drag reducing additives, (see references in the above references). In spite of considerable efforts, the physical mechanisms underlying the phenomenon remain poorly understood[28]. Perhaps, the common feature of all speculations is the belief that the effects observed are directly associated with extension of polymeric coils. This extension is caused by the field of strain, mostly by its fluctuative part. The existence of the threshold effect leads to a conjecture that only strain above some level is able to stretch the polymer coils. An important requirement is that the field of strain should be complicated enough, e.g. random in some sense, to be able to stretch the polymers effectively (see Chertkov, 2000; Groisman and Steinberg, 2000; and references therein). The reaction back of the polymer coils then is expected to change the field of strain in such a way (nobody seems to know/understand how precisely) as to cause the observed effects. Most probably, the presence of polymer molecules resists large strain. This is consistent with strong reduction of the occurrence of bursting events, and one can expect also reduction of small scale intermittency in other turbulent flows. In order to explain the effects with very low polymer concentrations, it is tempting to assume that the polymer molecules form associations which are preferentially located in regions of large strain, as happens with small particles possessing densities

---

[28]This is not surprising, since the problem is a combination of two poorly understood problems, which in the words of McComb (1990) comprise a *possible candidate for the title of 'most difficult problem in physics'*.

Figure 8.22.   Turbulent jet flow in water top) and in polymer solution (bottom) from Gadd (1965).

larger than that of the carrying fluid (Eaton and Fessler, 1994; Elperin et al., 2000; Vaillancourt and Yau, 2000; and references therein). This in turn results in the suppression of production of strain (and vorticity) in the 'hot spots', reduced dissipation and drag reduction, but not necessarily any suppression of turbulence. In such a case the fluid in a turbulent flow of dilute poymer solution is intermittently rheological just like, and because of, the turbulent flow is itself intermittent (Bewersdorff et al., 1993). The above is, however, just one more speculation among many. It seems, however, that in any case the intermittency, especially of strain and thereby of dissipation, is of central importance and hardly can be neglected.

There is no contradiction between the primary role of the interaction of strain with polymer coils and the relation (8.1), showing the importance of vorticity in maintenance of the Reynolds stress. The answer is that in the near-wall region vorticity and strain are strongly correlated, so that close to the wall ω^2 and $2s^2$ are (practically) instantaneously equal (e.g. Sandham and Tsinober, 1990). Hence suppression of strain (by whatever mechanism) results in suppression of vorticity and thereby of the Reynolds stresses, as follows from (8.1).

Drag reduction and other modification effects on turbulence are observed in turbulent flows with other additives. Turbulent flows with added surfactants, very large aspect ratio fibres and some non-fibrous additives exhibit clear drag reducing effects (Gyr and Bewersdorff, 1995; McComb, 1990). However, much less is known about the structure of these flows. For example, some authors report suppression of turbulent energy, while others observed the contrary. A similar situation concerns particle and bubble loaded flows (Crowe, 2000; Kaftori et al., 1998; Tsui, 1991).

It is noteworthy that, the effects of additives on turbulence is one of the manifestations of nonlocality of turbulence: the primary effect is in the smallest scales which results in changes at all levels.

The overview presented in this chapter attempted to give an exposition of turbulent flows in a variety of physical situations. There are many more. See for example, the paper by Gibson (1996) for an interesting account of turbulent flows in geophysical and astrophysical contexts. Turbulent flow phenomena in liquid helium at very low temperatures, at which it is a mixture of normal and quantum fluids, comprise another most fascinating field (Donnelli, 1999); and Liepmann and Laguna, 1984).

All the factors influencing turbulent flows (those mentioned above and many others) can be used in attempting to control turbulence (Gad-el-Hak, 2000; Soldati, 2000). Apart of practical importance such an approach may be useful in using these influences in a systematic way to study the basic mechanisms of turbulence.

CHAPTER 9

CONCLUSION/CLOSE

In spite of a huge number of papers and a large amount of research on turbulence, it remains an unsolved problem left for future generations (Sinai, 1999).

... with all due respect for the coherent structures of the seventies, the insight gained from the chaos theory in the eighties, the achievements of Direct Numerical Simulations (DNS) and Large Eddy Simulation in the nineties, we all hope in our heart of hearts to see the great breakthrough in turbulence, stirring Sir Horace Lamb in his eternal sleep to bring him the long awaited revelation (Wijngaarden, 2000).

The purpose of this last chapter is twofold. First, it aims to recapitulate some main points with somewhat different emphasis, and to discuss some issues of general nature not addressed above. The first issue is universality.

9.1. Universality

Since the Kolmogorov papers (1941ab), there exists almost a religious belief in some universal properties of turbulence. This belief was strengthened by achievements in dynamical chaos, such as the discovery of some universal numbers by Feigenbaum, etc.

On the other hand, with the exception of the 2/3 (5/3) and - 4/5, laws there appeared to exist no quantitative universality so far: the first doubt came from the famous remark by Landau in the first Russian edition of Fluid Mechanics by Landau and Lifshitz about the fluctuations of energy dissipation rate. These were followed by various 'universal' corrections, which did not appear to be universal either. These corrections were followed by the (multi) fractal approach using either the so-called $D(h)$ or $f(\alpha)$ formalisms, in which the *functions* $D(h)$ and/or $f(\alpha)$ are assumed to be universal. However, they do not seem to be universal either. There is quite solid evidence accumulated during the last fifty years against the most beautiful hypothesis on the restoring of the symmetries in the statistical sense of the Navier-Stokes equations locally in time and space, i.e. local isotropy together with scale invariance (see discussion and references in Chapters 5, 6 and 7). And so people started to look for some universality in the anisotropic properties of turbulent flows (see references in Kurien and Sreenivasan, 2000). In other words, it seems that our dream of quantitative universality of turbulence, i.e. universality of numbers, will never come true[1]. However, though there may not exist such a thing like quantitative universality of turbulence (i.e. universality of numbers), there seems to exist a qualitative one. It is naturally to include into the term 'qualitative universality' such general properties of turbulence as randomness, enhanced effective diffusivity and dissipation, rotational nature, and others as discussed in chapter 1. The question is whether there exist more *specific* qualitative universal properties of turbulent flows. The answer is positive. Moreover, these qualitative universal properties possess quantitative attributes, as will be seen in what follows. The likely reason for the qualitative universality is that the *nonlinear terms...remain active at surprisingly low Reynolds numbers*, as observed by Mansour and Wray (1994) in DNS of decaying turbulence at low Reynolds numbers. The resemblance of the flow patterns of turbulent flows in the same geometry, but very different Reynolds numbers also can be seen as one of the manifestations of the qualitative universality (see figure 1.9 and plate 1.3).

9.1.1. ON UNIVERSAL ASPECTS OF TURBULENCE STRUCTURE

In dynamical systems, one looks for structure in the *phase space* (Shlesinger, 2000; Zaslavsky, 1999), since it is relatively 'easy' due to low dimensional nature of the problems involved. In turbulence nothing is known about its properties in the corresponding infinite dimensional phase space[2]. Therefore, it is common to look for structure in the *physical space* with the hope that the structure(s) of turbulence - as we observe it in *physical space* - is (are) the manifestation of the generic structural properties of mathematical objects (*in phase space*), which are called (strange) attractors and which are invariant in some sense. In other words, the structure(s) is (are) assumed to be 'built in' in the turbulence independently of its (their) origin - hence universality. However, as mentioned, the expectation for universal numbers seems to be unjustified. It is more natural to expect universal qualitative statistical features in the physical space rather than universal numbers. Indeed, some of such features have been already observed, which are common for very different - essentially all known - turbulent flows. These are not only the general qualitative features of turbulent flows as described in chapter 1, but rather specific ones.

---

[1] For other negative statements about universality see, for example, Saffman (1978, p. 216) and Hunt and Carruthers (1990, pp. 497, 498).

[2] Hopf (1948) conjectured that the underlying attractor is finite dimensional due to presence of viscosity.

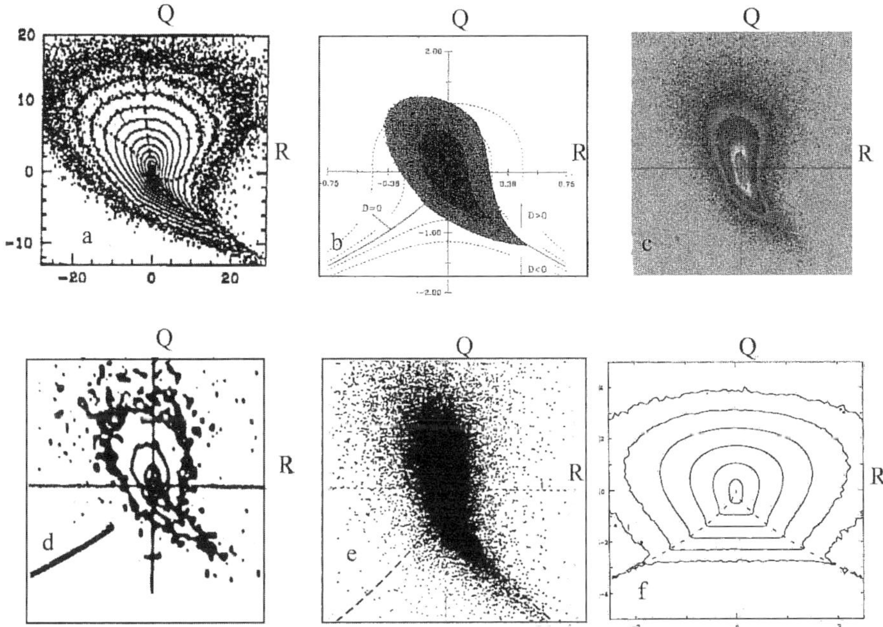

Figure 9.1. The 'tear drop' pattern in the Q-R plot (a) through e)) in different turbulent flows and the symmetric pattern, and (f) for a Gaussian velocity field (Chertkov et al., 1999). a) - turbulent flow in a periodic box (Borue and Orszag, 1998), b) - turbulent boudary layer (Chong et al., 1998), c) - compressible flow, courtesy of A. Pouquet, P. Woodward and D. Porter, d) - mixing layer (Soria et al., 1994), e) - turbulent grid flow (Tsinober et al., 1997). Note that the two invariants are not describing all the aspects of the structure of the field of velocity derivatives (see Chacin and Cantwell, 2000; and Tsinober, 2000a) for Q-R plots with a variety of additional information.

We bring two examples with features which are essentially the same for all known incompressible flows such as grid turbulent flow, periodic flow in a computational box, turbulent boundary layer and channel flow, mixing layer and compressible flows as well. Such features can be seen as universal statistical manifestations of the structure of turbulent flows.

The first example, is the so-called 'tearing drop' feature observed in the invariant map of the second invariant, $Q = \frac{1}{4}(\omega^2 - 2s_{ik}s_{ik})$, versus the third invariant $R = -\frac{1}{3}(s_{ik}s_{km}s_{mi} + \frac{3}{4}\omega_i\omega_k s_{ik})$ of the velocity gradient tensor $\partial u_i/\partial x_k$. This feature appears to be essentially the same for a great variety of flows, some of which are shown in figure 9.1.

We draw the attention to the 'tail' of the teardrop which is mainly located in the quadrant $Q < 0$, $R > 0$, in which most of turbulent activity happens in a variety of ways (Chacin and Cantwell, 2000; Chertkov et al. 1999; Tsinober, 2000). The important point is that this is the region dominated by strain as compared with enstrophy $(2s_{ik}s_{ik} > \omega^2)$

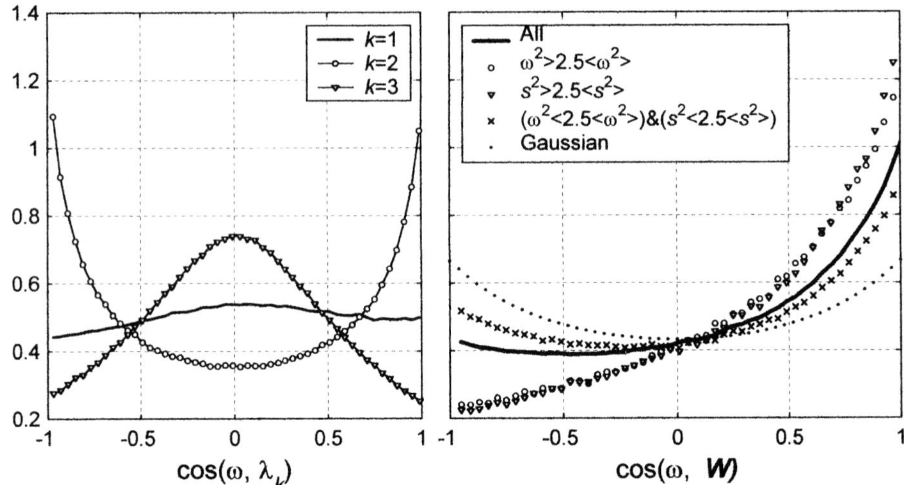

Figure 9.2.   Alignments in a turbulent boundary layer at $Re_\lambda = 10^4$. Right - PDFs of the cosine of the angle between vorticity, and the vortex stretching vector, $\cos(\omega, W)$. Left - PDFs of the cosine of the angle between vorticity, and the eigenframe of the rate of strain tensor, $\cos(\omega, \lambda_k)$ (Kholmyansky et al., 2001a).

and by *production* of strain as compared with production of enstrophy $(-s_{ik}s_{km}s_{mi} > \frac{3}{4}\omega_i\omega_k s_{ik})$. This is in full conformity with the behaviour of nonlinearities in these regions (see section 6.5 and section 9.1.4 below).

The second example is related to *geometrical statistics*. These are various alignments such as the alignments between vorticity and the eigenbasis of the rate of strain tensor, and between vorticity and the vortex stretching vector. It appears that these and many other similar properties are the same for all known flows and, moreover, for a broad range of Reynolds numbers. For example, the character of the above alignments is essentially the same at $Re_\lambda \sim 10^2$ and $Re_\lambda \sim 10^4$. A recent example for $Re_\lambda = 10^4$ is shown in figure 9.2.

It is seen that the qualitative[3] behaviour of the above mentioned alignments is precisely the same as at $Re_\lambda = 75$ shown in figures 6.11-6.13. This brings us to the issue of Reynolds number dependence and the (possible) asymptotic state of turbulent flows at very large Reynolds numbers.

9.1.2. REYNOLDS NUMBER DEPENDENCE

Some properties of turbulent flows become Reynolds number independent as the Reynolds number becomes large enough. An example is given in

---

[3]The quantitative difference is mostly due to the problems of underresolution in the case of $Re_\lambda = 10^4$

figure 1.8 - the drag coefficient of a disc is independent of Reynolds number beyond $Re \sim 10^3$. Another example is given in figure 5.1, showing the independence of dissipation of Reynolds number over three decades of Re. There are many other related examples (Idelchik, 1986), just like the friction factor for pipes with rough walls possessing large enough roughness (Schichting, 1979). It is this property that gives a special status to the scaling exponent 2/3. However, dissipation (energy input) or drag only are not sufficient to define the properties of a turbulent flow. For example, Bevilaqua and Lykoudis (1978) performed experiments on flows past a sphere and a porous disc with the same drag. However, other properties of these flows even on the level of velocity fluctuations were quite different; see also Wygnanski et al. (1986) who performed similar experiments with a larger variety of bodies with the same drag. It should be kept in mind that in both cases the flow was partially turbulent and most probably had different overall stability properties for different bodies. Similarly, many properties of turbulent flows with rough boundaries are not defined uniquely by their friction factor (Krogstad and Antonia, 1999).

There exists considerable evidence on Reynolds number dependence of various properties in different turbulent flows; for a partial list of recent references, see Belin et al. (1997), Ferchichi and Tavoularis (2000), Fisher et al. (2001), Kahaleras et al. (1998), Metzger et al. (2001), Moser et al. (1999), Östrelund et al. (2000), Shen and Warhaft (2000), Sreenivasan and Antonia (1997), Tsinober (1998b) and Zhou and Antonia (2000). For example, the flatness factor of the streamwise velocity derivative $\partial u_1/\partial x_1$ is increasing from $3 \div 4$ at $Re_\lambda \sim 10$ to about 40 at $Re_\lambda \sim 4 \cdot 10^4$, without showing any trend for saturation (see figure 6 in Sreenivasan and Antonia, 1997). There is no understanding of the reasons for such a strong Reynolds number dependence at large values of the Reynolds number. Another example is about the Reynolds number dependence of the relation between the solenoidal and irrotational 'components' of the nonlinearity as represented by $(\mathbf{u} \cdot \nabla)\mathbf{u}$ and shown in figure 9.3. There is a clear tendency of enhancement of solenoidality of the nonlinearity as the Reynolds number increases. Here too it is not clear what will happen when the Reynolds number will become very large.

These examples, along with other results, show that the issue of the asymptotic 'ultimate' regime/state of turbulent flows at very large Reynolds numbers remains open. However, in view of the qualitative universality of turbulent flows, it seems that in order to study the basic physics of turbulent flows - at least at present stage - one needs neither very high Reynolds numbers nor precise determination of scaling exponents[4] , etc.

---

[4]Scaling and related matters have not proven very useful (so far) in understanding the basic physics of turbulence or in justifying, e.g. enormous efforts in accurately measuring

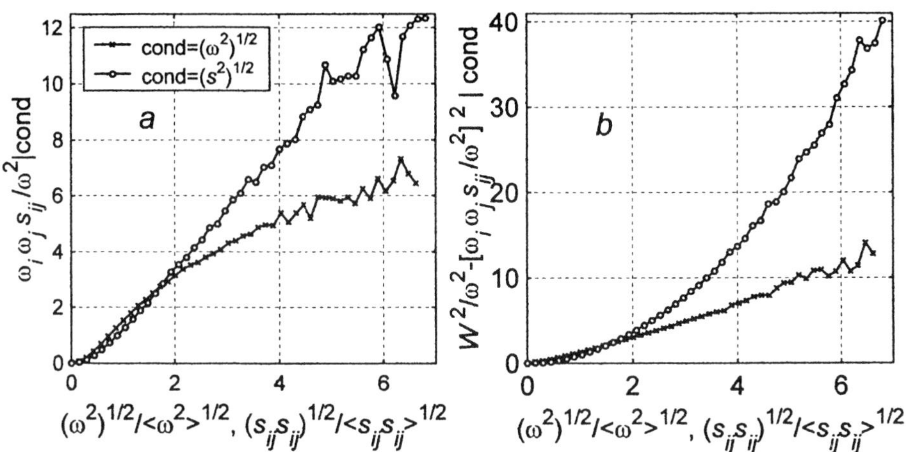

Figure 9.3. Reynolds number dependence of the ratio of the variances of the irrotational and solenoidal parts of the nonlinear term $(u \cdot \nabla)u$ in a DNS simulation of quasi-isotropic turbulence (Tsinober et al., 2001). This is a version of the bottom figure 6.29.

at such Reynolds numbers. Turbulent flows at the asymptotic regime at $Re \gg 1$ ($\rightarrow \infty$) - if such exists - do not seem to be simpler than those at very moderate Re, and are definitely much less accessible in every respect. Many will ask: but what about the (so-called) inertial range? As discussed in chapter 6, the pure inertial range is not well defined due to nonlocality of turbulence; *because of scale invariance breaking, the notion of inertial range is not well defined* (Arneodo et al., 1999). We wish to reiterate that independence of some (statistical) parameters or properties of viscosity at large Reynolds numbers does not mean that viscosity is unimportant. It means *only* that the effect of viscosity is Reynolds number independent.

9.1.3. SELF-AMPLIFICATION OF VELOCITY DERIVATIVES

As discussed in chapter 6 there is some evidence that the process of self-amplification of velocity derivatives, both vorticity and strain, is a universal phenomenon which occurs at Reynolds numbers as low as $Re_\lambda \sim 60$. This seems to be one of the key physical processes in all turbulent flows. The details of this process are not well understood, and apart from geometrical statistics and similar things, one needs much more. Since the whole flow is defined by the field of velocity derivatives (either vorticity or strain), proper understanding of the process of self-amplification of velocity derivatives in

exponents at very large values of Reynolds number. It is not at all clear why and to what extent accurate measurement and/or knowledge of exponents would aid understanding of turbulence. Moreover, the very existence of such exponents (with few exceptions) is quite problematic, e.g. Badii and Talkner (2001), Feigenbaum (1997), Tsinober (1996b).

turbulent flows would bring considerable progress in the understanding of the physics of turbulence as a whole.

It should be stressed that the process of self-amplification of strain is a specific feature of the dynamics of turbulence having no counterpart in the behaviour of passive objects. In contrast, the process of self-amplification of vorticity, along with essential differences, has a number of common features with analogous processes in passive vectors; in both the main factor is their interaction with strain, whereas the production of strain is much more 'self' (section 6.3).

9.1.4. DEPRESSION OF NONLINEARITY

We mention here one aspect of this problem, which seems to be universal in the sense that it is true for different flows and different Reynolds numbers, though the evidence is still quite limited. Namely, practically all nonlinearities appear to be much stronger in the strain dominated regions rather than in regions with concentrated vorticity, in contrast to the common expectation that, for example, the vorticity amplification process will be strongest where the vorticity already happens to be large. The regions with concentrated vorticity are in approximate equilibrium in the sense that the rate of enstrophy production is in approximate balance with the viscous destruction in these regions even at low Reynolds numbers, $Re_\lambda \sim 80$. Therefore, their life time is considerably larger than the life time of the regions dominated by strain, which are in strong disequilibrium in the sense that the rate enstrophy production is much larger than the destruction by viscosity in these regions. Here we show some relevant results from the latest observations at large Reynolds numbers (figure 9.4).

The general conclusion drawn from the whole experiment is that the basic physics of turbulent flow at high Reynolds number $Re_\lambda \sim 10^4$, at least qualitatively, is the same as at moderate Reynolds numbers, $Re_\lambda \sim 10^2$. This is true of such basic processes as enstrophy and strain production, geometrical statistics, the role of concentrated vorticity and strain, and depression of nonlinearity. In other words, as long as the flow is turbulent, there is no qualitative difference between flows with large and small Re. Dimotakis (2000) brings evidence that Re_λ should be larger than 10^2 for the flow to be 'truly' turbulent. Therefore, it seems not necessary to 'hunt' very large values of Reynolds number in studying and trying to understand the basic physics of turbulence.

9.2. Some mathematical and related aspects

As mentioned, there is reasonable evidence that the normalized mean dissipation $\varepsilon = U^3 L^{-1} \langle \epsilon \rangle$ tends to a finite limit as $Re \to \infty$. Since $\langle \epsilon \rangle =$

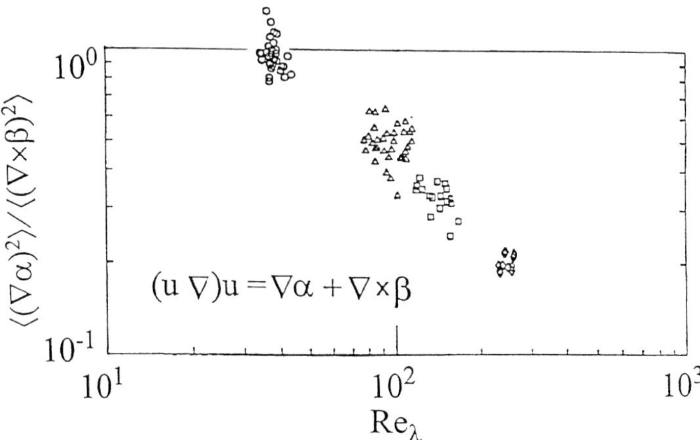

Figure 9.4. Examples of conditional averages showing the difference in the behaviour of nonlinearities in vorticity and strain dominated regions in a turbulent boundary layer at $Re_\lambda = 10^4$ (Kholmyansky et al., 2001a). The quantity $\eta^2 \equiv W^2/\langle\omega^2\rangle - \{\omega_i\omega_j s_{ij}/\langle\omega^2\rangle\}^2$ is responsible for vorticity tilting; see equation (6.4) and the text below.

$2\nu\langle s^2\rangle \approx \nu\langle\omega^2\rangle$ this means that at large Reynolds numbers both $\langle s^2\rangle$, $\langle\omega^2\rangle \approx \nu^{-1}$, i.e. the field of velocity derivatives is not only Reynolds number dependent, but also becomes very large. Due to the intermittent nature of the field of velocity derivatives one can expect that the maximal values of s^2, ω^2 (or $\max(|\partial u_i/\partial x_k|)$) increase even faster with the Reynolds number. This possible unboundness of the field of velocity derivatives as $Re \to \infty$ has an implication that the Newtonian approximation can break down as $\nu \to 0$, since the linear stress/strain relation is only the first term in the gradient expansion. So far, however, there seems to be no evidence that the Navier-Stokes equations are inadequate for describing turbulent flows[5]. Among the possible reasons that this (possible) violation is not so easy to detect is that, even if it happens, it will occur in rather small regions due to the strong intermittency of the field of velocity derivatives. We will not discuss here the possible breakdown of the NSE due to possible formation of singularities in finite time (see Constantin, 1995, 1996). There is a more general mathematical question: *Do the Navier-Stokes equations on a 3-dimensinal domain Ω have a unique smooth solution for all time?..* The belief is that *The solution of this problem might well be a fundamental step toward the very big problem of understanding turbulence* (Smale, 1998), and that *turbulence will be solved when well formulated mathematical statements describing the properties of Navier-Stokes systems are proven* (Sinai, 1999).

---

[5]But see Ladyzhenskaya (1975) and McComb (1990, pp. 401-403) on alternatives to NSE, and Tsinober (1993b) and references therein. It is safe to keep in mind that any equations are not Nature.

However, mathematics alone does not seem to help much (so far). Assume that the answer is known. So what does it add to understanding? Indeed, in the two-dimensional case the above problems are resolved (Doering and Gibbon, 1995), but this did not add much to the understanding of the basic physics of two-dimensional chaotic flows. *Nonuniqueness of the solutions per se thus cannot not play a role in turbulence* (Lumley, 1970).

There are many deep mathematical ideas behind the dynamical systems approach to the study of nonlinear phenomena which are aimed at understanding the origins and structures of complicated behaviour, not merely describing it (Mullin, 1993, p. xvii). However, so far these ideas appear to be very useful in low dimensional systems. It remains to hope that similar ideas will emerge in regard with highly dimensional systems, as well as with the PDEs

There is some evidence that the (statistical) properties of some turbulent flows with the same geometry at very modest Reynolds numbers are invariant of the boundary and initial conditions (BC and IC). For example, typical DNS computations of NSE of turbulent flows (e.g. in a circular pipe and a plane channel, in a cubic box, etc.) involve extensive use of periodic BC. The results of these agree very well with those obtained in laboratory experiments, in which the BC have nothing to do with periodicity[6] and in which the IC were totally different from those in DNS (De Bruyn Kops and Riley, 1998; Nikitin, 1995, 2000; and references in Tsinober, 1998b). No explanation of this kind of invariance is known so far, but it is natural to expect that it is related to some kind of hidden symmetry(ies) of the NSE. If such exist, this may be the reason for the similarity of results obtained via DNS of NSE in, e.g. periodic boxes by various forcing (different deterministic, random/stochastic, etc.).

9.3. On the goals of basic research in turbulence

The phenomenon of turbulence is still one of the least understood ones, so that many issues in the 'problem' of turbulent flows are unsettled. As a consequence, different, strongly disparate and even contradictory views are not a rarity. As mentioned, there is even no consensus on what is(are) the problem(s), just as there is no agreement on what are the aims/goals of turbulence research/theories. In our view the first priority should be given to study of basic physical mechanisms of turbulence with the emphasis on

---

[6]The correlation coefficient between two values of *any* quantity at the opposite such boundaries (i.e. the points separated at *maximal distance* in the flow domain) will be precisely equal to unity and close to unity for the points in the proximity of such boundaries, whereas in any *real* flow the correlation coefficient becomes very small for points separated by a distance on the order of (and larger than) the integral scale of turbulent flow.

qualitative aspects, keeping in mind a somewhat old-fashioned view that curiosity drives better science than 'strategies'. This priority includes the study of turbulence itself (*per se*), rather than multitudes of its models. The only exception seems to be the Navier-Stokes equations. Since there is no simple mathematical aid for what is usually called understanding, it is guileless to think that the 'problem of turbulence' would be resolved if one would have a super-hyper computer enabling to 'solve' the NSE or whatever at any Reynolds number. Suppose one can do this and also to measure whatever one wants. The real problem is to know what to do next with the really huge amounts of data, which is intimately related to the skill/art to ask the right and correctly posed questions. Only this knowledge will aid to expose the basic problems and will lead to real understanding and real 'solutions' and possibly will *produce a method of understanding the qualitative content of equations* [PDEs] (Feynmann, 1963). *Qualitative* is the key word, since there is a qualitative difference between being able to measure and/or compute/calculate all one wants and understanding. Perhaps the efforts of the turbulence community should be somewhat shifted to the qualitative aspects of the problem.

It is remarkable that in spite - or perhaps just because - of frustrated and unsuccessful attempts to construct a predictive theory of turbulent flows based on the first principles, the attraction of the turbulence problem is only growing. This is reflected in continuing and increasing efforts in most of the areas of the field, in spite of considerably reduced funding: curiosity drives better science than 'strategies'. And so one can be optimistic that in the end, the glorious enigma of turbulence as a physical phenomenon will be resolved. At that moment, the turbulence community will be not devided anymore into experimentalists who observe what still cannot be explained and theoreticians which try to explain what can not be observed[7]. And there will be no more attempts only to describe things and to replace the sought explanation by this mere description.

---

[7]Presently the turbulence community is not that simple and is devided to more than just two groups. To quote Lumley: *Turbulence is rent by factionalism. Traditional approaches in the field are under attack, and one hears intemperate statements against long time averaging, Reynolds decomposition, and so forth. Some of these are reminiscent of the Einstein-Heisenberg controversy over quantum mechanics, and smack of a mistrust of any statistical approach. Coherent structure people sound like* The Emperors's new Clothes *when they say that all turbulent flows consist primarily of coherent structures, in the face of visual evidence to the contrary. Dynamical systems theory people are sure that turbulence is chaos. Simulators have convinced many that we will be able to compute anything within a decade... The card-carrying physicists dismiss everything that has been done on turbulence from Osborne Reynolds until the last decade. Cellular Automata were hailed on their appearance as the answer to a màidens prayer, so far as turbulence was concerned* (Lumley, 1990, as quoted by Cantwell, 1990). There are more, see, for example, Giralt et al. (2000).

CHAPTER 10

APPENDIX A. WHAT IS TURBULENCE?

Attempts of definition(s)

This is an updated version of a collection of citations prepared by the author for the *First Monte Vertia Colloquium on Turbulence* held in Ascona, Switzerland, September 1991. Other citations are given in the main text.

* - Turbulence is the name given to imperfectly understood class of chaotic solutions to the Navier-Stokes equation in which many degrees of freedom are excited.

H. Aref 1999, Turbulent statistical dynamics of a system of point ortices, in A. Gyr, W. Kinzelbach and A. Tsinober, (eds.) Fundamental roblematic issues in turbulence, Birkhäuser.

* - It is a well-known fact that under suitable conditions, which normally amount to a requirement that the kinematic viscosity ν be sufficiently small, some of this motions are such that the velocity at any given time and position in the fluid is not found to be the same when it is measured several times under seemingly identical conditions. In these motions the velocity takes random values which are not determined by the ostensible, or controllable, or 'macroscopic', data of the flow, although we believe that the *average* properties of the motion are determined uniquely by the data. Fluctuating motions of this kind are said to be turbulent.

G. K. Batchelor 1953, The theory of homogeneous turbulence, 1953, CUP, p.1.

* - Turbulence is a three-dimensional time-dependent motion in which vortex stretching causes velocity fluctuations to spread to all wavelengths between a minimum determined by viscous forces and a maximum determined by the boundary conditions of the flow. It is the usual state of fluid motion except at low Reynolds numbers.

P. Bradshaw 1972, An introduction to turbulence and its measurement, Pergamon, p. 17.

* - The only short but satisfactory answer to the question "what is turbulence?" is that it is the general-solution of the Navier Stokes equation.

P. Bradshaw 1972, The understanding and prediction of turbulent flow, Aeronaut. J., p. 406.

* - The distinguishing feature of turbulent flow is that its velocity field appears to be random and varies unpredictably. The flow does, however, satisfy a set of differential equations, the Navier Stokes equations, which are not random. This contrast is the source of much of what is interesting in turbulence theory.

A. J. Chorin 1975, Lectures on turbulence theory, Publish or Perish, Berkeley, p. 1.

* - Creation of small scale activity and dissipation, is the principle of turbulence. Classical fluid dynamical instabilities play a role of the fuel, vortex stretching is the engine, and viscous dissipation is the breaks.

P. Constantin 1994, Geometric statistics in turbulence, SIAM Review, **36***, p. 73.*

* - The next great era of awakening of human intellect may well produce a method of understanding the *qualitative* content of equations. Today we cannot. Today we cannot see that the water flow equations contain such thing as the barber pole structure of turbulence that one sees between rotating cylinders. Today, we cannot see whether Schrödinger's equation contains frogs, musical composers, or morality - or whether it does not. We cannot say whether something beyond it like God is needed, or not. And we can all hold strong opinions either way.

Richard P. Feynman 1963, The Feynman lectures on physics II, Addison-Wesley, p. 41-12.

* - Turbulence with its limit of self-excitation, with the characteristic hysteresis in its appearance and disappearance as the velocity of flow producing it is increased or reduced and the primary role of nonlinearity in its developed (stationary) state, is, in fact, a *self-oscillation*. Its specific features are determined by the fact that it is self-oscillation of a continuous medium, i.e., a system with an infinite number of degrees of freedom.

G. S. Gorelik, 1956 as quoted by M.I. Rabiniovich, Stochastic self-oscillations and turbulence, 1978, Sov. Phys. Uspekhi, **21***, p. 444.*

* - Das "Turbulenzproblem" der Hydrodynamik ist ein Problem der energetischen, nicht der dynamischen Stabilität.

W. Heisenberg 1923, Über Stabilität und Turbulenz von Flüssigkeitströmen, Ph.D. Thesis, p. 37.

* - The following definition of turbulence can thus be tentatively proposed and may contribute to avoiding the somewhat semantic discussion on this matter:

a) Firstly, a turbulent flow must be unpredictable, in the sense that a small uncertainty as to its knowledge at a given initial time will amplify so as to render impossible a precise deterministic prediction of it evolution; b)

Secondly, it has to satisfy the increased mixing property defined above; c) Thirdly, it must involve a wide range of spatial wave lengths.

M. Lesieur, 1997 Turbulence in fluids, Kluwer, p. 2.

* - Turbulence can be defined by a statement of impotence reminiscent of the second law of thermodynamics: flow at a sufficiently high Reynolds number cannot be decelerated to rest in a steady fashion. The deceleration always produces vorticity, and the resulting vortex interactions are apparently so sensitive to the initial conditions that the resulting flow pattern changes in time and usually in stochastic fashion.

H. W. Liepmann 1979, The rise and fall of ideas in turbulence, American Scientist, 67, p. 221.

* - A body of fluid is a mechanical system with an infinite number of degrees of freedom. It may therefore be expected to execute a rather random motion comparable to that of the molecules in a gas. If one regards such a chaotic motion as analyzed into harmonic components of various scales, one recognizes that frictional forces tend to dissipate the small scale oscillations and keep the motion more or less regular. Thus, when viscous forces are sufficiently strong, i.e. at sufficiently low Reynolds numbers, the motion will become laminar. On the other hand, at sufficiently high Reynolds numbers the motion will tend to become random fluctuating, even when external conditions are steady.

C. C. Lin and W. H. Reid 1963, Turbulent flow, Theoretical aspects, in: Handbuch der Physik, Band VIII/2, Springer, p. 438.

* - Perhaps a satisfactory definition would be an ensemble of nonperiodic solutions of the Navier-Stokes equations. Ensembles of solutions of simplified or otherwise modified forms of the Navier-Stokes equations will not qualify as turbulence; we shall instead regard them as models of turbulence.

E.N. Lorenz 1972, Investigating the predictability of turbulent motion, In: M. Rosenblatt and C. van Atta, editors, Statistical models and turbulence, Springer, p. 195.

* - We have therefore defined turbulence as random fluctuations of the thermodynamic characteristics of *vortex* flows, thereby distinguishing it at the outset from any kind of whatever random irrotational i.e., *potential* flows, ...

A. S. Monin 1978, On the nature of turbulence, Sov, Phys. Uspekhi, 21, p. 430.

* - Definition of Randomness

Of special interest to us here are the strange attractors, on which phase trajectories display the following properties of randomness :

(1) An extremely sensitive dependence on initial conditions, due to *exponential divergence* of trajectories which are initially close together (and leading to their unpredictability for initial conditions which are given with arbitrarily high (but finite) precision). (2) The everywhere-denseness at the attractor of almost all trajectories, i.e., their arbitrarily close approach to any of the attractor's points (which implies that they return infinitely often to the attractor), and the property that any initial nonequilibrium probability distribution (measure) over the phase space (or, more precisely, over the region of attraction of the strange attractor) reduces to some limiting equilibrium distribution at the attractor (an invariant measure). (3) The mixing property : For any (measurable) subsets A and B of the attractor, the probability after emerging from A of arrival at B is proportional after a long time of measure of B :

$$\lim_{t \to \infty} P\{(F^t A) \cap B\} = P(A)\, P(B)$$

where the symbol $\cap$ denotes set intersection. A consequence of the mixing property is the fact that the time-averaged value $< \Phi[u(t)] >$ of any function $\Phi(u)$ defined on the strange attractor is independent of the initial conditions $\mathbf{u}_0$ (for almost all $\mathbf{u}_0$) and that this average value coincides with the average $\bar{\bar{\Phi}}(u)$ over the invariant measure (*ergodicity*) :

$$< \Phi > \equiv \lim_{T \to \infty} T^{-1} \int_0^T \Phi[u(t)]dt = \int \Phi(u)Pd\mathbf{u} \equiv \bar{\bar{\Phi}}$$

A characteristic of the mixing property is a rather *rapid decay of the correlation functions as* $\tau \to \infty$:

$$B^{jl}(\tau) = < [u^j(t) - < u^j >][u^l(t + \tau) - < u^l >] >$$

which is to say *continuity* of their Fourier transforms with respect to τ, i.e., their *spectral functions.*

It appears expedient to have the term *turbulence* refer to the random evolution [in the sense of (1) - (3) above] of the flow of a (viscous) fluid which possesses *vorticity.* Stochastic *potential* flows of a fluid are by preference referred to as *random wave fields*, while for nonhydrodynamic systems one should preferably restrict oneself, where necessary, to the adjective *stochastic.*

Monin, A. S. 1978, Hydrodynamic instability, Sov.Phys.Uspekhi, 29, 856-857.

* - Turbulence is a phenomenon which sets in in a viscous fluid for small values of the viscosity coefficient ν (reckoning ν in significant units, that is, as the reciprocal Reynolds' number $1/Re$), hence its purest, limiting form

may be interpreted as the asymptotic, limiting behavior of a viscous fluid for $\nu \to 0$...

The circumstances described above made it very plausible that turbulence is a phenomenon of instability...

A complete theory of the general solutions of the Navier-Stokes equations are called for. ... nothing less than a thorough understanding of the system of all their solutions would seem to be adequate to elucidate the phenomenon of turbulence.

Turbulence proper is tied ... to 3-dimensionality.

John von Neumann, Recent theories of turbulence - A report to office of Naval Research, 1949, in: Collected works, vol. 6, pp. 439,441,448, 462, ed. A.H. Taub, Pergamon

* - One of the best definitions of turbulence is that it is a field of random chaotic vorticity.

P. G. Saffman 1981, Vortex interactions and coherent structures in turbulence, in: Transition and turbulence, ed. R. Meyer

* - ...the turbulence syndrome includes the following symptoms: The velocity field is such a complicated function of space and time that a statistical description is easier than a detailed description; it is essentially three-dimensional, in the sense that the dynamical mechanism responsible for it (the stretching of vorticity by velocity gradients) can only take place in three dimensions; it is essentially nonlinear and rotational, for the same reasons; a system of partial differential equations exists, relating the instantaneous velocity field to itself at every time and place.

R. W. Stewart, as quoted by J.L. Lumley 1972, Stochastic tools in turbulence, Academic Press, p. ix.

* - Turbulence is an irregular motion which, in general, makes its appearance in fluids, gaseous or liquid, when they flow past solid surfaces or even when neighboring streams of the same fluid flow past or over one another.

G. I. Taylor and Th. von Karman, in: von Karman, Th., Twenty-fifth Wilbur Wright memorial lecture - Turbulence, 1937, J. Roy. Aeronaut. Soc., 41, p. 1109.

* - In many cases, it is allowable to deal with a system of finite dimensions as a model of a continuous fluid and this is particularly the case at the stage of generation of "turbulence", at which only a limited number of degrees of freedoms of motion have been excited. The approximation of a fluid in terms of a model system of finite dimensions provides us with a powerful means of analysis and it is by this reason that the recent progress in the theory of "chaos" has enabled us to look straight at the fundamental

mechanism of "turbulence".

On the other hand, it is generally recognized that "turbulence" in its fully developed state has a *singular* structure in space and time and that the singularity is closely connected with the peculiar property of "turbulence" such as the nonzero viscous dissipation in the limit of vanishing viscosity. Such a singular behaviour of the fluid cannot be described correctly by means of a model system of finite dimensions which remains *regular* in the inviscid limit. Thus, in this restricted area, "chaos in fluids" covers only a part of "turbulent" phenomena.

T. Tatsumi 1984, Irregularity, regularity and singularity in turbulence, in: Turbulence and chaotic phenomena in fluids, ed. T. Tatsumi, Elsevier, p. 1.

* - Before 1970, I would not have dreamt of putting the words turbulence and predictability side by side, as in the title of this summer course. To me, turbulence was unpredictable by definition. Turbulence was the chaos that arises in fluids because of the innumerable instabilities associated with vortex stretching.

These days, I tend to think of turbulent flow as flow in which deterministic calculations become useless in a finite time interval.

H. Tennekes 1985, A comparative pathology of atmospheric turbulence in two and three dimensions, in: Turbulence and predictability in geophysical fluid dynamics and climate dynamics, eds. M. Ghil, R. Benzi and G. Parisi, North-Holland, p. 45.

* - Everyone who, at one time or another, has observed the efflux from a smokestack has some idea about the nature of turbulent flow. However, it is very difficult to give a precise definition of turbulence. All one can do is list some of the characteristics of turbulence flows:

Irregularity ... Diffusivity ... Large Reynolds numbers ... Three-dimensional vorticity fluctuations ... Dissipation ... Continuum ... Turbulent flows are flows ...

H. Tennekes and J. L. Lumley 1972, A first course to turbulence, MIT Press, pp. 1-3.

CHAPTER 11

APPENDIX B. ABOUT THE 'SNAGS' OF THE PROBLEM

$\diamond$ - *I had less difficulty in the discovery of the motion of heavenly bodies in spite of their astonishing distances, than in the investigation of the movement of flowing water before our very eyes.* Galileo as cited by R. Narasimha 1983 The turbulence problem: a survey, *J. Indian Inst. Sci,* **64(A)** Jan. p.1. (1-59)

$\diamond$ - *As a doctorate I proposed to Heisenberg no theme from Spectroscopy but the difficult problem of Turbulence, in the hope, that WENN IRGEN-DEINER (if anybody), would solve this problem. However, the problem is until now not solved.* A. Sommerfeld (1942), Scientia, Nov./Dez 1942.

$\diamond$ - *The universal similarity theory of the small-scale components of the motion stands out in this rather grey picture as a valuable contribution, of which an increasing number of applications is being made...* Batchelor G. K. (1962) The dynamics of homogeneous turbulence: introductory remarks, In A. Favre, editor, Mécanique de la Turbulence, *Colloques Interntionaux du CNRS, No. 108, (Marseille 28 août-2 septembre 1961), p. 96.*

$\diamond$ - *It remains to call attention to the chief outstanding difficulty (i.e. turbulence) of our subject. H. Lamb 1927 Hydrodynamics, p.651.*

$\diamond$ - *I am an old man now, and when I die and go to Heaven there are two matters on which I hope for enlightenment. One is quantum electrodynamics and the other is the turbulent motion of fluids. And about the former I am rather optimistic. Sir Horace Lamb 1932 as quoted by S. Goldstein 1969, ARFM, 1, 23.*

$\diamond$ - *I soon understood that there was little hope of developing a pure, closed theory, Selected works of A. N. Kolmogorov, I, ed. Tikhomirov, p. 487, Kluwer, 1991.*

$\diamond$ - *It is at this point that the study of turbulence does prove to be an exception: the applied physics involvement is almost completely absent. In view of the extraordinary practical importance of turbulence ..., this is quite astonishing. Yet the reason for such apparent neglect is easily found. Quite simply the fundamental problems of turbulence are still unresolved.* D. McComb 1990 The physics of turbulence, Oxford Univ. Press p. vii.

⋄ - *The entire experience with the subject indicates that the purely analytical approach is beset with difficulties, which at this moment are still prohibitive. The reason for this is probably as was indicated above: That our intuitive relationship to the subject is still to loose – not having succeeded at anything like deep mathematical penetration in any part of the subject, we are still quite disoriented as to the relevant factors, and as to the proper analytical machinery to be used. Under these conditions there might be some hope to "break the deadlock" by extensive , but well planned computational efforts. J. von Neumann (1949) Recent theories of turbulence"* - A report to Office of Naval Research. Collected works, **6** *(1963), 469, ed. Taub., A. H., Pergamon.*

⋄ - *...the absence of a sound theory is one of the most disturbing aspects of the turbulence syndrome. R. W. Stewart.*

⋄ - *Turbulence is the graveyard of theories. H. W. Liepmann as cited by S. J. Klein, in: Self-sustaining mechanisms of wall turbulence, ed. R. L. Panton, p.4, Comp. Mech Publ.*

⋄ - *... we should not altogether neglect the possibility that there is no such as thing 'turbulence'. That is to say, it is not meaningful to talk of the properties of a turbulent flow independently of the physical situation in which it arises. In searching for a theory of turbulence, perhaps we are looking for a chimera, P. G. Saffman 1978 Problems and progress in the theory of turbulence,* Lect Notes in Phys., **76** *(II), 276.*

⋄ - *... I just cannot think of anything where a genuine prediction for the dynamics of turbulent flow has been confirmed by an experiment. So we have a big vast empty field. P. G. Saffman (1991) in The Global Geometry of Turbulence, NATO ASI Ser.* **B 268**, *ed. J. Jimenez, p. 349, Plenum.*

⋄ - *Turbulence was probably invented by the Devil on the seventh day of Creation when the Good Lord wasn't looking. P. Bradshaw 1994* Experiments in Fluids, **16,** *203.*

⋄ - *... less is known about the fine scale turbulence ... than about the structure of atomic nuclei. Lack of basic knowledge about turbulence is holding back progress in fields as diverse as cosmology, meteorology, aero-*

nautics and biomechanics. *U. Frisch and S. Orszag 1990 Phys Today*, **43** , *32.*

$\diamond$ - ... *a fundamental theoretical understanding is still lacking. M. Nelkin,* *1994* Adv. Phys., **43**, *143.*

$\diamond$ - Turbulence is the last great unsolved problem of classical physics. *Remarks of this sort have been variously attributed to Sommerfeld, Einstein, and Feynman, although no one seems to know precise references, and searches of some likely sources have been unproductive. Of course, the allegation is a matter of fact, not much in need of support by a quotation from a distinguished author. However, it would be interesting to know when the matter was first recognised. P.J. Holmes G. Berkooz and J.L. Lumley 1996 'Turbulence, coherent structures, dynamical systems and symmetry, Cambridge University Press.*

CHAPTER 12

APPENDIX C. GLOSSARY OF ESSENTIAL FLUID MECHANICS

This appendix contains basic information on fluid mechanics, in general, and turbulent flows, in particular, with some specific relevant items. This includes flow kinematics, equations of motion, and some of their consequences for velocity derivatives, and basic relations for description of turbulent flows.

12.1. Kinematics

The evolution of vector, l_i, connecting two material points, $\mathbf{x}$ and $\mathbf{x} + \mathbf{l}$, follows the equation $\frac{Dl_i}{Dt} = u_i(\mathbf{x} + \mathbf{l}) - u_i(\mathbf{x})$. If the vector l_i is infinitesimal this equation becomes

$$\frac{Dl_i}{Dt} = l_j \frac{\partial u_i}{\partial x_j} = l_j s_{ij} + l_j a_{ij} = l_j s_{ij} + \frac{1}{2}\varepsilon_{ijk}\omega_j l_k, \qquad (C.1a)$$

or in vector notations

$$\frac{D\mathbf{l}}{Dt} = (\boldsymbol{\nabla} \cdot \mathbf{u})\mathbf{l} = \mathbf{s} \cdot \mathbf{l} + \frac{1}{2}\boldsymbol{\omega} \times \mathbf{l}, \qquad (C.1b)$$

where $s_{ij} = \frac{1}{2}\left(\frac{\partial u_i}{\partial x_j} + \frac{\partial u_j}{\partial x_i}\right)$, is the rate of strain tensor, $a_{ij} = \frac{1}{2}\left(\frac{\partial u_i}{\partial x_j} - \frac{\partial u_j}{\partial x_i}\right)$, is the rotation tensor, and $a_{ij} = -\frac{1}{2}\varepsilon_{ijk}\omega_k$, where $\omega_i = \varepsilon_{ijk}\frac{\partial u_k}{\partial x_j} = \varepsilon_{ijk}a_{kj}$, $\boldsymbol{\omega} = curl\,\mathbf{u}$ is the vorticity vector and $\frac{D}{Dt} = \frac{\partial}{\partial t} + u_j\frac{\partial}{\partial x_j}$.

The equivalent to (1) statement is that the velocity field in a small region surrounding the position $\mathbf{x}$ consists, to the first order in the linear dimensions of this region, of the superposition of a uniform translation with velocity $\mathbf{u}(\mathbf{x})$, a pure straining motion characterized by the rate-of-strain tensor s_{ij}, and a rigid-body rotation with an angular velocity $\frac{1}{2}\boldsymbol{\omega}$:

$$u_i(\mathbf{x} + \mathbf{dx}) = u_i(\mathbf{x}) + s_{ij}(\mathbf{x})dx_j + \frac{1}{2}\,\varepsilon_{ijk}\omega_j(\mathbf{x})dx_k, \qquad (C.1c)$$

The equation for infinitesimal area, identified by its vector normal, N_i, follows from the conservation of fluid volume, $N_i l_i$, i.e. $\frac{DN_i l_i}{Dt} = 0$ which together with (1) gives

$$\frac{DN_i}{Dt} = -N_k \frac{\partial u_k}{\partial x_i}, \qquad (C.2)$$

In compressible fluids this should be replaced by $\frac{D\rho N_i l_i}{Dt} = 0$ with the resulting equation $\frac{DN_i}{Dt} = -N_k \frac{\partial u_k}{\partial x_i} + N_i \frac{\partial u_k}{\partial x_k}$, where ρ is the fluid density.

Note that, generally, neither vector $\mathbf{l}$ is solenoidal, nor vector $\mathbf{N}$ is potential.

12.2. Dynamics

12.2.1. BASIC EQUATIONS AND THEIR CONSEQUENCES

In the sequel $\frac{D}{Dt} = \frac{\partial}{\partial t} + u_i \frac{\partial}{\partial x_i}$ is the material (Lagrangian) derivative.

Incompressibility

$$div\ \mathbf{u} \equiv \frac{\partial u_i}{\partial x_i} = 0 \qquad (C.4)$$

Euler Equations (EE)

$$\frac{Du_i}{Dt} = -\frac{\partial (p/\rho)}{\partial x_i}. \qquad (C.5)$$

Navier-Stokes equations (NSE)

$$\frac{Du_i}{Dt} = -\frac{\partial (p/\rho)}{\partial x_i} + \nu \nabla^2\ u_i + F_i. \qquad (C.6)$$

The Lamb's form of the NSE in vector notations

$$\frac{\partial \mathbf{u}}{\partial t} + \boldsymbol{\omega} \times \mathbf{u} = \boldsymbol{\nabla} \left(p/\rho + \frac{1}{2}u^2 \right) + \nu \nabla^2\ \mathbf{u} + \mathbf{F}. \qquad (C.7)$$

Three typical examples of real forces, $\mathbf{F}$, are as follows. The first one is the force due to buoyancy, $\mathbf{F}_b = \frac{\Delta \rho}{\rho_o}\mathbf{g}$, in fluids with density variations, represented by the difference, $\Delta \rho = \rho - \rho_o$, in respect with some reference density ρ_o, where $\mathbf{g}$ is the gravitational acceleration. The second example is the Coriolis force in rotating systems, $\mathbf{F}_c = -2\boldsymbol{\Omega}_{syst} \times \mathbf{u}$, where $\boldsymbol{\Omega}_{syst}$ is the angular velocity of the system. Finally, the third example is the electromagnetic force in electrically conductive fluids, $\mathbf{F}_{em} = \frac{1}{\rho}(\mathbf{j} \times \mathbf{B})$, where $\mathbf{j} = curl\mathbf{B} = \sigma(\mathbf{E} + \mathbf{u} \times \mathbf{B})$, $\mathbf{B}$ – magnetic field, $\mathbf{E}$ – electrical field.

It is important that the force $\mathbf{F}$ can be prescribed in any desirable way in direct numerical simulations of the Navier-Stokes equations. For example, one can realize the 'unrealistic' (quasi) isotropic turbulent flow by choosing an appropriate force.

Note that NSE are *integro*-differential equations, since the pressure field at some point in space is defined by the velocity in the whole flow domain, due to the nonlocality of the inverse Laplace operator, because $\nabla^2 p = \{\rho(\omega^2/2 - s^2)\} \equiv 2Q$, $s^2 \equiv s_{ij}s_{ij}$. In an unbounded fluid flow $p(\mathbf{x}) = -(2\pi)^{-1} \int Q(\mathbf{x})\frac{d\mathbf{y}}{|\mathbf{x}-\mathbf{y}|}$. This can be represented as a sum of local and nonlocal terms, $p(\mathbf{x},t) = -\frac{1}{3}u^2(\mathbf{x},t)+N(\mathbf{x},t)$, $N(\mathbf{x},t) = \frac{1}{4\pi} \int_{P.V.} \{3\frac{y_i y_j}{y^2} - \delta_{ij}\} R_{ij}(\mathbf{x} - \mathbf{y},t)\frac{d\mathbf{y}}{|y|^3}$, $R_{ij} = u_i u_j$, (Constantin and Fefferman, 1994). Here $\int_{P.V.}$ stands for the Cauchy's principal value.

Equation for the kinetic energy, $\frac{1}{2}u^2$,

$$\frac{D}{Dt}\left(\frac{1}{2}u^2\right) = -\frac{\partial}{\partial x_j}\left\{u_j\left(p/\rho + \frac{1}{2}u^2\right) - 2\nu u_i s_{ij}\right\} - 2\nu s_{ij}s_{ij} + u_i F_i. \quad (C.8)$$

Note that – as follows from the NSE in Lamb's form (C.7) – vorticity does not contribute *directly* to the local (i.e. without integration over the whole flow domain) energy balance/transfer, since $\mathbf{u}\cdot(\boldsymbol{\omega} \times \mathbf{u}) \equiv 0$.

Equation(s) for vorticity, ω_i,

$$\frac{D\omega_i}{Dt} = \omega_j s_{ij} + \nu\nabla^2\omega_i + \varepsilon_{ijk}\frac{\partial F_k}{\partial x_j}, \quad (C.9)$$

or in vector notations

$$\frac{D\boldsymbol{\omega}}{Dt} = (\boldsymbol{\omega} \cdot \boldsymbol{\nabla})\mathbf{u} + \nu\nabla^2\boldsymbol{\omega} + curl\mathbf{F}. \quad (C.10)$$

The vortex stretching vector $W_i \equiv \{(\boldsymbol{\omega} \cdot \boldsymbol{\nabla})\mathbf{u}\}_i = \omega_j\frac{\partial u_i}{\partial x_j} = \omega_j s_{ij}$ reflects the interaction between vorticity and rate of strain tensor and is responsible for stretching (compressing) and tilting of vorticity.

If the fluid density is not constant, the equation for vorticity becomes

$$\frac{D\boldsymbol{\omega}}{Dt} = (\boldsymbol{\omega} \cdot \boldsymbol{\nabla})\mathbf{u} + \boldsymbol{\omega}div\mathbf{u}+\frac{1}{\rho^2}\boldsymbol{\nabla}\rho \times \boldsymbol{\nabla}p + \nu\nabla^2\boldsymbol{\omega}+curl\mathbf{F}, \quad (C.11)$$

This equation can be rewritten for the potential vorticity, $\boldsymbol{\omega}/\rho$, as

$$\frac{D\boldsymbol{\omega}/\rho}{Dt} = \{(\boldsymbol{\omega}/\rho) \cdot \boldsymbol{\nabla}\}\mathbf{u}-\frac{1}{\rho}\boldsymbol{\nabla}\left(\frac{1}{\rho}\right) \times \boldsymbol{\nabla}p + \frac{\nu}{\rho}\nabla^2\boldsymbol{\omega}+\frac{1}{\rho}curl\mathbf{F}, \quad (C.12)$$

where use was made of mass conservation $div\mathbf{u} = -\frac{1}{\rho}\frac{D\rho}{Dt}$.

A useful equation for the dilatation, $\Theta = div\mathbf{u}$, is

$$\frac{D\Theta}{Dt} = \frac{1}{2}\omega^2 - s^2 - \frac{1}{\rho}\nabla^2 p + \frac{1}{\rho^2}\boldsymbol{\nabla}\rho \cdot \boldsymbol{\nabla}p + \nu\nabla^2\Theta + div\mathbf{F}, \quad (C.13)$$

Note, that the equation for vorticity is *integro*-differential, since the velocity field at some point in space is defined by the vorticity in the whole flow domain, for example, by the Biot-Savart law in case of the whole space,

$$u_i(\mathbf{x},t) = \int \alpha_{ij}(\mathbf{r})\omega_j(\mathbf{y},t)d\mathbf{y}, \ \alpha_{ij}(\mathbf{r}) = \frac{1}{4\pi}\varepsilon_{ijk}\frac{r_k}{r^3}, \ r_i = x_i - y_i, \qquad (C.14)$$

and consequently the rate of strain tensor

$$s_{ij}(\mathbf{x}) = \int_{P.V.} \beta_{ijk}(\mathbf{r})\omega_j(\mathbf{y},t)d^3\mathbf{y}, \ \beta_{ijk} = -\frac{3}{8\pi}\frac{\varepsilon_{ijl}r_l r_k + \varepsilon_{kjl}r_l r_i}{r^5}, \qquad (C.15)$$

where $\int_{P.V.}$ stands for the Cauchy principal value. A similar expression for ω_i can be obtained in terms of strain, s_{ij}, and both relations can be cast in a symmetric form using the rotation tensor $a_{ij} = \frac{1}{2}\left(\frac{\partial u_i}{\partial x_j} - \frac{\partial u_j}{\partial x_i}\right)$ (Ohkitani, 1994). It is important that this symmetry is only kinematic. Dynamically the two quantities are very different in many respects. For example, pressure is present in the equation for the rate of strain tensor, but it is absent in the equation for vorticity. Another difference is seen in the absence of viscosity. In this case vorticity is frozen in the fluid motion. There no such a property associated with the rate of strain.

In an arbitrary flow domain the velocity field is defined from a Poisson equation $\nabla^2\mathbf{u} = -curl\boldsymbol{\omega}$, with appropriate boundary conditions. This is true also of the rate of strain tensor: just like in the case of vorticity the whole flow field is determined entirely by the field of strain. This is seen from the equation $\nabla^2 u_i = 2\partial s_{ik}/\partial x_k$, which together with boundary conditions defines uniquely the velocity field. In particular, it is straightforward to obtain an analogue of the Biot-Savart law for the whole space under the same conditions the Biot-Savart law is valid

$$u_i(\mathbf{x},t) = \int \gamma_j(\mathbf{r})s_{ij}(\mathbf{y},t)d\mathbf{y}, \ \gamma_j(\mathbf{r}) = -\frac{1}{2\pi}\frac{r_j}{r^3}, \ r_i = x_i - y_i, \qquad (C.14')$$

We stress these simple purely kinematic relations with the emphasis that the velocity field is completely defined by the field of vorticity or strain. Thus it is the first and the simplest indication that the 'small' scales (represented by velocity derivatives, i.e. vorticity and strain) and the 'large' scales (represented by velocity) are not that separate for whatever Reynolds number. Note that the small scales, generally, are not 'integrated out' in (C.14) and (C.14') due to the singular nature of the kernel.

Equation for enstrophy $\frac{1}{2}\omega^2$

$$\frac{1}{2}\frac{D\omega^2}{Dt} = \omega_i\omega_j s_{ij} + \nu\omega_i\nabla^2\omega_i + \varepsilon_{ijk}\omega_i\frac{\partial F_k}{\partial x_j}. \qquad (C.16)$$

The term $\omega_i\omega_j s_{ij}$ is responsible for the enstrophy production.

The enstrophy production and its rate, $\alpha = \frac{\omega_i\omega_j s_{ij}}{\omega^2}$ can be expressed in terms of vorticity only for an infinite domain in the following nonlocal form (Constantin, 1994)

$$\alpha(\mathbf{x}) = \frac{3}{4\pi}\int_{P.V.} D\{\tilde{\mathbf{y}},\tilde{\boldsymbol{\omega}}(\mathbf{x}+\mathbf{y},t),\tilde{\boldsymbol{\omega}}(\mathbf{x},t)\}|\boldsymbol{\omega}(\mathbf{x}+\mathbf{y},t)|\frac{d\mathbf{y}}{|y|^3}, \qquad (C.15')$$

where $D\{\mathbf{a}_1,\mathbf{a}_2,\mathbf{a}_3\} = (\mathbf{a}_1\cdot\mathbf{a}_2)Det\{a_{ij}\}$ and $\tilde{\mathbf{a}} = \mathbf{a}/a$.
Equation for the rate of strain tensor, s_{ij}, (Yanitski, 1982),

$$\frac{Ds_{ij}}{Dt} = -s_{ik}s_{kj} - \frac{1}{4}(\omega_i\omega_j - \omega^2\delta_{ij}) - \frac{\partial^2 p}{\partial x_i \partial x_j} + \nu\nabla^2 s_{ij} + F_{ij}, \qquad (C.17)$$

where $F_{ij} = \left(\frac{\partial F_i}{\partial x_j} + \frac{\partial F_j}{\partial x_i}\right)$.
Equation for the total strain, $s_{ij}s_{ij} \equiv s^2$,

$$\frac{1}{2}\frac{Ds^2}{Dt} = -s_{ij}s_{jk}s_{ki} - \frac{1}{4}\omega_i\omega_j s_{ij} - s_{ij}\frac{\partial^2 p}{\partial x_i \partial x_j} + \nu s_{ij}\nabla^2 s_{ij} + s_{ij}F_{ij}. \quad (C.18)$$

Just like the term $\omega_i\omega_j s_{ij}$ in (16) is called enstrophy production the term $-s_{ij}s_{jk}s_{ki} - \frac{1}{4}\omega_i\omega_j s_{ij} - s_{ij}\frac{\partial^2 p}{\partial x_i \partial x_j}$ in (18) can be called as (inviscid) production of total strain. Note that the mean $\langle -s_{ij}s_{jk}s_{ki} - \frac{1}{4}\omega_i\omega_j s_{ij}\rangle = \frac{1}{2}\langle\omega_i\omega_j s_{ij}\rangle$ is strictly positive in homogeneous flows and comprises the generation of strain in such (incompressible) flows, since $s_{ij}\frac{\partial^2 p}{\partial x_i \partial x_j} = \frac{\partial}{\partial x_k}\{\cdots\}$ and therefore $\langle s_{ij}\frac{\partial^2 p}{\partial x_i \partial x_j}\rangle = 0$.

If the fluid density is not constant the equation (18) for s^2 contains two more terms in its RHS: $-\frac{1}{\rho}s_{ij}\frac{\partial^2 p}{\partial x_i \partial x_j} + \frac{s_{ij}}{2\rho^2}\left\{\frac{\partial\rho}{\partial x_i}\frac{\partial p}{\partial x_j} + \frac{\partial\rho}{\partial x_j}\frac{\partial p}{\partial x_i}\right\}$.

There is a following *qualitative* difference between the equation for the kinetic energy, $\frac{1}{2}u^2$, and the equations for the enstrophy, $\frac{1}{2}\omega^2$, and total strain, $\frac{1}{2}s^2$. Integrate these over a domain with homogeneous, periodic or/and other boundary conditions causing the surface integrals to vanish. The nonlinear terms do not contribute to the equation for energy due to their conservative nature - they can be written in the form $\frac{\partial}{\partial x_j}(\cdots)$.

$$\frac{d}{dt}\int_V \left(\frac{1}{2}u^2\right)dV = -\int_V \epsilon dV + \int_V u_i F_i dV. \qquad (C.19)$$

That is kinetic energy is an inviscid invariant, and therefore it is meaningful to speak of its dissipation[1].

On the contrary the nonlinear term, $\int \omega_i \omega_j s_{ij} dV$, corresponding to the net enstrophy production is (empirically) known to be strictly positive both from laboratory and numerical experiments (Taylor 1938, Betchov 1976, Tsinober et al., 1997), i.e. its contribution does not vanish

$$\frac{1}{2}\frac{d}{dt}\int_V \omega^2 dV = \int_V \omega_i\omega_j s_{ij} dV - \int_V \epsilon_\omega dV + \int_V \varepsilon_{ijk}\omega_i \frac{\partial F_k}{\partial x_j} dV. \qquad (C.20)$$

That is enstrophy is not an inviscid invariant and therefore the expression $\epsilon_\omega = -\nu\omega_i\nabla^2\omega_i$, as any other, cannot be termed as dissipation of enstrophy, since there is no way to find a unique expression for ϵ_ω (as in the case of ϵ): it can be written as $(curl\boldsymbol{\omega})^2$, $\frac{\partial\omega_i}{\partial x_j}\frac{\partial\omega_i}{\partial x_j}$ and many other forms all of them differing by terms in the form $\frac{\partial}{\partial x_j}(\cdots)$.

A similar equation can be written for the total strain

$$\frac{1}{2}\frac{d}{dt}\int_V s^2 dV = -\int_V (s_{ij}s_{jk}s_{ki} + \frac{1}{4}\omega_i\omega_j s_{ij}) dV - \int_V \epsilon_s dV + \int_V s_{ij}F_{ij} dV.$$
$$(C.21)$$

where $\epsilon_s = -\nu s_{ij}\nabla^2 s_{ij}$, and the $-\int(s_{ij}s_{jk}s_{ki}+\frac{1}{4}\omega_i\omega_j s_{ij})dV = \frac{1}{2}\int \omega_i\omega_j s_{ij}dV$, corresponding to the net strain production, is also positive.

[1]See Loitsyanskii (1966), also Serrin (1959), for a simple demonstration that dissipation rate of kinetic energy, i.e. the rate at which mechanical energy is turned *locally* into heat in incompressible Newtonian fluid, is $\epsilon = 2\nu s_{ij}s_{ij}$. It should be emphasized that this is the true local energy dissipation rate in incompressible flows. The stress is made here, since this expression can be written for example, as $\epsilon = 2\nu s_{ij}s_{ij} = \nu\omega_i\omega_i + \frac{\partial^2}{\partial x_i\partial x_j}\{u_i u_j\}$. That is with appropriate boundary conditions $2\nu\int s_{ij}s_{ij}dV = \nu\int \omega_i\omega_i dV$. Also for homogeneous turbulence $2\nu\langle s_{ij}s_{ij}\rangle = \nu\langle\omega_i\omega_i\rangle$, and at large Reynolds numbers $2\nu\langle s_{ij}s_{ij}\rangle \approx \nu\langle\omega_i\omega_i\rangle$. These are nothing more than *kinematic* relations. The *true physical causal* relation is between dissipation and strain both in Newtonian and non-Newtonian fluids. Therefore it is a misconception to associate dissipation directly with vorticity. Dissipation is a quantity essentially associated with strain.

For a viscous Newtonian compressible fluid

$$\epsilon_c = 2\nu\left(s_{ij} - \frac{1}{3}\delta_{ij}\frac{\partial u_k}{\partial x_k}\right)\left(s_{ij}s_{ij} - \frac{1}{3}\delta_{ij}\frac{\partial u_k}{\partial x_k}\right) + \frac{\zeta}{\rho}\left\{\frac{\partial u_k}{\partial x_k}\right\}^2$$

where ζ is the so called second or bulk viscosity. Even if $\zeta \approx 0$ (an assumption valid in many cases) compressibility effects reduce the dissipation.

It is customary to rearrange the terms in the expression for ϵ_c and to represent it as

$$\epsilon_c = \left(\frac{4}{3}\nu + \frac{\zeta}{\rho}\right)\left\{\frac{\partial u_k}{\partial x_k}\right\}^2 + \nu\omega_i\omega_i + \nu\frac{\partial}{\partial x_i}\left\{\frac{\partial u_i u_j}{\partial x_j} + u_j\frac{\partial u_i}{\partial x_j} - 3\frac{\partial}{\partial x_j}\left(u_j\frac{\partial u_k}{\partial x_k}\right)\cdots\right\}$$

with the first term called compressible or dilatational dissipation and the second term called solenoidal or incompressible dissipation (Lele, 1994; Friedrich and Bertolotti; 1997 and references therein). Though this is done for their means only from the physical point the latter is still misleading.

In summary, the nonlinearity does not contribute to the rate of change of energy, but only redistributes it in space. On the contrary, the nonlinearity makes a positive contribution to the rate of change of the total enstrophy and strain[2].

In two-dimensional flows the 'source' terms $\omega_i\omega_j s_{ij}$ and $s_{ij}s_{jk}s_{ki}$ vanish and both enstrophy and total strain are inviscid invariants. Moreover, vorticity, and consequently enstrophy (but not the total strain) is a pointwise Lagrangian inviscid invariant, i.e. vorticity of any infinitesimal material fluid element does not change. However, the quantity of the next level, the palinstrophy, $\boldsymbol{\xi}^2, \boldsymbol{\xi} = curl\boldsymbol{\omega}$, is not an inviscid invariant, since it obeys the equation

$$\frac{1}{2}\frac{D\boldsymbol{\xi}^2}{Dt} = \xi_i\xi_j s_{ij} + \nu\xi_i\nabla^2\xi_i. \qquad (C.22)$$

and its net production $\int \xi_i\xi_j s_{ij}dV$ is again a positive quantity.

Equivalently, the equation for ζ^2, where $\boldsymbol{\zeta} = \nabla\omega \ (= \partial\omega/\partial x_i)$ is the vorticity gradient, takes the form

$$\frac{1}{2}\frac{D\zeta^2}{Dt} = -\zeta_i\zeta_j s_{ij} + \nu\zeta_i\nabla^2\zeta_i, \qquad (C.22')$$

and the net production of ζ^2 $(-\int \zeta_i\zeta_j s_{ij}dV = \int \xi_i\xi_j s_{ij}dV)$ is a positive quantity as well.

---

[2]This has interesting implications in the context of stability theory. Namely, the Reynolds-Orr equation for the total energy of a disturbance, u_i, of an undisturbed shear flow U_i (assumed to be a solution of NSE) does not contain cubic terms in the disturbance (corresponding to the nonlinear terms in NSE)

$$\frac{d}{dt}\int_V \left(\frac{1}{2}u^2\right)dV = -\int u_iu_j\frac{\partial U_i}{\partial x_j}dV - \int_V \epsilon dV,$$

This means that the rate of change of the energy of the disturbance $E^{-1}dE/dt$ does not depend on the disturbance amplitude, i.e. in some sense, is the same for infinitesimal and finite amplitude disturbances. This was interpreted (see references in Henningson, 1996) in the sense that the disturbance energy produced by linear mechanisms is the only disturbance energy available. In contrast to the energy equation the corresponding equation for enstrophy (and a similar equation for strain)

$$\frac{d}{dt}\int_V \left(\frac{1}{2}\omega^2\right)dV = \int \left(-\omega_iu_j\frac{\partial\Omega_i}{\partial x_j} + \omega_i\omega_j S_{ij} + \omega_i s_{ij}\Omega_j + \omega_i\omega_j s_{ij}\right)dV - \int_V \varepsilon_\omega dV.$$

does contain the cubic term, $\omega_i\omega_j s_{ij}$ corresponding to the self-amplification of vorticity. Hence the rate of change of the enstrophy of the disturbance $\mathcal{E}_\omega^{-1}d\mathcal{E}_\omega/dt$ does depend on the disturbance amplitude, and is different for infinitesimal and finite amplitude disturbances. A similar statement is true of the total strain.

This shows the advantages of using velocity derivatives (vorticity and strain) in elucidating the essential aspects of physics.

12.2.2. SOME ADDITIONAL CONSEQUENCES FROM THE NSE AND INVARIANT QUANTITIES

Equation for enstrophy production $\omega_i\omega_j s_{ij}$[3]

$$\frac{D}{Dt}\omega_i\omega_j s_{ij} = W^2 - \omega_i\omega_j\frac{\partial^2 p}{\partial x_i\partial x_j} + VT, \qquad (C.23)$$

Equation for $s_{ij}s_{jk}s_{ki}$

$$\frac{D}{Dt}s_{ij}s_{jk}s_{ki} = -3\left\{s^4 + \frac{1}{4}(W^2 - s^2\omega^2) + s_{ik}s_{kj}\frac{\partial^2 p}{\partial x_i\partial x_j}\right\} + VT, \quad (C.24)$$

where $W^2 = \omega_j s_{ij}\omega_k s_{ik}$, $s^4 = s_{ik}s_{kj}s_{il}s_{lj}$ and VT stands for viscous terms.

Writing such equations allows us to identify in a natural way the dynamically significant geometrical invariant quantities and relations between such invariants of different order via dynamical equations, all of which are the consequence of the Navier-Stokes equations. Namely the quantities of the second order are ω^2 and s^2; the invariants of the third order are $\omega_i\omega_j s_{ij}$, $s_{ij}s_{jk}s_{ki}$ and $s_{ij}\frac{\partial^2 p}{\partial x_i\partial x_j}$. All of them appear in the equations for enstrophy (C.16) and total strain (C.18). The invariants of the fourth order are

$$\mathcal{I}_1 = s_{ik}s_{kj}s_{il}s_{lj}; \quad \mathcal{I}_2 = \omega^2 s^2; \quad \mathcal{I}_3 = \omega_i s_{ij}\omega_k s_{ik} \equiv W^2; \quad \mathcal{I}_4 = \omega^4, \quad (C.25)$$

The right hand side of the equations (23) and (24) contain three of the four invariants. The fourth invariant, $\mathcal{I}_4 = \omega^4$, appears in the equations for the higher order quantities. Thus the invariants $\mathcal{I}_1 - \mathcal{I}_4$ are *pointwise* quantities of dynamical significance. Siggia (1981) identified the means $\langle\mathcal{I}_i\rangle$, $i = 1-4$, in the kinematical context as invariants determining all the 105 terms of the tensor of the fourth rank $T^{(4)}_{i,j,k,l...} = \langle\frac{\partial u_i}{\partial x_j}\frac{\partial u_k}{\partial x_l}...\rangle$ for an isotropic velocity field.

An important aspect is that the equations (18), (23) and (24) contain three additional invariant quantities containing the pressure Hessian, $h_{ij} = \frac{\partial^2 p}{\partial x_i\partial x_j}$: one of third order, $s_{ij}\frac{\partial^2 p}{\partial x_i\partial x_j}$ and two of fourth order

$$\mathcal{I}_5 = \omega_i\omega_j\frac{\partial^2 p}{\partial x_i\partial x_j}; \quad \mathcal{I}_6 = s_{ik}s_{kj}\frac{\partial^2 p}{\partial x_i\partial x_j}, \qquad (C.26)$$

All the three reflect the nonlocal *dynamical* effects due to interaction of pressure Hessian with vorticity and strain.

---

[3]In the equations of this subsection the terms associated with the external force, F_i, are dropped.

Geometrical invariants remain unchanged under the full group of rotations (i.e. rotations plus reflections)[4] in contradistinction with other *noninvariant* combinations of velocity derivatives. For this reason the geometrical invariants are mostly appropriate for studying physical processes in (turbulent) fluid flows, their structure and universal properties (Tsinober, 1995, 1996a; and references therein).

Another way of choosing invariant quantities is to look directly at the invariants of the velocity gradient tensor $\frac{\partial u_i}{\partial x_j}$: the first invariant $- P = \frac{\partial u_i}{\partial x_i}$, vanishing for incompressible flow; the second invariant $- Q = \frac{1}{4}(\omega^2 - 2s^2)$, and the third invariant $R = -\frac{1}{3}(s_{ij}s_{jk}s_{ki} + \frac{3}{4}\omega_i\omega_j s_{ij})$, both written for incompressible flows (see Chacin and Cantwell, 2000; Martin et al., 1998; Ooi et al., 1999; and references therein). These invariants arise naturally as coefficients in the equation of the characteristic equation for the eigenvalues of $\frac{\partial u_i}{\partial x_j}$. It is convenient and useful way to study some of the local flow properties in the $R - Q$ plane. This analysis should be complemented by looking at the invariants of s_{ij} (e.g. its eigenvalues, Λ_i), the invariant quantities mentioned above as well as some other discussed below.

It is noteworthy that some of the invariant quantities allow useful geometrical interpretation. For example,

$$\omega_i\omega_j s_{ij} = \omega^2\Lambda_i\cos^2(\boldsymbol{\omega},\boldsymbol{\lambda_i}) \equiv \boldsymbol{\omega}\cdot\mathbf{W} = \omega W\cos(\boldsymbol{\omega},\mathbf{W}), \qquad (C.27)$$

where $\boldsymbol{\lambda}_i$ is the eigenframe of the rate of strain tensor, s_{ij}, and Λ_i are its eigenvalues. The vector $W_i = \omega_j s_{ij}$ is the vortex stretching vector. Another example, $\omega_i s_{ij}\omega_k s_{ik} \equiv W^2 = \omega^2\Lambda_i^2\cos^2(\boldsymbol{\omega},\boldsymbol{\lambda_i})$.

Note a useful relation

$$\cos(\boldsymbol{\omega},\mathbf{W}) = \frac{\Lambda_i\cos^2(\boldsymbol{\omega},\boldsymbol{\lambda_i})}{\{\Lambda_i^2\cos^2(\boldsymbol{\omega},\boldsymbol{\lambda_i})\}^{1/2}}, \qquad (C.28)$$

which shows that $\cos(\boldsymbol{\omega},\mathbf{W})$ is independent both of the magnitude of vorticity and total strain. As expected it depends only on geometrical properties of the velocity gradients: the mutual orientation of vorticity vector, $\boldsymbol{\omega}$, and the eigenframe, $\boldsymbol{\lambda}_i$, of the strain tensor, s_{ij}, and of the shape of the latter, e.g. the ratios Λ_2/Λ_1 and Λ_3/Λ_1.

The cosines $\cos(\boldsymbol{\omega},\boldsymbol{\lambda_i})$ and $\cos(\boldsymbol{\omega},\mathbf{W})$ are examples of invariant quantities of the kind discussed here. As we shall see in chapters 6-9 they allow us to study an important aspect of the essential dynamics of 3-D turbulence associated with the geometrical properties of the field of velocity derivatives.

[4]They are also invariant under space/time translations and the Galilean transformation (see below).

Inviscid invariants

Here by invariants are meant quantities which remain unchanged under the inviscid dynamics, i.e. invariants of the Euler equations.

Kinetic energy

$$\frac{dE}{dt} = - \oint_A pu_n dA - \int_V \epsilon dV.$$

Helicity

$$\frac{d\mathcal{H}}{dt} = \oint_A \omega_n C\, dA - 2\nu \int_V \boldsymbol{\omega} \cdot curl\ \boldsymbol{\omega}\, dV,$$

where, $E = \int_V \frac{1}{2}u^2 dV$, $\mathcal{H} = \int_V h dV$, $h = \boldsymbol{\omega} \cdot \mathbf{u}$, $C = (1/2)\mathbf{u}^2 - p/\rho$, and A is the surface bounding some volume V, p - pressure, ρ - fluid density, and the fluid is assumed incompressible.

Thus in the absence of viscosity and external forces, and with vanishing surface integrals, both kinetic energy and helicity of a fluid (and space) volume are conserved. The essential difference is that - unlike kinetic energy - helicity is a non-positively defined quantity. This makes it more difficult to use this important quantity (see Moffatt and Tsinober, 1992; and Droegemeier et al., 1993). Another difference is that it is not invariant under reflections - it is pseudoscalar, so that it changes its sign under reflections.

We did not mention here other invariants such as those linear in velocity (e.g. angular momentum) and several others (see Monin and Yaglom, 1971, 1975), as their application to turbulent flows is quite limited and is beyond the scope of this small book.

12.2.3. SYMMETRIES OF EULER AND NAVIER-STOKES EQUATIONS

The Euler and the Navier Stokes equations are invariant under the following transformations (see Frisch, 1995; also Oberlack, 1999).

- translations in space and time
- full group of rotation including rotations and reflections
- Galilean transformation $\mathbf{u}(\mathbf{x}, t) \Rightarrow \mathbf{u}(\mathbf{x} - \mathbf{U}t, t) + \mathbf{U}$, $\mathbf{U} = \mathbf{const}$

The Euler equation is in addition invariant under

- time reversal $t \Rightarrow -t$, $\mathbf{u} \Rightarrow -\mathbf{u}$, $p \Rightarrow p$.
- scaling transformation

$$\mathbf{r} \Rightarrow \lambda\mathbf{r}; \quad t \Rightarrow \lambda^{1-h}t; \quad \mathbf{u} \Rightarrow \lambda^h\mathbf{u}; \quad p \Rightarrow \lambda^{2h}p, \quad \lambda > 0, \qquad (C.29)$$

for any h.

The Navier-Stokes equations obey the scaling transformation for $h = -1$ only. However, it is a common belief that it *may be justified at very high Reynolds number...that there are infinitely many scaling groups, labeled by their scaling exponent h, which can be any real number, i.e. in the inviscid limit, the Navier-Stokes equation is invariant under infinitely many scaling groups, labeled by an arbitrary real scaling exponent h* (Frisch, 1995, p. 18, 144) just like in the case of Euler equation. However, it is not at all clear why one can ignore the singular nature of the limit $Re \to \infty$ $(\nu \to 0)$ when handling the issue of scaling exponents and/or related matters.

12.3. Passive objects

12.3.1. PASSIVE SCALARS

The behaviour of a passive scalar with concentration, θ, is described by the (deceptively simple, but still linear) advection – diffusion equation[5]

$$\frac{D\theta}{Dt} = \mathcal{D}\nabla^2\theta + \Phi, \qquad (C.30)$$

where Φ is the external source/forcing[6]. The corresponding 'energy' equations are

$$\frac{D}{Dt}\left(\frac{1}{2}\theta^2\right) = -\mathcal{D}G^2 + \nabla^2\left(\frac{1}{2}\theta^2\right) + \Phi\theta, \qquad (C.31)$$

$$\frac{d}{dt}\int_V\left(\frac{1}{2}\theta^2\right)dV = -\mathcal{D}\int_V G^2(t)dV + \int_V \Phi\theta dV, \qquad (C.32)$$

where $G^2 = G_iG_i, G_i = \frac{\partial\theta}{\partial x_i}$ and in the last equation the surface integrals are assumed vanishing.

12.3.2. PASSIVE VECTORS

There are roughly two kinds of passive vectors.

Gradient of a Passive Scalar
It is a passive vector, $G_i = \frac{\partial\theta}{\partial x_i}$, governed by the equation

$$\frac{\partial G_i}{\partial t} + u_k\frac{\partial G_i}{\partial x_k} = -G_k\frac{\partial u_k}{\partial x_i} + \mathcal{D}\nabla^2 G_i + \frac{\partial\Phi}{\partial x_k}, \qquad (C.33)$$

---

[5]In compressible fluids $\frac{\partial\theta}{\partial t} + u_k\frac{\partial\theta}{\partial x_k} + \theta\frac{\partial u_k}{\partial x_k} = \mathcal{D}\nabla^2\theta$

[6]The external forcing can be roughly of two kinds. In the case of homogenous and isotropic flow, the only way to sustain a statistically stationary state is to apply an isotropic forcing in the RHS of the corresponding equations. In the case of homogeneous flow, the forcing term may have its origin in the mean gradient. For example, one of the simplest cases - which is homogeneous, but not isotropic, - is when a constant mean gradient, $\mathbf{g}$, is imposed on the scalar, i.e. $\Theta = \theta + \mathbf{g}\cdot\mathbf{x}$, and $\frac{\partial\theta}{\partial t} + u_k\frac{\partial\theta}{\partial x_k} = \chi\nabla^2\theta - u_kg_k$, so that the forcing is dependent on the velocity field.

with the energy equations in the form

$$\frac{D}{Dt}\left(\frac{1}{2}G^2\right) = -G_iG_k s_{ik} - \epsilon_G + \nabla^2\left(\frac{1}{2}G^2\right) + \frac{\partial\Phi}{\partial x_i}G_i, \qquad (C.34)$$

$$\frac{d}{dt}\int_V\left(\frac{1}{2}G^2\right)dV = -\int_V G_iG_k s_{ik}dV - \int_V \epsilon_G dV + \int_V \frac{\partial\Phi}{\partial x_i}G_i dV, \quad (C.35)$$

where $\epsilon_G = \mathcal{D}\frac{\partial G_i}{\partial x_k}\frac{\partial G_i}{\partial x_k}$. In the absence of diffusivity, $\mathcal{D} = 0$, and external forcing the vector G_i is proportional to the (infinitesimal) vector representing a material surface element (its normal), N_i, thus satisfying the equation

$$\frac{\partial N_i}{\partial t} + u_k\frac{\partial N_i}{\partial x_k} = -N_k\frac{\partial u_k}{\partial x_i}, \qquad (C.2)$$

'Frozen' Passive Vectors
Such vectors obey the equation[7]

$$\frac{\partial B_i}{\partial t} + u_k\frac{\partial B_i}{\partial x_k} = B_k\frac{\partial u_i}{\partial x_k} + \chi\nabla^2 B_i + F_i, \qquad (C.36)$$

with an 'energy' equation in the form

$$\frac{D}{Dt}\left(\frac{1}{2}B^2\right) = B_iB_k s_{ik} - \epsilon_B + \chi\nabla^2\left(\frac{1}{2}B^2\right) + F_iB_i, \qquad (C.37)$$

$$\frac{d}{dt}\int_V\left(\frac{1}{2}B^2\right)dV = \int_V B_iB_k s_{ik}dV - \int_V \epsilon_B dV + \int_V F_iB_i dV, \qquad (C.38)$$

where $\epsilon_B = \chi\frac{\partial B_i}{\partial x_k}\frac{\partial B_i}{\partial x_k}$.

Again such a vector, in the absence of molecular diffusive effects, $\chi = 0$, and external forcing is frozen into the fluid motion, i.e. it's lines consist of the same material particles during the evolution, and it is proportional to the (infinitesimal) material line elements, l_i, thus satisfying the equation[8]

$$\frac{\partial l_i}{\partial t} + u_k\frac{\partial l_i}{\partial x_k} = l_k\frac{\partial u_i}{\partial x_k}. \qquad (C.1)$$

Just like in the dynamical problem the quantities $G_iG_k s_{ik}$, $N_iN_k s_{ik}$, $B_iB_k s_{ik}$, $l_il_k s_{ik}$; $\cos(\mathbf{G}, \boldsymbol{\lambda_i})$, $\cos(\mathbf{N}, \boldsymbol{\lambda_i})$, $\cos(\mathbf{B}, \boldsymbol{\lambda_i})$, $\cos(\mathbf{l}, \boldsymbol{\lambda_i})$; $\cos(\mathbf{G}, \mathbf{W}^G)$, $\cos(\mathbf{G}, \mathbf{W}^N)$,

---

[7]Provided that $\mathbf{B}$ is solenoidal, for example, magnetic field. Otherwise an additional term $u_i\frac{\partial B_k}{\partial x_k}$ enters the RHS with one more term, $-B_i\frac{\partial u_k}{\partial x_k}$, in the case of compressible fluids.

[8]The equation for vorticity in an inviscid fluid has a *similar* form $\frac{\partial\omega_i}{\partial t} + u_k\frac{\partial\omega_i}{\partial x_k} = \omega_k\frac{\partial u_i}{\partial x_k}$. Moreover, vorticity is also frozen in the fluid motion. However, since $\omega_i = \epsilon_{ijk}\frac{\partial u_j}{\partial x_k}$, vorticity is *not* a passive vector – it 'reacts back', see Chapter 6.

Figure 12.1.   Studies of an Old Man Seated and of Swirling Water, Leonardo da Vinci, Windsor, RL, No. 12,579r, The Royal Collection ©2001, Her Majesty Queen Elizabeth II.

$\cos(\mathbf{B}, \mathbf{W}^B)$, $\cos(\mathbf{l}, \mathbf{W}^l)$ all are geometrical invariants in the sense as discussed above with $W_i^G = -G_j s_{ij}, W_i^N = -N_j s_{ij}, W_i^B = B_j s_{ij}, W_i^l = l_j s_{ij}$.

12.4. Some basic relations for the statistical description of turbulent flows

From now on (as in the main text) capital letters will be used for the mean/average quantities (if not indicated otherwise), whereas the low case letters – for the fluctuations, i.e. the instantaneous value of some quantity, $\tilde{z}$, is equal to $Z + z$. The mean/average in some sense (ensemble, time, space) will be denoted by $\langle \cdots \rangle$, i.e. $\langle \tilde{z} \rangle = \langle Z + z \rangle = Z$, $\langle z \rangle = 0$, and where necessary specified. This is called Reynolds decomposition, which was described in words by Leonardo da Vinci on one of his drawing, Figure C.1 (Richter, 1970): *Observe the motion of the surface of the water which resembles that of hair, which has two motions, of which one depends on the weight of the hair, the other on direction of curls; thus the water forms eddying whirlpools, one part of which is due the impetus of the principal current and the other of incidental motion and return flow.*

12.4.1. SCALING, SCALES AND RELATED MATTERS

> *Scaling laws are at the heart of turbulence research.*
> (Tennekes and Lumley, 1972).
> *The wonderful thing about scaling is that you can get everything right without understanding anything.*
> (Kraichnan as cited by Kadanoff, 1990).

There is no contradiction between the two views[9], since the first is mainly about the parameters determining the order of magnitude of some quantity of interest, whereas the second concerns primarily the scaling in the sense of power laws and their scaling exponents. Both are closely related to the use of dimensional analysis, similarity and related things.[10]

However,

...it is clear that if a result can be derived by dimensional analysis alone... then it can be derived by almost any theory, right or wrong, which is dimensionally correct and uses the right variables. (Bradshaw, 1994). That is the correspondence with the experimental results may occur for the wrong reasons as happens from time to time in the field of turbulence.

Use of similarity and dimensional analysis in turbulence is more an art than science. (Ya.B. Zeldovich, 1971 as quoted by G.S. Golitsyn, 1999, private communication).

In other words a prerequisite for the use of dimensional analysis, similarity and symmetries (group theoretical methods) in turbulence is at least some minimal understanding of the basic physics of turbulent flows.

Scales

> *What is meant by "scale" is hard to define precisely.*
> (Chorin, 1994).
> *Fully turbulent flows consist of a wide range of scales, which are classified somewhat loosely as either large or small scales.*
> (Sreenivasan and Antonia, 1997).
> *An eddy eludes a precise definition, but it is conceived to be turbulent motion, localized within a region of size ℓ.*
> (Pope, 2000).

It is rather common to admit that the term 'scale' is not a well defined thing. However, it is much more common to use the term scale(s) in a great variety of contexts and meanings in turbulence. For example, it is frequently used in physical space and in Fourier space without much distinction, though a narrow band in Fourier space involves a broad range

---

[9]There is a third view: *It is increasingly clear that deterministic chaos and universal scaling theories can explain everything* (Normal, 1993).

[10]See Corrsin (1953) for a short and very instructive exposition of dimensional analysis and similarity in the fluid dynamical contexts including turbulence.

of scales in physical space and vice versa due to the integral nature of the Fourier transform. This ambiguity of language is one (among many) of the reflections of the inherent problems in turbulence 'theories'[11]. One of the main difficulties is due to the nonlinearity and nonlocality of the turbulence phenomenon. That is different 'scales' are not so separated as usually assumed, and therefore the division of turbulent flow into large and small scales is quite problematic, though useful from the technical point. Speaking about interaction, dependence, and coupling of scales it is meant that some quantities 'residing' on these scales are involved and characterize such interaction. Therefore, it is important to specify these dynamically relevant quantities, which adequately represent the 'scales'. We will use the term scale mainly in two meanings: i - in its simplest direct geometrical meaning in the physical space and in a similar way in time, and ii - implying the quantities representing these scales in some sense. For instance, velocity (fluctuations) are representative of large scales, whereas the field of velocity derivatives (vorticity and strain) is appropriate to represent the small scales (chapter 5). Note that using such a 'definition' of small scales does not specify the small scales uniquely: the small scales associated with vorticity are different from the scales related to strain.

Some characteristic statistical scales of turbulent flows
There are several useful scales used in turbulence, which belong to its simplest statistical characteristics, i.e. they are not some specific scales, but they are statistically defined quantities. Some of these scales are given below, others, more specific appear in the main text.
Kolmogorov scales. The Kolmogorov length scale is defined on dimensional grounds as

$$\eta = (\nu^3/\epsilon)^{1/4}, \tag{C.39}$$

where ν is the fluid kinematic viscosity and ϵ is the mean rate of energy pumping into the system, which is equal to the mean dissipation in statistically stationary flows (see below). Similarly, the corresponding time, velocity and acceleration scales are defined as $\tau_\eta = (\nu/\epsilon)^{1/2}$, $u_\eta = (\nu\epsilon)^{1/4}$ and $a = \epsilon^{3/2}\nu^{-1/2}$. The Kolmogorov scale, η, can be defined in various ways. For example, imposing the condition $Re = \frac{u_\eta \eta}{\nu} \approx 1$ and writing $\epsilon \approx \nu(u_\eta/\eta)^2$ with $u_\eta \approx \nu/\eta$ one arrives again at (39). It is known that adequate space/time resolution (in laboratory, field or numerical experiments) is achieved if the smallest resolved scales are of the order of the space/time Kolmogorov scales as defined above. For instance, η is considered as the smallest spatial relevant scale in turbulence. This, however, is not obvious,

---

[11]The ambiguity of Fourier decomposition was addressed by Liepmann, 1962; and Tennekes, 1976; see also Lohse and Müller-Groeling, 1996.

since the instantaneous dissipation ε is not narrow banded around its mean ϵ, but is distributed with a rather long tail, so that values as large as, $10^2\epsilon$ are not that rare. This corresponds to scales more than three times smaller than η. It is due to the rather weak dependence of η on ϵ, $\eta \sim \epsilon^{1/4}$, that the resolution as mentioned above appears to be reasonably adequate.

Integral scales. These are in some sense the largest relevant scales of the system. Regions separated by scales much larger than the integral ones both in space and time do not 'know' much about each other. For a homogeneous turbulent field this is defined, e.g., as

$$\mathcal{L} = \frac{1}{R_{uu}(0)} \int R_{uu}(\mathbf{r})d\mathbf{r}, \qquad (C.40)$$

where $R_{uu}(\mathbf{r}) = \langle u_r(\mathbf{x}+\mathbf{r},t)u_r(\mathbf{x},t)\rangle$ is the longitudinal correlation function of the velocity component along $\mathbf{r}$. An integral scale which is associated with the transverse velocity correlation is of the same order of magnitude. Similarly the integral time scale $\mathcal{T}$ is defined via the analogous time correlation. One of the definitions of the integral velocity scale is $\mathcal{U} = \frac{1}{\mathcal{L}^3} \int R_{uu}(\mathbf{r})d\mathbf{r}$, which is of the order of the variance of velocity fluctuations $R_{uu}(0) = \langle u_r^2 \rangle$, and the integral time scale $\mathcal{T} \sim \mathcal{L}/\mathcal{U}$. For practical purposes $\mathcal{L}$ and $\mathcal{T}$ are usually taken to be of the order of (but smaller than) the external scales of the system.

Taylor microscale. The Taylor microscale, λ, is defined from the relation[12]

$$\epsilon \sim \nu \frac{\mathcal{U}^2}{\lambda^2}, \qquad (C.41)$$

where $\mathcal{U}$ is the integral scale of turbulent velocity fluctuations. That is, λ, a mixed hybrid scale, since $\mathcal{U}$ is an integral scale and ϵ occurs mostly in much smaller scales. The Taylor microscale, λ, is not the smallest relevant scale in turbulent flow, and it is different from the Kolmogorov scale. Though, λ is not a physically representative length scale it is used as a convenient length scale in evaluating the Reynolds number associated with the field of velocity derivatives.

Relation between the statistical scales. This is obtained using the estimate $\epsilon \sim \mathcal{U}^3/\mathcal{L}$, Taylor (1935). It follows, that

$$\frac{\eta}{\mathcal{L}} \sim Re^{-3/4}, \ \frac{\eta}{\lambda} \sim Re^{-1/2}; \ \frac{\lambda}{\mathcal{L}} \sim Re^{-1/2};$$

---

[12]This scale was originally defined by Taylor (1935) as $\epsilon \sim 15\nu \langle u_1^2 \rangle /\lambda^2$ where u_1 is the x_1- component of velocity fluctuations. The motivation for the coefficient 15 is that in isotropic turbulence $\epsilon = 15\nu \langle (\partial u_1/\partial x_1)^2 \rangle$, so that $\lambda^2 = \langle u_1^2 \rangle / \langle (\partial u_1/\partial x_1)^2 \rangle$ – the most frequently used definition. The Taylor microscale is associated with the correlation function $R_{uu}(\mathbf{r})$ and can be defined as $\lambda_T^2 = -R_{uu}(0)/2R''(0)$. It is a technical matter to show that $\lambda_T = 2\lambda$ in (41).

$$\tau_\eta = \tau_\lambda = \left(\frac{\nu}{\epsilon}\right)^{1/2} \sim \mathcal{T} Re^{-1/2}, \qquad (C.42)$$

These relations allow to estimate the number of degrees of freedom (Landau and Lifshits, 1959; first (!) edition) as $(\mathcal{L}/\eta)^3 \sim Re^{9/4}$, see the first (!) edition of Landau and Lifshits (1959), since this estimate was removed from the subsequent editions.

12.4.2. REYNOLDS AVERAGED NAVIER-STOKES EQUATIONS AND RELATED

Introducing the quantities decomposed into a sum of averages and fluctuations into the Navier-Stokes equations, and averaging results in
The Reynolds averaged Navier-Stokes equations (RANS) for the mean flow[13]

$$\frac{D_U U_i}{Dt} = \frac{\partial}{\partial x_j}\left(-\frac{1}{\rho}P\delta_{ij} + 2\nu S_{ij} - \langle u_i u_j\rangle\right), \qquad (C.43)$$

here $-\rho\langle u_i u_j\rangle = \tau_{ij}^{\text{Reynolds}}$ is the Reynolds stress tensor and $\frac{D_U}{Dt} = \frac{\partial}{\partial t} + U_j\frac{\partial}{\partial x_j}$. Substracting this equations from the NSE gives an equation for the field of fluctuations

$$\frac{\partial u_i}{\partial t} + U_j\frac{\partial u_i}{\partial x_j} + u_j\frac{\partial U_i}{\partial x_j} + \frac{\partial}{\partial x_j}\{u_i u_j - \langle u_i u_j\rangle\} = \frac{\partial}{\partial x_j}\{-\frac{1}{\rho}p\delta_{ij} + 2\nu s_{ij}\}. \quad (C.44)$$

Using a similar equation for u_j one can obtain an equation for the instantaneous tensor $u_i u_j$, which after averaging becomes
The equation for the Reynolds stress tensor [14]

$$\frac{D_U\langle u_i u_j\rangle}{Dt} + \langle u_j u_k\rangle\frac{\partial U_i}{\partial x_k} + \langle u_i u_k\rangle\frac{\partial U_j}{\partial x_k} =$$

$$-\frac{\partial\langle u_i u_j u_k\rangle}{\partial x_k} - \frac{1}{\rho}\left\{\frac{\partial}{\partial x_i}\langle pu_j\rangle + \frac{\partial}{\partial x_j}\langle pu_i\rangle\right\} + \frac{2}{\rho}\langle ps_{ij}\rangle +$$

$$2\nu\frac{\partial}{\partial x_k}\{\langle u_j s_{ik}\rangle + \langle u_i s_{jk}\rangle\} - 2\nu\left\{\langle s_{ik}\frac{\partial u_j}{\partial x_k}\rangle + \langle s_{jk}\frac{\partial u_i}{\partial x_k}\rangle\right\}. \qquad (C.45)$$

This equation contains third order quantities (in the second line). This process can be continued producing the (infinite) Keller-Friedman (1925) chain

---

[13]We subsequently dropped the external force. It is explicitly noted when the force is present.

[14]Similar equations can be written for *two point* correlations in space, such as $\langle u_i(\mathbf{x})u_j(\mathbf{x}+\mathbf{r})\rangle$, and in space/time $\langle u_i(\mathbf{x};t)u_j(\mathbf{x}+\mathbf{r};t+t')\rangle$ (see Hinze, 1975; Favre et al., 1976). Such quantities are particularly useful in the case of homogeneous and isotropic turbulence.

of equations. Each finite set of this chain has more unknowns than equations. Roughly, this comprises the essence of the famous 'closure' problem; see Kraichnan (1962b) for a detailed exposition of the closure problem. On the simplest level one tries to introduce some (statistical) hypotheses on the relation between the Reynolds stress tensor $\langle u_i u_j \rangle$ and the mean flow U_i. In order to do this in an intelligent way, one has to understand - at least qualitatively - the essential physical processes of turbulent flows under consideration. It should be emphasized, that in the strict sense such a relation does not exist due to the nonlocal nature relation between the mean flow and the fluctuations: it is not a function, but a functional as can be seen from the equations (C.43), (C.44) and a similar equation for u_j, see section 6.6.

The equation for the kinetic energy of the mean flow $E_T = \frac{1}{2} U_i U_i$

$$\frac{D_U E_T}{Dt} = \frac{\partial}{\partial x_j} \left\{ -\frac{1}{\rho} P U_j + 2\nu U_i S_{ij} - \langle u_i u_j \rangle U_i \right\} + \langle u_i u_j \rangle S_{ij} + 2\nu S^2. \quad (C.46)$$

The equation for turbulent kinetic energy $e_T = \frac{1}{2} u_i u_i$

Instantaneous

$$\frac{D_U e_T}{Dt} = u_i \frac{\partial \langle u_i u_j \rangle}{\partial x_j} - \frac{\partial}{\partial x_j} \left\{ u_j e_T + \frac{1}{\rho} p u_j - 2\nu u_i s_{ij} \right\} - u_i u_j S_{ij} - 2\nu s^2. \quad (C.47)$$

Mean

$$\frac{D_U \langle e_T \rangle}{Dt} = -\frac{\partial}{\partial x_j} \left\{ \langle u_j e_T \rangle - 2\nu \langle u_i s_{ij} \rangle + \frac{1}{\rho} \langle p u_j \rangle \right\} - \langle u_i u_j \rangle S_{ij} - 2\nu \langle s^2 \rangle. \quad (C.48)$$

The total turbulent energy balance for the whole flow domain is

$$\frac{d\mathcal{E}_T}{dt} = \mathcal{P} - \mathcal{D} \quad (C.49)$$

where $\mathcal{E}_T = \int e_T dV$ - is the total kinetic energy of turbulent fluctuations, $\mathcal{P} = \int - <u_i u_j> S_{ij} dV$ - is the total rate of production/ destruction of energy turbulent fluctuations by the mean strain (velocity gradients), and $\mathcal{D} = 2\nu \int s_{ij} s_{ij} dV$, - is the total rate of dissipation (simply dissipation) of energy of turbulent fluctuations by viscosity. If the flow is statistically stationary, the dissipation equals production $\mathcal{P} = \mathcal{D}$, i.e. $\mathcal{P} > 0$.

If there is no mean shear (more precisely mean strain) to supply the energy to the field of fluctuations, the statistically stationary state is still possible in the presence of some external forcing, as it happens in DNS of NSE. Then the energy balance equation takes the form $\mathcal{W}_F = \mathcal{D}$, where $\mathcal{W}_F = \int U_i F_i dV$ is the total rate of production of energy of turbulent fluctuations by the external forces. It is noteworthy that, in the presence of some energy supply other than the mean strain, the total rate of production/destruction of energy turbulent fluctuations by the mean strain, $\mathcal{P}$, does not have to be positive even in the case of a statistically stationary turbulent flow, since in this case the balance is $\mathcal{W}_F + \mathcal{P} - \mathcal{D} = 0$. For example, there is some evidence that in this way the fluctuative motions produced by the supply of solar energy to the Earth atmosphere are feeding such a 'mean' flow as the famous jet stream (see chapter 8).

A similar procedure can be applied to the vorticity equation giving
The equation for the enstrophy of the mean flow $\frac{1}{2}\Omega^2$

$$\frac{1}{2}\frac{D_U\Omega^2}{Dt} = -\frac{\partial}{\partial x_j}\{\Omega_j\langle\omega_i u_j\rangle\} + \langle u_j\omega_i\rangle\frac{\partial\Omega_i}{\partial x_j} + \Omega_i\Omega_j S_{ij} + \nu\Omega_i\nabla^2\Omega_i. \quad (C.50)$$

The equation for the mean enstrophy of the turbulent fluctuations $\frac{1}{2}\langle\omega^2\rangle$

$$\frac{1}{2}\frac{D_U\langle\omega^2\rangle}{Dt} = -\langle u_j\omega_i\rangle\frac{\partial\Omega_i}{\partial x_j} - \frac{1}{2}\frac{\partial}{\partial x_j}\{\langle u_j\omega_i\omega_i\rangle\} + \langle\omega_i\omega_j s_{ij}\rangle +$$

$$+\langle\omega_i\omega_j\rangle S_{ij} + \langle\omega_i s_{ij}\rangle\Omega_j + \nu\langle\omega_i\nabla^2\omega_i\rangle.; \quad (C.51)$$

One of the possible interpretations of the terms in this equation for turbulent enstrophy is given in Tennekes and Lumley (1972). Here we draw attention to the two kinds of terms associated with the production of enstrophy of fluctuations $\langle\omega^2\rangle$: i – the RDT[15]-type terms involving the mean velocity gradients, $-\langle u_j\omega_i\rangle\frac{\partial\Omega_i}{\partial x_j}$, $\langle\omega_i\omega_j\rangle S_{ij}$ and $\langle\omega_i s_{ij}\rangle\Omega_j$, and ii – the term containing the fluctuative quantities only $\langle\omega_i\omega_j s_{ij}\rangle$.
A similar procedure can be applied to the equation for strain, s_{ij}, giving
The equation for the strain of the mean flow $\frac{1}{2}S_{ij}S_{ij}$

$$\frac{D_U\frac{1}{2}S_{ij}S_{ij}}{Dt} = -\frac{\partial}{\partial x_k}\{S_{ij}\langle u_k s_{ij}\rangle\} + \langle u_k s_{ij}\rangle\frac{\partial S_{ij}}{\partial x_k} - S_{ij}S_{jk}S_{ki}$$

$$-\frac{1}{4}\{\langle\omega_i\omega_i\rangle S_{ij} + \Omega_i\Omega_j S_{ij}\} - S_{ij}\frac{\partial^2 P}{\partial x_k\partial x_k} + \nu S_{ij}\nabla^2 S_{ij}. \quad (C.52)$$

The equation for the mean total strain of the turbulent fluctuations $\langle s_{ij}s_{ij}\rangle$

$$\frac{D_U\frac{1}{2}\langle s_{ij}s_{ij}\rangle}{Dt} = -\langle u_k s_{ij}\rangle\frac{\partial S_{ij}}{\partial x_k} - \frac{1}{2}\frac{\partial}{\partial x_k}\{\langle u_k s_{ij}s_{ij}\rangle\} - 2\langle s_{ij}s_{ik}\rangle S_{kj}$$

$$-\frac{1}{2}\langle\omega_i s_{ij}\rangle\Omega_j - \langle s_{ij}s_{jk}s_{ki}\rangle - \frac{1}{4}\langle\omega_i\omega_j s_{ij}\rangle - \langle s_{ij}\frac{\partial^2 p}{\partial x_i\partial x_j}\rangle + \nu\langle s_{ij}\nabla^2 s_{ij}\rangle. \quad (C.53)$$

[15]Rapid distortion theory (Savill, 1987; Hunt and Carruthers, 1990).

Like in (C.51), there are two kinds of terms in the equation (C.53) associated with the production of the total strain of fluctuations $\langle \omega^2 \rangle$: i – the RDT-type terms involving the mean velocity gradients, $-\langle u_k s_{ij} \rangle \frac{\partial S_{ij}}{\partial x_k}$, $-2\langle s_{ij} s_{ik} \rangle S_{kj}$ and $-\frac{1}{2}\langle \omega_i s_{ij} \rangle \Omega_j$, and ii – the terms containing the fluctuative quantities only $-\langle s_{ij} s_{jk} s_{ki} \rangle$, $-\frac{1}{4}\langle \omega_i \omega_j s_{ij} \rangle$, and $-\langle s_{ij} \frac{\partial^2 p}{\partial x_i \partial x_j} \rangle$.

Equations similar to those given above can be written for passive objects as well.

12.4.3. FILTER DECOMPOSITION

In this case instead of some average as above, one defines a large-scale or 'resolvable' component of each 'raw' variable, $\widetilde{f}(\mathbf{x},t))$ as its convolution with some filter function[16]

$$\prec \widetilde{f}(\mathbf{x},t) \succ \equiv F(\mathbf{x},\mathbf{t}) \;=\; \int \mathcal{G}(\mathbf{x} - \mathbf{x}', t - t'; \Delta, \theta) f(\mathbf{x}',t') d\mathbf{x}' dt', \quad (C.54)$$

where Δ, θ are the widths of the spatial and temporal filters respectively, so that $\widetilde{f} = F + f$, where f represents the subgrid scale (SGS) or the unresolved part, which is an analogue of the fluctuations in the Reynolds decomposition.

The equations for the filtered ('resolved') quantities are found in a similar way as in the case of RANS applying the operation $\prec \cdots \succ$ to the NSE[17]

$$\frac{\partial U_i}{\partial t} + \frac{\partial U_i U_j}{\partial x_j} = \frac{1}{\rho} \frac{\partial}{\partial x_j} \left(-P\delta_{ij} - \tau_{ij}^{\mathrm{SGR}} + 2\rho\nu S_{ij} \right), \qquad (C.55)$$

where the *subgrid-scale stress*[18]

$$\tau_{ij}^{\mathrm{SGR}} = -\rho \left\{ \prec \widetilde{u}_i \widetilde{u}_j \succ - U_i U_j \right\} \qquad (C.56)$$

should be modelled in some way, i.e. expressed in terms of the filtered field $\prec u_i \succ$ on the basis of some hypothesis, though, again, in the strict sense such a relation does not exist for the same reason as in RANS.

Substracting the equation (C.55) from the NSE gives the equation for the unresolved part of the flows, which is analogous to the field of fluctua-

---

[16]This approach is based on Leonard (1974); see also Germano (1999); Piomelli (1999); Meneveau and Katz (2000); and references therein.

An example of a spatial filter function is $\mathcal{G}(\mathbf{x}; \Delta) = 6/(\pi\Delta^2)^{3/2} \; exp(-6\mathbf{x}^2/l^2)$; - $\int \mathcal{G}(\mathbf{x}; \Delta) d\mathbf{x} = 1$.

[17]That is, the quantities $\prec \widetilde{u}_i \succ$, etc., are analogous to U_i in RANS.

[18]Generally, $\tau_{ij}^{\mathrm{SGR}} = -\rho \left\{ U_i u_j + U_j u_i + \prec u_i u_j \succ \right\} \neq - \prec u_i u_j \succ$. Only for filters such that $\prec\prec f \succ\succ = \prec f \succ$ the subgrid stress takes the form of a usual Reynolds stress $\tau_{ij}^{\mathrm{SGR}} = -\rho \prec u_i u_j \succ$.

tions in the Reynolds decomposition

$$\frac{\partial u_i}{\partial t} + U_j\frac{\partial u_i}{\partial x_j} + u_j\frac{\partial U_i}{\partial x_j} = \frac{1}{\rho}\frac{\partial}{\partial x_j}\left\{-p\delta_{ij} + 2\rho\nu s_{ij} - \rho u_i u_j - \tau_{ij}^{\mathrm{SGR}}\right\} \qquad (C.57)$$

The filter decomposition is formally more general than the Reynolds decomposition. However, the former is one among many decompositions, so to say, of a technical nature, whereas the latter is physically more natural provided that the means/averages do exist in some sense. It is noteworthy that equations such as (45) for correlations, or (55) for filtered quantities by their very nature (apart from the 'closure problem') contain much less information on the turbulent flow than the Navier- Stokes equations. The main problem is the lack of clarity about how much and which physics is retained in the RANS or LES equations.

12.4.4. EQUATIONS GOVERNING THE DYNAMICS OF 'ERROR'

Since in many cases one is unable to reproduce precisely the initial (and boundary) conditions, it is of interest to follow the dynamics of an 'error', e.g. in initial conditions[19]. That is, one looks at the behaviour of the difference $\boldsymbol{\Delta}^u$ of some undisturbed flow realization $\mathbf{u}$ and the one with a disturbance $\mathbf{u} + \boldsymbol{\Delta}^u$. This behaviour is governed by the equation similar to (C.44)

$$\frac{D\Delta_i^u}{Dt} = \frac{\partial}{\partial x_j}\{-\frac{1}{\rho}\Delta^p\delta_{ij} + 2\nu\Delta_{ij}^s - \Delta_i^u\Delta_j^u\} - \Delta_j^u\frac{\partial u_i}{\partial x_j}, \qquad (C.58)$$

where $\frac{D}{Dt} = \frac{\partial}{\partial t} + u_j\frac{\partial}{\partial x_j}$, and Δ^p, Δ_{ij}^s are the pressure and the rate of strain errors.

The corresponding equation for the energy of the error, $e_{\boldsymbol{\Delta}^u} = \frac{1}{2}\Delta_i^u\Delta_i^u$ is analogous to the equation (C.47)

$$\frac{De_{\boldsymbol{\Delta}^u}}{Dt} = -\frac{\partial}{\partial x_j}\left\{u_j e_{\boldsymbol{\Delta}^u} + \frac{1}{\rho}p\Delta_j^u - 2\nu\Delta_i^u\Delta_{ij}^s\right\} - 2\nu\Delta_{ij}^s\Delta_{ij}^s - \Delta_i^u\Delta_j^u s_{ij}$$
$$(C.59)$$

The 'source' term, $-\Delta_i^u\Delta_j^u s_{ij}$, in the RHS of (C.59) is analogous to the turbulent energy production term, $-u_i u_j S_{ij}$, in (C.47). The analogy is not only formal. Namely, just like the mean, $-\langle u_i u_j\rangle S_{ij}$, is positive in turbulent shear flows, the integral of the production of the energy of error, $P_{\boldsymbol{\Delta}^u} = -\int \Delta_i^u\Delta_j^u s_{ij}dV$, over the flow domain at any time moment is positive, because the production of the error energy , $-\Delta_i^u\Delta_j^u s_{ij}$, appears to be a positively skewed quantity (Tsinober and Ortenberg, 2001).

---

[19]This approach takes its beginning from the predictability problem in meteorology (Holloway and West, 1984; Lorenz, 1985; Novikov, 1959; but is of more general importance. See also Bohr et al. (1998) and Lesieur (1997) and references therein.

Similar equations can be easily written also for the errors of vorticity, Δ_i^ω, and strain, Δ_{ij}^s. For example,

$$\frac{D\Delta_i^\omega}{Dt} = \omega_j \Delta_{ij}^s + \Delta_j^\omega s_{ij} - \Delta_j^u \frac{\partial \omega_i}{\partial x_j} + \Delta_j^\omega \Delta_{ij}^s - \Delta_j^u \frac{\partial \Delta_i^\omega}{\partial x_j} + \nu \nabla^2 \Delta_i^\omega. \quad (C.60)$$

$$\frac{De_{\Delta^\omega}}{Dt} = \Delta_i^\omega \Delta_{ij}^s \omega_j + \Delta_i^\omega \Delta_j^\omega s_{ij} - \Delta_j^u \Delta_i^\omega \frac{\partial \omega_i}{\partial x_j}$$

$$+ \Delta_i^\omega \Delta_j^\omega \Delta_{ij}^s - \Delta_j^u \Delta_i^\omega \frac{\partial \Delta_i^\omega}{\partial x_j} + \nu \Delta_i^\omega \nabla^2 \Delta_i^\omega. \quad (C.61)$$

where $e_{\Delta^\omega} = \frac{1}{2}\Delta_i^\omega \Delta_i^\omega$ is the energy of the vorticity error. It is seen from (C.61) that apart of $\Delta_i^\omega \Delta_j^\omega s_{ij}$ several other terms are involved in the production of e_{Δ^ω}

CHAPTER 13

APPENDIX D. IT IS A MISCONCEPTION THAT

— 'Statistical' and 'structural' contrapose each other.

— Turbulence possesses a random (quasi)-Gaussian background.

— Kolmogorov picture is structureless and quasi-Gaussian.

— Large scales and small scales are decoupled.

— Turbulence can be described adequately by equations 'simpler' than the Navier-Stokes equations, e.g. by a low dimensional system.

— 'Eddy viscosity' and 'eddy diffusivity' explain the enhanced transfer rates of momentum, energy and passive objects.

— Spatial fluxes represent 'cascade' in physical space.

— Vorticity amplification is a result of the kinematics of turbulence.

— Vortex lines are on average stretched rather than compressed because two particles on average move apart from each other.

— The vorticity intensification process is the strongest where vorticity already happens to be large.

— Vorticity is stretched only. Hence inadequate representation of turbulent field by a collection of purely stretched (or other 'simple') objects.

— Strain rate in turbulent flows is irrotational.

— Enhanced dissipation in turbulent flows is due to vortex stretching.

— The difference between quasi-two-dimensional and pure-two-dimensional turbulent flows is always small.

CHAPTER 14

APPENDIX E. ON METHODS OF STUDYING TURBULENT FLOWS

The methods of studying turbulent flows are usually divided into theoretical and experimental. As discussed in chapters 1 and 3, in fact, there exist no adequate theoretical methods[1]. However, most valuable is the language and terminology which comes from numerous theoretical approaches. The experimental methods are subdivided into physical (laboratory and field), and numerical. Both are extensively described in the literature with thousands of references, though most books on turbulence contain very little (if at all) material on the physical methods, whereas the numerical methods, including modelling, are covered extensively, the latest examples being the books by Pope (2000) and Mathieu and Scott (2000).

The aim of the remarks below is to give some general comments and to list the main methods found in the references of some major sources of information.

14.1. Direct numerical simulations of the Navier Stokes equations

The ability of a computer code to simulate flow that is difficult to realize in the laboratory has its unfortunate extension the ability of a computational solution to be altogether unphysical. (Aref, 1986).
It is not sufficient to set up the code and let the computer zip along. It zips all right, but to where? (Kadanoff, 1997).

The above citations serve as a warning about taking special care in using numerical simulations. The problem arises due to the extreme sensitivity to very small variations in the initial and boundary conditions and possible different behaviour of the chaotic system approximating the original one[2]. Therefore it is is not clear how to interpret numerical experiments designed

---

[1]Since our main concern is with basic aspects of turbulent flows various methods of modelling are not mentioned here. The only exception are the Navier-Stokes equations.

[2]This problem is hoped to be less serious when handling statistics, but can be acute when attempting to follow a particular realization.

to test the accuracy of DNS (hence the term numerical *experiments*)[3]. Nevertheless it is impossible to underestimate the importance of direct numerical simulations of the Navier-Stokes equations (Moin and Manesh, 1998; Mathieu and Scott, 2000; and Pope, 2000). This is especially true regarding most of the issues in basic research, since one does need neither high Reynolds numbers nor very complex geometry. By its very nature numerical simulation allows not only to simulate flow that is difficult to realize in the laboratory and to get access to quantities which are not accessible in the laboratory, but also to realize situations which are not reproducible in the laboratory. The immediate examples are the 'unphysical' situations, such as pure two-dimensional turbulence, and three-dimensional turbulence with any desirable forcing.

14.2. Physical experiments

In view of the above, it is natural to ask: why bother *measuring* especially difficult quantities such as velocity derivatives in the age of supercomputers? Indeed, many numerical fluid dynamists seem to be inclined to consider physical experiments as superfluous. Yet not every problem in turbulence can be studied by a numerical approach. First, it is useful to remember that Nature is far richer than the Navier-Stokes equations, so that one can encounter surprises at large Reynolds numbers[4] Turbulent processes in very large systems such as the atmosphere and the ocean are not and never will be accessible to direct numerical computation. Today, it is possible to perform direct numerical simulations of the Navier-Stokes equations at rather moderate Reynolds numbers in simple geometries and the prospects for higher Reynolds numbers are rather modest and not only for the immediate future. Though the scale resolution problem is serious, both in laboratory and numerical experiments, there is an essential difference between the two: inadequate resolution in numerical experiments leads usually to erroneous results, whereas in laboratory/field experiments one has the *true* flow and correct results for the scales resolved even when some range of scales is not resolved. Second, even at small Reynolds numbers, it is not always possible to use the numerical approach. An important example is represented by flows in complex geometries such as flows with rough

---

[3] *Whenever they fail in their predictions, scientists tend to blame the poor accuracy of the observations, the lack of computer power and the inadequate parametrization in their numerical models, rather than their own lack of skill in computing the accuracy that can be obtained with present resources. Sloppy reasoning of this kind is responsible for much of the thoughtless expansion and escalation numerical modellers in all branches of science indulge in... A calculation that does not include a calculation of its predictive skill is not a legitimate scientific product.* (Tennekes, 1993).

[4] Apart from things like aliasing, numerical viscosity, inadequate BC and other 'small' problems.

walls and flows in plant canopies. Similar problems arise in handling flows involving sediment transport and other additives (various particles, bubles), and especially polymers and surfactants for which even no adequate equations seem to be known. This is a partial list only of the limitations of numerical approaches and the reasons for the importance of physical experiments. One more reason is that it is the physical experiment (dealing with real turbulent flows) that provides the final verdict to the results of both numerical simulations and theories.

The experimental methods of studying turbulent flows are quite elaborate. There is a host of methods of flow visualization providing mostly qualitative information; see references mentioned in section 1.1. Of special interest are methods providing local-in-space and instantaneous-in-time values of various quantities. The main methods include hot-wire anemometry, laser doppler anemometry and ultasonic anemometry, electromagnetic methods, particle image velocimetry and stereoscopic particle image velocimetry, particle tracking velocimetry, laser induced velocimetry and other methods of local tagging of fluid elements, holographic velocimetry and nuclear magnetic resonance. Selected references can be found in Bonnet et al. (1998), Bruun (1995), Dracos (1996), Fukushima (1999), Kholmyansky et al. (2001c), Sanford et al. (1999), Tao et al. (2000). The review by Corrsin (1963) deserves special mentioning for everybody intending to perform any experiments in turbulence. There exist also numerous methods to measure temperature, salinity, moisture and other species.

CHAPTER 15

APPENDIX F. GLOSSARY OF SOME TERMS

Anisotropy - Lack of isotropy. There are many kinds and different manifestations of anisotropy of turbulent flows, depending on quantities in question. These may be of different order, consist of velocity components, their derivatives of various orders taken at the same or different space/time points. See Monin and Yaglom (1971.1975), Hinze (1975), Favre et al. (1976) and Lumley(1978).

Attractor - A set in the phase space that the system approaches at large times. Invariant under evolution (phase flow).

Bifurcation - topological change of the phase flow as some parameter, say Reynolds number, changes.

Degrees of freedom - Proportional to (say, 1/2 of) the number of first order ODE's adequately describing the system in question. The effective number of degrees of freedom is smaller due to the conservation laws and constraints imposed on the system.

Ergodicity - For statistically stationary flows roughly equivalence of 'true' statistical properties (not only means/averages, but 'almost' all statistical properties) of an ensemble to those obtained using time series in one very long realization. A similar property is defined in space by replacing time by space coordinate(s) in which the flow domain has an infinite extension, at least in one direction. Turbulent flows are known (empirically) to be ergodic. Other chaotic systems may be non-ergodic; see Shlesinger (2000), Zaslavsky (1999).

Homogeneity - Invariance of all statistitical properties/parameters of turbulent flow to translations in space. All real flows are inhomogeneous.

Isotropy - Invariance of all statisitical properties/parameters of turbulent flow to rotations and reflections, i.e. to the full rotation group. This is essentially the definition given by Taylor (1935) and Kolmogorov (1941a). Note the *statistical* nature of isotropy, homogeneity and stationarity. Isotropic flows are necessarily homogeneous. All real flows are anisotropic.

Phase flow - All trajectories originating from all possible initial conditions in a given flow geometry (boundary conditions), i.e. all fluid flows in a given geometry for all possible initial conditions comprise the phase flow.

Phase space - Set of all possible (instantaneous) states of a system. For incompressible fluid flows, a 'point' in the phase space is a solenoidal velocity vector in the flow domain, satisfying the boundary conditions. A time dependent fluid flow comprises a trajectory in the phase space.

Stationarity - Invariance of all statistical properties/characteristics of turbulent flow to translations in time.

A glossary of basic terms related to turbulence is found in Bradshaw (1971).

BIBLIOGRAPHY

Acrivos, A., editor (1971) Symposium on Fluid Mechanics of Stirring and Mixing, *Phys. Fluids,* **A5**, Part 2.

Adrian, R.J., Meinart, C.D. and Tomkins, C.D. (2000) Vortex organization in the outer region of the turbulent boundary layer, *J. Fluid Mech.,* **422**, 1–54.

Anderson P.W. (1972) More is different, *Science,* **177**, 393–396.

Anderson P.W. (1991) Is complexity physics? Is it science? What it is?, *Physics today,* **44**, 9,11.

Anderson P.W. (1995) Physics: The opening to complexity *Proc. Natl. Acad. Sci. USA*, **192**, 6653–4.

Andreopulos, Y., Agui, J.H. and Briassulis, G. (2000) Shock wave–turbulence interation, *Annu. Rev. Fluid. Mech.,* **32**, 309–345.

Anselmet, F., Gagne, Y., Hopfinger, E. G. and Antonia, R. A. (1984) High-order velocity structure functions in turbulent shear flows, *J. Fluid Mech.,* **140**, 63–89.

Antonia, R.A. (1981) Conditional sampling in turbulence measurement, *Annu. Rev. Fluid. Mech.,* **13**, 131–156.

Antonia, R.A., Hopfinger, E.G., Gagne, Y and Anselmet, F. (1984) Temperature structure functions in turbulent shear flows, *Phys. Rev.,* **A30**, 2704 - 2707.

Antonia, R.A., Ould-Rouis, M. Anselmet, F. and Zhu, Y. (1997) Analogy between predictions of Kolmogorov and Yaglom, *J. Fluid Mech.,* **332**, 395–409.

Aref, H. (1984) Stirring by chaotic advection, *J. Fluid Mech.,* **143**, 1–21.

Aref, H. (1986) The numerical experiment in fluid mechanics, *J. Fluid Mcch.,* **173**, 15 41.

Aref, H. and El Naschie, M.S. (1994) *Chaos applied to fluid mixing,* Pergamon.

Aref, H. and Gollub, J.P. (1996) Application of dynamical systems theory to fluid mechanics, in: J. Lumley, A. Acrivos, G. Leal and S. Leibovich, editors, *Research trends in Fluid Mechanics,* AIP Press, pp. 15–30.

Armi, L. and Flament, P. (1987) Cautionary remarks on the spectral interpretation of turbulent flows, *J. Geophys. Res,* , **90**, C6, 11,779–11,782.

Arneodo, A., Manneville, S. Muzy, J.F. and Roux, S. (1999) Revealing a log-normal cascading process in turbulent velocity statistics with wavelet analysis, *Phil. Trans. Proc. Roy. Soc. London*, **A 357**, 2415–38.

Arnold, V.I. (1991) Kolmogorov's hydrodynamics attractors, *Proc. Roy. Soc. London*, **A434,** 19–22.

Aronson, D., Johansson, A.V. and Löfdal (1997) Shear-free turbulence near a wall, *J. Fluid Mech.*, **338,** 363–385.

Ashurst, Wm.T., Kerstein, A. R., Kerr, R. A. and Gibson, C. H. (1987) Alignment of vorticity and scalar gradient with strain rate in simulated Navier-Stokes turbulence, *Phys. Fluids*,**30**, 2343–2353.

Atlas of Visualization III (1997), edited by The Visualization Society of Japan, CRC Press.

Avelius, K. (1999) Random forcing of three-dimensional homogeneous turbulence, *Phys. Fluids*, **11**, 1880–1889.

Babiano, A., Basdevant, C., Legras, B. and Sadourny, R. (1987) Vorticity and passive-scalar dynamics in two-dimensional turbulence, *J. Fluid Mech*, **183**, 379–397.

Badii, R. and Talkner, P. (2001) Refined scaling hypothesis for anomalously diffusing processes, *Physica*, **A291**, 229–243.

Batchelor, G.K. and Townsend, A.A. (1949) The nature of turbulent motion at large wave-numbers, *Proc. Roy. Soc. London*, **A199,** 238–255.

Belin, F., Maurer, J., Tabeling, P. and Williame, H. (1996) Observation of intense filaments in fully developed turbulence, *J. Phys. II France*, **6**, 573 - 583.

Belin, F., Maurer, J., Tabeling, P. and Willaime, H. (1997) Velocity gradient distributions in fully developed turbulence: and experimental study, *Phys. Fluids*, **9,** 3843–3850.

Betchov, R. (1956) An inequality concerning the production of vorticity in isotropic turbulence, *J. Fluid Mech.*, **1**, 497–503.

Betchov, R. (1961) Semi-isotropic turbulence and helicoidal flows, *Phys. Fluids*, **4,** 925–926.

Betchov, R. (1974) Non-Gaussian and irreversible events in isotropic turbulence, *Phys. Fluids*, **17,** 1509–1512.

Betchov, R. (1976) On the non-Gaussian aspects of turbulence, *Archives of Mechanics (Archiwum Mechaniki Stosowanej)*, **28**, Nos. 5-6, 837–845.

Betchov, R. (1993) in: T. Dracos and A. Tsinober, editors, *New Approaches and Turbulence*, Birkäuser, p.155.

Bevilaqua, P.M and Lykoudis, P.S. (1978) Turbulence memory in self-preserving wakes, J. Fluid. Mech., **89**, 589–606.

Bewersdorff, H.-W., Gyr, A., Hoyer, K. and Tsinober, A. (1993) An investigation on possible mechanism of heterogeneous drag reduction in pipe and channel flows, *Rhelogica Acta*, **32**, 140–149.

Biferale, L., Blank, M. and Frisch, U. (1994) Chaotic cascades with Kolmogorov 1941 scaling, *J. Stat. Phys.*, **75**, 781–795.

Billant, P. and Chomaz, J.-M. (2000) Experimental evidence for a new instability of vertical columnar vortex pair is strongly stratified fluid, *J. Fluid Mech.*, **418**, 167–188.

Birkhoff, G. (1960) *Hydrodynamics*, Princeton Univ. Press.

Biswas, G. and Eswaran, V. (2000) *Turbulent flows*, Narosa Publ. House, New Dehli.

Bisset, D.K., Hunt, J.C.R., Cai, X. and Rogers, M.M. (1998) Interfaces at the outer boundaries of turbulent motions, *CTR- Ann. Res. Briefs*, 125–135.

Bisset, D.K., Hunt, J.C.R., and Rogers, M.M. (2001) The turbulent/non-turbulent interface, *J. Fluid Mech.*, to be published.

Bjerknes, V., Bjerknes, J., Solberg, H. and Bergeron, T. (1933) *Physikalishe Hydrodynamik*, Springer, Berlin.

Boffetta, G., Celani, A. and Vergassola, M. (2000) Inverse energy cascade: deviations from Gaussian behaviour, *Phys. Rev.*, **E61,** R29–R32.

Bohr, T., Jensen, M. H., Paladin, G. and Vulpiani, A. (1998) *Dynamical systems approach to turbulence*, Cambr. Univ. Press.

Bonnet, J. P. (ed.) (1996) *Eddy structure identification*, Springer.

Bonnet, J.P., Gesilon, D. and Taran, J.P. (1998) Nonintrusive measurements for high-speed, supersonic, and hypersonic flows, *Annu.Rev. Fluid Mech.*, **30**, 231–273.

Boratav, O.N. and Pelz, R.B. (1997) Structures and structure functions in the inertial range of turbulence, *Phys. Fluids*, **9**, 1400–1415.

Borue, V. and Orszag, S.A. (1996) Numerical study of three-dimensional Kolmogorov flow at high Reynolds numbers, *J. Fluid Mech.*, **306**, 293–323.

Borue, V. and Orszag, S.A. (1998) Local energy flux and subgrid scale statistics in three-dimensional turbulence, *J. Fluid Mech.*, **336**, 1-31.

Boussinesq, J. (1877) Essai théorique sur les lois eaux courantes, Mém. prés. par div. savants à l'Acad. Sci., Paris **23,** No. 1, 1-680.

Bracco, A., McWilliams, J.C., Murante, G. and Weiss, J.B. (2000) Revising freely decaying two-dimensional turbulence at millennial resolution, *Phys. Fluids*, **12**, 2931–2941.

Brachet, M.E., Meneguzzi, M., Vincent, A. Politano, H. and Sulem, P.L. (1992) Numerical evidence of smooth self-similar dynamics and possibility of subsequent collapse for three-dimensional ideal flows, *Phys. Fluids*, **A4**, 2845–2854.

Bradshaw, P. (1994) *An introduction to turbulence and its measurement*, Pergamon.

Bradshaw, P. (1994) Turbulence: the chief outstanding difficulty of our subject, *Experiments in fluids* **16,** 203–216.

Brandenburg, A. (1995) Flux tubes and scaling in MHD dynamo simulations, *Chaos, solitons and fractals*, **5** (10), 2203–2245.

Brandenburg, A., Jennings, R.L., Nordlund, A., Rieutord, M., Stein, R.F. and Touminen, I. (1996) Magnetic structures in a dynamo simulation", *J. Fluid Mech.*, **306,** 325–352.

Briassulis, G., Agui, J.H. and Andreopulos, Y. (2001) The structure of weakly compressible grid-generated turbulence, *J. Fluid Mech.* **,433**, 219–283.

Browand, F.K., Guyomar, D. and Yoon, S.C. (1987) The behaviour of a turbulent front in a stratified fluid: experiments with an oscillating grid, *J. Geophys. Res.*, **92**, C5, 5329–41.

Brown, R.A. (1980) Longitudinal instabilities and secondary flows in the planetary boundary layer: a review, *Rev. Geophys. Space Phys.*, **18** (3), 683–697.

Brun, C. and Pumir, A. (2001) Statistics of Fourier modes in a turbulent flow, *Phys. Rev.*, **E63,** in press (056313 – 1-12).

Bruun, H.H. (1995) *Hot-wire anemometry–Principles and Signal analysis.* Oxford Univ. Press, London.

Buffon, G. (1777) Essai d'Arithmétique Morale, *Suppl. à l'Histoire.* Nat. 4; see also BUFFON 88, *Actes Coll. Intern. Bicentenaire de la mort de Buffon, 14-22, Juin 1988*, Publ. de l'Inst. Interdisc. d'Etudes Epistém.

Burr, U., Barleon, L., Jochmann, P. and Tsinober, A. (2000) MHD-convection in a vertical slot with horizontal magnetic field, submitted.

Busse, F.H. (1983) Generation of mean flows by thermal convection, *Physica*, **D9**, 287-299.

Cadot, O., Douady, S. and Couder, Y. (1995) Characterization of the low-pressure filaments in three-dimensional turbulent shear flow, *Phys. Fluids*, **7**, 630–646.

Cadot, O., Couder, Y., Daerr, A., Douady, S. and Tsinober, A. (1997) Energy injection in closed turbulent flows: stirring through boundary-layers versus inertial stirring, *Phys. Rev.*, **E56,** 427–433.

Cadot, O., Bonn, D. and Douady, S. (1998) Turbulent drag reduction in a closed flow system: boundary layer versus bulk effects, *Phys. Fluids*, **10**, 426–436.

Cambon, C. (1994) Turbulent flows undergoing distortion and rotation — A report on EUROMECH 288, *Fluid Dyn. Res.*, **13**, 281–298.

Cantwell, B.J. (1990) Future directions in turbulence research and the role of organized motion, in J.L.Lumley (ed)., *Whither turbulence?*, pp. 97–131, Springer.

Cantwell, B.J. (1992) Exact solution of a restricted Euler equation for the velocity gradient tensor, *Phys. Fluids*, **A4,** 782–793.

Celani, A., Lanotte, A., Mazzino, A. and Vergassola, M. (2001) Fronts in passive scalar turbulence, *Phys. Fluids*, **13**, 1768–1783.

Chacin, J., and Cantwell, B.J. (2000) Dynamics of a low Reynolds number turbulent boundary layer, *J. Fluid Mech.*, **404**, 87–115.

Chacin, J., Cantwell, B.J. and Kline, S.J. (1996) Study of turbulent boundary layer using invariants of the velocity gradient tensor, *Expl. Thermal Fluid Sci.*, **33**, 308–317.

Chasnov, J.R. (1997) On the decay of two-dimensional homogeneous turbulence, *Phys. Fluids*, **9**, 171-180.

Chaté, H., Villermaux, E. and Chomaz, J.M., editors, *Mixing: Chaos and Turbulence*, *ASI NATO Ser.*, **B373,** Plenum Press.

Chavanis, P.-H. and Sire, C. (2000) Statistics of velocity fluctuations arising from a random distribution of point vortices: the speed of fluctuations and the diffusion coefficient, *Phys. Rev.*, **E62**, 490–506.

Chen, H., Herring, J. R., Kerr, R. and Kraichnan, R.H. (1989) Non-Gaussian statistics in isotropic turbulence, *Phys. Fluids.*, **A1**, 1844–1854.

Cheklov, A. and Yakhot, V. (1995) Kolmogorov turbulence in a random-force-driven Burgers equation, *Phys. Rev.*, **E51**, R2739–R2744.

Chen, S., and Cao, N. (1997) Anomalous scaling and structure instability in three-dimensional passive scalar turbulence, *Phys. Rev. Lett*, **78**, 3459–3462.

Chen, S., Sreenivasan, K.R. and Nelkin, M. (1997) Inertial range scalings of dissipation and enstrophy in isotropic turbulence, *Phys. Rev. Lett.*, **79**, 1253–1256.

Chertkov, M. (2000) Polymer stretching by turbulence, *Phys. Rev. Lett.*, **84**, 4761–4764.

Chertkov, M., Pumir, A. and Shraiman, B. I., (1999), Lagrangian tetrad dynamics and the phenomenology of turbulence, *Phys. Fluids*, **11**, 2394–2410. Childress, S. and Gilbert, A.D. (1995) *Stretch, Twist, Fold: The fast dynamo*, Springer.

Chong, M.S., Soria, J., Perry, A.E., Chacin, J., Cantwell, B.J. and Na, Y. (1998) Turbulence structures of wall-bounded shear flows found using DNS data, *J. Fluid Mech.*, **357**, 225–247.

Chorin, A.J. (1982) The evolution of a turbulent vortex, *Commun. Math. Phys.*, **83**, 517–535.

Chorin, A.J. (1994) *Vorticity and Turbulence*, Springer.

Chorin, A.J. (1996) Turbulence cascades across equilibrium spectra, *Phys. Rev.*, **54**, 2616–2619.

Ciliberto, S., Cioni, S. and Laroche, C. (1996) Large-scale flow properties of turbulent thermal convection, *Phys. Rev.*, **E54**, R5901–R5904.

Cimbala, J.M., Nagib, H. M and Roshko, A. (1988) Large structures in the far wakes of two-dimensional bluff bodies, *J. Fluid Mech.*, **190**, 265–298.

Cocke, W.J. (1969) Turbulent hydrodynamic line stretching: consequences of isotropy, *Phys. Fluids*, **12**, 2488–2492.

Constantin, P. (1994) Geometrical statistics in turbulence, *SIAM Rev.*, **36**, 73 - 98.

Constantin, P. (1995) A few results and open problems regarding incompressible fluids, *Notices of the AMS*, **42**, 658–663.

Constantin, P. (1996) Navier-Stokes equations and incompressible fluid turbulence, *Lectures in Applied Mathematics*, **31**, 219–234.

Constantin, P. and Fefferman, C., (1994), Scaling exponents in fluid turbulence: some analytic results, *Nonlinearity*, **7**, 41–57.

Corrsin, S. (1943) Investigation of flow in an axially symmetric heated jet in air, *NACA Wartime Reports*, W-94.

Corrsin, S. (1953) Dimensional analysis and similarity, in, R. Long (ed.) *Fluid models in geophysics*, 1–17.

Corrsin, S. (1959) Lagrangian correlations and some difficulties in turbulent diffusion experiments, *Adv. Geohys.*, **6**, 441–448.

Corrsin, S. (1961) Turbulent flow, *American Scientist*, **49** (3), 300–325.

Corrsin, S. (1972) Random geometric problems suggested by turbulence, in Rosenblatt, M. and C. van Atta, C., editors, *Statistical models and turbulence, Lect. Notes Phys.*, **12**, 300–316.

Corrsin, S. (1963) Turbulence: experimental methods, In S. Flugge and C. Truesdell, editors, *Handbuch der Physik*, Band **VIII/2**, Stromungsmechanik, pp. 525–590, Springer.

Corrsin, S. and Kistler, A.L. (1954, 1955), The free-stream boundaries of turbulent flows, NACA, TN-3133, TR-1244, pp. 1033–1064.

Čolak-Antić, P. (1964) "Dreidimensionale Instabilitätserscheinungen des laminarturbulenten Umschlags bei freier Konvektion längs einer vertikaler geheizten Platte", in *Sitzungsberichte der Heidelberger Akademie der Wissenschaften, Mathem.-Naturw. Klasse. Jahrg. 1962/64, 6. Abhandlung*, Springer, Berlin.

Crowe, T.C., (2000) On models for turbulence modulation in fluid particle flows, *Int. J. Multiphase Flow*, **26**, 719–727.

Curry, J.H., Herring, J.R., Loncaric, J. and Orszag, S.A. (1984) Order and disorder in two and three-dimensional Bénard convection, *J. Fluid Mech.*, **147**, 1–38.

Dalziel, S.B., Linden, P.F. and Youngs, D.L. (1999) Self-similarity and internal structure of turbulence induced by Rayleigh-Taylor instability, *J. Fluid Mech.*, **399**, 1–48.

Danaila, L., Anselmet, F., Zhou, T. and Antonia, R.A. (1999) A generalization of Yaglom's equation which accounts for the large-scale forcing in heated decaying turbulence, *J. Fluid Mech.*, **391**, 359–372.

De Bruyn Kops, S.M. and Riley, J.J. (1998) Direct numerical simulation of laboratory experiments in isotropic turbulence, *Phys. Fluids*, **10**, 2125–2127.

Deissler, R.G. (1978) On the localness of the spectral energy transfer in turbulence, *Appl. Sci. Res.*, **34**, 379-392.

Dernoncourt, B., Pinton, J.-F. and Fauve, S. (1998) Experimental study of vorticity filaments in a turbulent swirling flow, *Physica*, **D117**, 181-190.

Dimotakis, P.E. (2000) The mixing transition in turbulent flows, *J. Fluid Mech*, **409**, 69–98.

Doering, C.R. (1999) Does turbulence saturate global transport estimates?, in: A. Gyr, W. Kinzelbach and A. Tsinober, editors, *Fundamental Problematic Issues in Turbulence*, pp. 3–16, Birkhäuser.

Doering, C.R. and Gibbon, J.D. (1995) *Applied analysis of the Navier-Stokes Equations*, Cambr. Univ. Press.

Dolzhansky, F.V. (1999) Transverse structures of quasi-two-dimensional geophysical and magnetohydrodynamic flows, *Izvestiya, Atmosperic and Oceanic Phys.*, **35** (2), 147–156.

van Doorn, E., White, C.M. and Sreenivasan, K. (1999) The decay of grid turbulence in polymer and surfactant solutions, *Phys. Fluids*, **11**, 2387–2393.

Donnelli, R. (1999) Cryogenic fluid dynamics, *J. Phys., Condens. Matter*, **11**, 7783-7834.

Douady, S. Couder, Y. and Brachet, M.E. (1991) Direct observation of the intermittency of intense vorticity filaments in turbulence, *Phys. Rev. Lett.*, **67**, 983-986.

Dracos, Th., editor, (1996) *Three-dimensional velocity and vorticity measuring and image analysis techniques*, Kluwer.

Drazin, P.G. and Reid, W.H. (1981) *Hydrodynamic stability*, Cambr. Univ. Press.

Dritcshel, D.G., Juarez, M.T. and Ambaum, M.H.P. (1999) The three-dimensional vortical nature of atmospheric and oceanic turbulent flows, *Phys. Fluids*, **11**, 1512–1519.

Droegemeier, K.K., Lazarus, S.M. and Davies-Jones, R. (1993) The influence of helicity on numerically simulated convective stroms, *Month. Weather Rev.*, **121**, 2006–2029.

Drummond, I.T. (1993) Stretching and bending of line elements in random flows, *J. Fluid Mech.*, **252**, 479–498.

Drummond, I.T. and Münsch, W. (1990) Turbulent stretching of line and surface elements, *J. Fluid Mech.*, **215**, 45–59.

Dryden, H. (1948) Recent Advances in Boundary Layer Flow, *Adv. Appl. Mech.*, **1**, 1-40.

Dubs, W. (1939) Über den Einfluss laminarer and turbulenter Strömung auf das Röntgenstreubild von Wasser und Nitrobenzol. Ein röntgenographischer Beitrag zum Turbulenzproblem, *Helv. Phys. Acta,* **12**, 169–228.

Durbin, P. and Pettersson, R. (2000) *Theory and modelling of turbulent flows*, Wiley.

Eaton, J.K. and Fessler, J.R. (1994) Preferential concentration of particles by turbulence, *Int. J. Multiphase Flow*, **20**, 169–209.

Eggels, J.G.M., Unger, F., Weiss, M.H., Westerweel, J., Adrian, R.J., Friedrich, R. and Nieuwstadt, F.T.M. (1994) Fully developed turbulent pipe flow: A comparison between direct numerical simulation and experiment, *J. Fluid Mech.*, **268**, 175–209.

Elliott, F.W. and Majda A.J. (1995) A new algorithm with plane waves and wavelets for random velocity fields with many spatial scales, *J. Comput. Phys.*, **117**, 146–162.

Elliott, F. W. and Majda, A. J., 1996 Pair dispersion over an inertial range spanning many decades, *Phys. Fluids.*, **8**, 1052-1060.

Elperin, T., Kleeorin, N., Rogachevskii, I. and Sokoloff, D.(2000) Turbulent Transport of Atmospheric Aerosols and Formation of Large-Scale Structures, *Physics and Chemistry of Earth*, **A25**, 797-803.

Emmons, H.W. (1951) The laminar-turbulent transition in a boundary layer – part I, *J. Aero. Sci*, **18**, 490-498.

Falkovich, G. (1999) Universal and nonuniversal properties of the passive scalar statistics, in: A. Gyr, W. Kinzelbach and A. Tsinober, editors, *Fundamental Problematic Issues in Turbulence*, pp. 419-425, Birkhäuser.

Fantasy of Flow (1993), edited by The Visualization Society of Japan, Ohmsha, Ltd.

Farge, M. and Guyon, E. (1999) A philosophical and historical journey through mixing and fully-developed turbulence, in: Chaté, H., Villermaux, E. and Chomaz, J.M., editors, *Mixing: Chaos and Turbulence*, NATO ASI Series, **B 373**, pp. 11–36, Plenum Press.

Farge, M., Schneider, K. and Kevlahan, N. (1999) Non-Gaussianity and coherent vortex simulation for two-dimensional turbulence using an adaptive orthogonal wavelet basis, *Phys. Fluids*, **11**, 2187–2201.

Farge, M., Schneider, K. and Pelegrino, G. (2000) Vortex tube extraction in three-dimensional turbulence using orthogonal wavelets, *Advances in turbulence*, **8**, 979-800.

Farrell, B.F. and Ioannou, P.J. (1998) Perturbation structure and spectra in turbulent channel flow, *Theoret. Comput. Fluid Dynamics*, **11**, 237–250.

Fasel, W.F. and Saric, W.S. (2000) *Laminar-Turbulent Transition*, Springer, Proceedings of the *IUTAM Symposium on Laminar-Turbulent Transition*, September 13-17, 1999, Sedona, Arizona.

Favre, A., Kovaszny, L.S.G., Dumas, R., Gaviglio, J., Coantic, M., (1976) *La turbulence et mécanique des fluides*, Gauthiers-Villars.

Feigenbaum, M. (1997) Where will the future go? in E. Infeld, R. Zelazny and A. Galkovski, editors, *Nonlinear dynamics, chaotic and complex systems*, pp. 321–326.

Ferchichi, M. Tavoularis, S. (2000) Reynolds number dependence of the fine structure of uniformly sheared turbulence, *Phys. Fluids*, **12**, 2942-2953.

Feynmann R., *1963 Lectures on Physics*, **2**.

Fincham, A.M., Maxworthy, T., Spedding, G.R. (1996) Energy dissipation and vortex structure in freely decaying, stratified grid turbulence, *Dyn. Atm. Oceans*, **23**, 155–169.

Fisher, M., Jovanovich, J. and Durst, F. (2001) Reynolds number effects in the near-wall region of turbulent channel flows, *Phys. Fluids*, **13**, 1755–1767.

Flohr, P. (1999) Small-scale flow structure in turbulence: fundamentals and numerical models, Ph.D. Thesis, Cambridge, Wolfson College.

Foias, C. (1997) What do the Navier-Stokes equations tell us about turbulence?, *Contemporary Mathematics*, **208**, 151–180.

Fornberg, B. (1977) A numerical study of 2-D turbulence, *J. Comp. Phys.*, **25**, 1–31.

Foss, J.F. and Klewicki, C.L. (1984) Vorticity and velocity measurements in the intermittent region of large plane shear layer, *Fourth Symposium on turbulent shear flows, Karlsruhe*, pp 8.1–8.8.

Fountain, G.O., Kharkar, D.V., Mezic, I. and Ottino, J.M. (2000) Chaotic mixing in a bounded three-dimensional flow, *J. Fluid Mech.*, **417**, 265–301.

Frederiksen, R.D., Dahm, W. and Dowling, D.R. (1997) Experimental assessment of fractal scale similarity in turbulent flows. Part 2. Higher-dimensional intersections and non-fractal inclusions, *J. Fluid Mech.*, **338**, 89–126.

Frederiksen, R.D., Dahm, W. and Dowling, D.R. (1998) Experimental assessment of fractal scale similarity in turbulent flows. Part 4: Effects of Reynolds and Schmidt numbers, *J. Fluid Mech.*, **377,**, 169–187.

Frenkiel, F.N., Klebanoff, P.S. and Huang, T.T (1979) Grid turbulence in air and water, *Phys. Fluids*, **22**, 1606–1617.

Frick, P. and Zimin, V. (1993) Hierarchical models of turbulence, in M. Farge, J. C. R. Hunt and J. C. Vassilicos (editors), *Wavelets, Fractals, and Fourier Transforms*, Clarendon Press, Oxford, pp. 265–283.

Friedlander, S.K. and Topper, L., editors, (1961) *Turbulence: Classic papers on statistical theory*. Interscience.

Friedrich, R. and Bertolotti, F.P (1997) Compressibility effects due to turbulent fluctuations, *Appl. Sc. Res.*, **57**, 165–194.

Frisch, U. (1995) *Turbulence: The legacy of A.N. Kolmogorov*. Cambridge University Press.

Frisch, U. and Orszag, S.A. (1990) Turbulence: challenges for theory and experiment, *Phys. Today*, **9**, 24–32.

Fukushima, E. (1999) Nuclear magnetic resonance as a tool to study flow, *Annu. Rev. Fluid Mech.*, **31**, 95–123.

Fursikov, A.V. and Emanuilov, O.Yu. (1995) The rate of convergence of approximations for the closure of the Friedman-Keller chain in the case of large Reynolds numbers", *Russian Acad. Sci. Sb. Math*, **81**, 235–259.

Gadd, G.E. (1965) Turbulence damping and drag reduction produced by certain additives in water, *Nature*, **206**, 463–467.

Gad-el-Hak, M. (2000) *Flow control: Passive, active and reactive flow management*, Cambridge University Press.

Gagne, Y. (1987) Etude expérimentale de l'intermittence et des singularités dans le plan complexe en turbulence développée, Thése de Docteur és-Sciences Physiques, Université de Grenoble.

Galanti, B. and Tsinober, A. (2000) Self-amplification of the field of velocity derivatives in quasi-isotropic turbulence, *Phys. Fluids*, **12**, 3097–3099; erratum (2001) *Phys. Fluids*, **13**, 1063.

Galanti, B., Sulem, P.-L. and A. Pouquet, A. (1992) Linear and nonlinear dynamos associated with ABC flows, *Geophys. Astrophys. Fluid Dynamics*, **66**, 183-208.

Gampert, B. and Yong, C.K. (1990) The influence of polymer additives on the coherent structure of turbulent channel flow, in A. Gyr, editor, *Structure of turbulence and drag reduction*, pp. 223–232.

Garg, S. and Warhaft, Z. (1998) On the small scale structure of simple shear flow, *Phys. Fluids*, **10**, 662–673.

Gawedzki, K. and Vergassola, M. (2000) Phase transition in the passive scalar advection, *Physica*, **D138**, 63–90.

Germano, M. (1999) Basic issues of turbulence modelling, in: A. Gyr, W. Kinzelbach and A. Tsinober, editors, *Fundamental Problematic Issues in Turbulence*, 213–219, Birkhäuser.

Gerz, T. and Schumann, U. (1996) A possible explanation of countergradient fluxes in homogeneous turbulence, *Theoret. Comput. Fluid Dynamics*, **8**, 169–181.

Gibson, C.H. (1996) Turbulence in the ocean, atmosphere, galaxy, and universe, *Appl. Mech. Rev.*, **49** (5), 299–315.

Gibson, C.H., Stegen, G. S. and Williams, R. B. (1970) Statistics of the fine structure of turbulent velocity and temperature fields measured at high Reynolds numbers, *J. Fluid Mech*, **41**, 153–167.

Gibson, C.H., Friehe, C.A. and McConnell, S.O. (1977) Structure of sheared turbulent fields, *Phys. Fluids*, **20**, pt II, S156–S167.

Giralt, F., Arenas, A., Ferre-Giné, J., Rallo, R. and Kopp, G.A. (2000) The simulation and intrepetation of free turbulence with cognitive neural system, *Phys. Fluids*, **12**, 1826–1835.

Girimaji, S.S. and Pope, S.B. (1990) Material-element deformation in isotropic turbulence, *J. Fluid Mech.*, **220,** 427–458.

Godeferd, F.S. and Lollini, L. (1999) Direct numerical simulations of turbulence with confinement and rotation, *J. Fluid Mech,* **393**, 257–308.

Groisman, A. and Steinberg, V. (2000) Elastic turbulence in a polymer solution flow, *Nature*, **405**, 53–55.

Grossman, S. and Lohse, D. (2000) Scaling in thermal convection *J. Fluid Mech.*, **407,** 27–56.

Guckenheimer, J. (1986), Strange attractors in fluids: another view, *Ann. Rev. Fluid. Mech.*, **18,** 15–31.

Gyr, A. and Bewersdorff, H.-W. (1995) *Drag reduction of turbulent flows by additives*, Kluwer.

Gyr, A. and Tsinober A. (1996) On some local aspects of turbulent drag reducing flows of dilute polymers and surfactants, *Advances in Turbulence*, **6**, 449–452.

Hancock, R., Glezer A. and Trump, D. (1992) Manipulation of a square jet by piezoelectric actuators, *Phys. Fluids*, **A4**, 1877.

Harris, V.G., Graham, J.A.H. and Corrsin, S. (1977) Further experiments in nearly homogeneous turbulent shear flow, *J. Fluid Mech.*, **81,** 657–687.

Harvey, E.N. (1952) *Bioluminescence*, AP, NY.

Henningson, D.S. (1996) Comment on transition in shear flows. Nonlinear normality versus non-normal linearity, *Phys. Fluids*, **8**, 2257–6.

Herring, P.J. (1998) Dolphins glow with the flow, *Nature*, **393**, 731, 733.

Herring, J.R. and Métais, O. (1989) Numerical experiments in forced stably stratified turbulence *J. Fluid Mech.*, **202,** 97–115.

Herring, J.R. and Orszag, S.A., Kraichnan, R.H. and Fox, D.G. (1974) Decay of two-dimensional homogeneous turbulence *J. Fluid Mech.*, **66,** 417–444.

Hidenaru, M., Takao, I. and Akiyoshi, I. (1988) An experimental study of axisymmetic turbulence, *Trans. Jap. Soc. Mech. Engn.*, **B54**, No. 505, 2408–2415.

Hibberd, M.F. and Dohmann, J. (1988) The anisotropy of grid-generated turbulence in dilute polymer solutions, in *Xth International Congress on Rheology, Sydney*, **1**, 404–6.

Hill, R.J. (1997) Applicability of Kolmogorov's and Monin's equations to turbulence, *J. Fluid Mech.*, **353**, 67–81.

Hill, R.J. and Thoroddsen, S.T. (1997) Experimental evaluation of acceleration correlations for locally isotropic turbulence, *Phys. Rev.*, **E55**, 1600–1606.

Hinze, O. (1975) *Turbulence*, McGraw-Hill.

Holloway, G. and West, B.J. (1984) *Predictability of fluid motions*, AIP, New York.

Holmes, P.J., Berkooz, G. and Lumley, J.L. (1996) *Turbulence, coherent structures, dynamical systems and symmetry*, Cambridge University Press.

Holmes, P.J., Lumley, J.L., Berkooz, G., Mattingly, J.C. and Wittenberg, R.W., (1997) *Low-dimensional models of coherent structures in turbulence*, *Phys. Reports*, **287**, 337–384.

Holzer, M. and Siggia, E. (1993) Skewed, exponential pressure distributions from Gaussian velocities, *Phys. Fluids*, **A5**, 2525-2532.

Holzer, M. and Siggia, E. (1994) Turbulent mixing of a passive scalar, *Phys. Fluids*, **6**, 1820–1837.

Hopf, E. (1948) A mathematical example displaying features of turbulence, *Comm. Pure Appl. Math.*, **1**, 303–322.

Hopf, E. (1952) Statistical hydromechanics and functional calculus, *J. Rat. Mech. Anal.*, **1**, 87–123.

Hopf, E. (1962) Remarks on functional-analytic approach to turbulence, in G. Birkhoff, R. Bellman, and C.C. Lin, editors (1962) *Hydrodynamic instability, Proceedings of Symposia in Applied Mathematics*, **XIII**, Amer. Math. Soc., pp. 157–164.

Hopfinger, E.J. (1987) Turbulence in stratified fluids: a review, *J. Geophys. Res.*, **92**, C5, 5287–5303.

Hopfinger, E.J. (1989) Turbulence and vortices in rotating fluids, in A. Germain, M. Piau and D. Cailerie, editors, *Theoretical and applied mechanics*, pp. 117–138.

Hosokawa, I. (2000) Statistics of "worms" in isotropic turbulence treated on the multifractal basis, *J. Stat. Phys.*, **99**, 783–798.

Huang M.-J. (1996) Correlations of vorticity and material line elements with strain in decaying turbulence, *Phys. Fluids*, **8**, 2203–2214.

Huang Y.-n. and Huang, Y-d. (1989) On the transition to turbulence in pipe flow, *Physica*, **D37**, 153–159.

Huerre, P. and Rossi, M. (1998) Hydrodynamic instabilities in open flows, in: C. Godréche and P. Manneville, editors, *Hydrodynamics and nonlinear instabilities*, pp. 81–294, Cambr. Univ. Press.

Hunt, J.C.R. and Carruthers, D.J. (1990) Rapid distortion theory and 'problems' of turbulence, *J. Fluid Mech.*, **212**, 497-532.

Hunt, J.C.R., Vassilicos, J.C. and Kevlahan, N.K.R. (1994) Turbulence: A state of nature or a collection of phenomena? *Progr. Astron. Aeronaut.*, **162**, 1–18.

Hussein, H.J. (1994) Evidence of local axisymmetry in small scales of a turbulent planar jet, *Phys. Fluids*, **6**, 2508–2070.

Ibbetson, A. and Tritton, D.J. (1975) Experiments in turbulence in a rotating fluid, *J. Fluid Mech.*, **68**, 639–672.

Idelchik, I.E. (1996) *Handbook of hydraulic resistance*, Springer.

Jimenez, J. (1990) Transition to turbulence in two-dimensional Poiseuille flow, *J. Fluid Mech.*, **218**, 265–297.

Jimenez, J. (1991) The minimal flow unit in near-wall turbulence, *J. Fluid Mech.*, **225**, 213–240.

Jimenez, J. (1999) The physics of wall turbulence, *Physica*, **A263**, 252–262.

Jimenez, J. (2000) Intermittency and cascades, *J. Fluid Mech.*, **409**, 99–120.

Jimenez, J. and Moin, P., (2000) The minimal flow unit in near-wall turbulence, *J. Fluid Mech.*, **225**, 213–240.

Jimenez, J. and Pinelli, A, (1999), The autonomous cycle of a near-wall turbulence, *J. Fluid Mech.*, **389**, 335–359.

Jimenez, J. and Wray, A.A. (1998) On the characteristics of vortex filaments in isotropic turbulence, *J. Fluid Mech.*, **373**, 255–285.

Jimenez, J., Wray, A. A., Saffman, P. G. and Rogallo, R. S. (1993) The structure of intense vorticity in homogeneous isotropic turbulence, *J. Fluid Mech.*, **255**, 65-91.

Kadanoff L.P. (1990) Scaling and structures in the hard turbulence region of Rayleigh-Bénard convection, in L. Sirovich, editor, *New perspectives in turbulence*, p.265, Springer.

Kadanoff, L.P. (1997) Singularities and blowups, *Phys. Today*, **50**, No.9 (Sept.), 12–13.

Kaftori, D., Hetsroni, G. and Banerjee, S. (1998) The effect of particles on wall turbulence, *Int. J. Multiphase Flow*, **24**, 359–386.

Kahaleras, H., Malecot, Y. and Gagne, Y. (1998) Intermittency and Reynolds number, *Phys. Fluids*, **10**, 910–921.

Kambe, T. and Hatakeyama, N. (2000) Statistical laws and vortex structures in fully developed turbulence, *Fluid Dyn. Res.*, **27**, 247–267.

Karman, Th. von (1937) The fundamentals of the statistical theory of turbulence, *J. Aeronaut. Sci.*, **4**, 131–138.

Katul, G.G., Parlange, M.B. and Chu, C.R. (1994) Intermittency, local isotropy, and non-Gaussian statistics in atmospheric surface layer, *Phys. Fluids*, **6**, 2480–2492.

Keefe, L. (1990a) Connecting coherent structures and strange attractors, in *Near wall turbulence - 1988 Zaric memorial conference, editors, S. J. Kline and H. N. Afgan*, Hemisphere, pp. 63–80.

Keefe, L. (1990b) in J.L.Lumley (ed)., *Whither turbulence?*, Springer, p. 189.

Keefe, L., Moin, P. and Kim, J. (1992) The dimension of attractors underlying periodic turbulent Poiseulle flow, *J. Fluid Mech.*, **242**, 1–29.

Keller, L.V and Fridman, A.A. (1925) Differentialgleichung für die turbulente Bewegung einer kompressiblen Flüssigkeit, *Proc. First Int. Congr. Appl. Mech.*, ed. C.B. Biezeno and J. M. Burgers, Waltman, Delft.

Kennedy, D.A. and Corrsin, S. (1961) Spectral flatness factor and "intermittence" in turbulence and noinlinear noise, *J. Fluid Mech.*, **10**, 366–370.

Kerswell, R.R. (1999) Variational principle for the Navier-Stokes equations, *Phys. Rev.*, **E59,** 5482–5494.

Kevlahan, N.K.-R. and Hunt, J.C.R. (1997) Nonlinear interactions in turbulence with strong irrotational straining, *J. Fluid Mech.*, **337**, 333–364.

Kholmyansky, M. and Tsinober, A. (2001) On the origins of intermittency in real turbulent flows, in Vassilicos, J.C., editor, *Intermittency in Turbulent Flows*, pp. 183-192, Cambridge University Press.

Kholmyansky, M. Tsinober, A. and Yorish, S. (2001a) Velocity derivatives in the atmospheric turbulent flow at $Re_\lambda = 10^4$, *Physics of Fluids* , **13**, 311–314.

Kholmyansky, M. Tsinober, A. and Yorish, S. (2001b) Velocity derivatives in an atmospheric surface layer at $Re_\lambda = 10^4$. Further results. in Proceedings of the Second International Symposium on *Turbulence and Shear Flow Phenomena*, Stockholm, 27-29 June 2001, in press.

Kholmyansky, M. Shapiro-Orot, M. and Tsinober, A. (2001c) Experimental observations of spontaneous breaking of reflexional symmetry in turbulent flows, *Proc. Roy. Soc. London*, in press.

Kida, S. and Takaoka, M. (1994) Vortex reconnection, *Annu.Rev. Fluid Mech.*, **26,** 169–189.

Kiya, M. and Ishii, H. (1991) Vortex interaction and Kolmogorov spectrum, *Fluid Dyn. Res,* **8**(11), 73–83.

Kim, J. (1989) On the structure of pressure fluctuations in simulated turbulent channel flow, *J. Fluid Mech.*,**205,** 421–451.

Kim, H.T., Kline, S.J. and Reynolds, W.C. (1971) The production of turbulence near smooth wall in a turbulent boundary layer, *J. Fluid Mech.*, **50,** 133–160.

Kim, J.-H. and Stinger, J. (1992), editors, *Applied Chaos*, Wiley.

Kit, L.G., Turuntaev, S.V. and Tsinober, A. (1970) Investigation with conduction anemometer of the effect of magnetic field on disturbances in the wake of a cylinder, *Magnetohydrodynamics*, **5**, 331–335.

Klimontovich, Yu. L. (1996) Is turbulent motion chaos or order? Is the hydrodynamic or the kinetic description of turbulent motion more natural? *Physica*, **B228**, 51–62.

Kline, S.J., Reynolds, W.C., Schraub, F.A. and Runstadler, P.W. (1967) The structure of turbulent boundary layers, *J. Fluid Mech.*, **30**, 741–733.

Knobloch, E. and Moehlis, J. (2000) Burst mechanisms in hydrodynamics, in Debnath, L. and Riahi, D.N., editors, *Nonlinear Instability, Chaos and Turbulence*, vol. 2, pp. 237-287, WIT Press.

Kolmogorov, A.N. (1941a) The local structure of turbulence in incompressible viscous fluid for very large Reynolds numbers, *Dokl. Akad. Nauk SSSR*, **30**, 299–303; for English translation see *Selected works of A. N. Kolmogorov*, **I**, ed. V.M. Tikhomirov, pp. 321–318, Kluwer, 1991.

Kolmogorov, A.N. (1941b) Dissipation of energy in locally isotropic turbulence, *Dokl. Akad. Nauk SSSR,* **32**, 19–21; for English translation see *Selected works of A. N. Kolmogorov,* **I,** ed. V.M. Tikhomirov, pp. 324-327, Kluwer, 1991.

Kolmogorov, A.N. (1962) A refinement of previous hypotheses concerning the local structure of turbulence is a viscous incompressible fluid at high Reynolds number", *J. Fluid Mech.*, **13**, 82–85.

Kolmogorov, A.N. (1978) Remarks on statistical solutions of Navier-Stokes equations, *Uspekhi Matematicheskikh Nauk,* **33**, 124, in Russian. This section was omitted in the English translation.

Komori, S. and Nagata, K. (1996) Effects of molecular diffusivities on counter-gradient scalar and momentum transfer in strongly stable stratification, *J. Fluid Mech.,***326,** 205–237.

Konrad, J.H. (1976) An experimental investigation of mixing in two-dimensional turbulent shear flows with applications to diffusion-limited chemical reactions, *Project SQUID*, Techn. Rep. CIT-8-PU.

Kraichnan, R.H. (1962a) Turbulent thermal convection at arbitrary Prandtl number, *Phys. Fluids,* **5**, 1374–1389.

Kraichnan, R.H. (1962b) The closure problem in turbulence, in Birkhoff, G., Bellman, R. and Lin, C.C., editors (1962) *Hydrodynamic instability, Proceedings of Symposia in Applied Mathematics,* **XIII,** Amer. Math. Soc., pp. 199–232.

Kraichnan, R. H. (1967) Intermittency in the very small scales of turbulence, *Phys. Fluids,* **10**, 2080–2082.

Kraichnan, R.H. (1968) Small-scale structure of a scalar field convected by turbulence, *Phys. Fluids,* **11**, 945–963.

Kraichnan, R.H. (1972) Some modern developments in the statistical theory of turbulence, in: S.A. Rice, K.F. Freed, and J.C. Light, editors, *Statistical mechanics. New concepts, new problems, new applications,* The Univ. Chicago Press, pp. 201–228.

Kraichnan, R.H. (1974) On Kolmogorov's inertial-range theories, *J. Fluid Mech.*, **62**, 305–330.

Kraichnan, R.H. (1988) Reduced descriptions of hydrodynamic turbulence, *J. Stat. Phys.*, **51**, 949–963.

Kraichnan, R.H. (1987) Eddy viscosity and diffusivity: exacts formulas and approximations, *Complex systems*, **1**, 805–820.

Kraichnan, R.H. and Chen, S. (1989) Is there a statistical mechanics of turbulence?, *Physica*, **D37**, 160–172.

Kraichnan, R.H. and Kimura, Y. (1994) Probability distributions in hydrodynamic turbulence, *Progr. Astron. Aeronaut.*, **162**, 19–27.

Kraichnan, R.H. and Montgomery, D. (1980) Two dimensional turbulence, *Rep. Prog. Phys.*, **43**, 571–619.

Kraichnan, R.H. and Panda, R. (1988) Depression of nonlinearity in decaying isotropic turbulence, *Phys. Fluids*, **31**, 2395–2397.

Krasilnikov, E.Yu., Luschchik, V.G., Nikolaenko, V.S. and Panevin, I.G., (1971) Experimental study of the flow of an electrically conducting liquid in a circular tube in an axial magnetic field, *Fluid Dynamics*, **6**, 317–320.

Krishnamurti, R. and Howard, L.N. (1981) Large scale flow generation in turbulent convection, *Proc. Natl. Acad. Sci. USA*, **78**, 1981–1985.

Krishnamurti, R. and Howard, L.N. (1983) Large scale flow in turbulent convection: laboratory experiments and a mathematical model, *Papers in Meteorological Research, A Journal of the Meteorological Soc. of the Republic of China*, **6**(2), 143–159.

Krogstad, P.-Å. and Antonia, R.A. (1999) Surface effects in turbulent boundary layers, *Experiments in Fluids*, **27**, 450-460.

Kurien, S. and Sreenivasan, K.R. (2000) Anisotropic scaling contributions to high-order structure functions in high-Reynolds-number turbulence, *Phys. Rev.*, **E62**, 2206–2212.

Kush, H.A. and Ottino, J.M. (1992) Experiments in continuous chaotic flows, *J. Fluid Mech.*, **236**, 319–348.

Kuo, A.Y.-S. and Corrsin S. (1971) Experiments on internal intermittency and fine-structure distribution functions in fully turbulent fluid, *J. Fluid Mech.*, **50**, 285–319.

Kuznetsov, V.R., Praskovsky, A.A. and Sabelnikov, V.A. (1992) Finescale turbulence structure of intermittent shear flows, *J. Fluid Mech.*, **243**, 595–622.

Ladyzhenskaya, O.A. (1975) Mathematical analysis of NSE for incompressible liquids, *Ann. Rev. Fluid Mech.*, **7**, 249–272.

Landau, L.D. (1944) On the problem of turbulence, *Doklady Akad. Nauk SSSR*, **44**, 339-343 (in Russian). English translation in: D. Ter Haar, editor, *Collected papers of L.D. Landau*, Pergamon, Oxford, pp. 387–391.

Landau, L.D. (1960) Fundamental problems, in M. Fierz and V. F. Weisskopf, editors, *Theoretical Physics in the Twentieth Century, a Memorial Volume to Wolfgang Pauli*, Interscience, New York, pp. 245–247.

Landau, L.D. and Lifshits, E. M. (1944) *Fluid Mechanics*. First Russian edition.

Landau, L.D. and Lifshits, E.M. (1959) *Fluid Mechanics*, Pergamon

Landau, L.D. and Lifshits, E.M. (1987) *Fluid Mechanics*, Pergamon.

Lapeyre, G., Hua, B.L. and Klein, P. (2001), Dynamics of the orientation of active and passive scalars in two-dimensional turbulence, *Phys. Fluids*, **13**, 251–264.

Laws, E.M. and Livesey, J.L. (1978) Flow through screens, *Annu. Rev. Fluid Mech.*, **10**, 247–266.

Le Dizes, S., Rossi, M. and Moffatt, H.K. (1996) On the three-dimensional instability of elliptical vortex subjected to stretching, *Phys. Fluids*, **8**, 2084–2090.

Lele, S.K. (1994) Compressibility effects in turbulence, *Annu. Rev. Fluid Mech.*, **26**, 211–254.

Lele, S.K. (1997) Flows with density variations and compressibility: similarities and differences, in Fulachier L., Lumley, J.L. and Anselmet, F., editors, *Variable density low-speed turbulent flows*, pp. 279–301, Kluwer.

Leonard, A. (1974) Energy cascade in large-eddy simulations of turbulent fluid flows, *Adv. Geophys.*, **18A**, 237–248.

Leray, J. (1934) Sur le mouvement d'un liquide visqueux emplissant l'espace, *Acta Math.*, **63**, 193–248.

Lesieur, M. (1997) *Turbulence in fluids*, Kluwer.

Lesieur, M. and Métais, O. (1996) New trends in large eddy simulations, *Annu. Rev. Fluid Mech.*, **28**, 45–82.

Lewalle, J., Delville, J. and Bonnet, J.-P. (2000) Decomposition of mixing layer turbulence into coherent structures and background fluctuations, *Flow, Turulence and Combustion*, **64**, 301-328.

Libby, P.A. (1996) *Introduction to turbulence*, Taylor & Francis, Washington.

Lighthill, M.J. (1978) Acoustic streaming, *J. Sound Vib.*, **61**, 391–418.

Liepmann, H.W. (1962) Free turbulent flows, in Favre, A., (ed.), *Mécanique de la turbulence*, Proceedings of the *Colloques Internationaux du CNRS, Marseille, 28 Aug. - 2 Sept. 1961*, Publ. CNRS N° 108, Paris, pp. 211–227.

Liepmann, H.W. (1979) The rise and fall of ideas in turbulence, *Amer. Scientist*, **67**, 221–228.

Liepmann, H.W. and Laguna, G.A. (1984) Nonlinear interactions in the fluid mechanics of Helium II, *Annu. Rev. Fluid Mech.*, **16**, 139–177.

Lindborg, E. (1996) A note on Kolmogorov's third-order structure-function law, the local isotropy hypothesis and the pressure-velocity correlation, *J. Fluid Mech.*, **326**, 343–356.

Lindborg, E. (1996) Correction to the four-fifths law due to variations of the dissipation, *Phys. Fluids*, **11**, 510–512.

Lindborg, E. (1999) Can atmospheric kinetic energy spectrum be explained by two-dimensional turbulence?, *J. Fluid Mech.*, **388**, 259–288.

Lohse, D., Müller-Groeling, A. (1996) Anisotropy and scaling corrections in turbulence, *Phys. Rev.*, **54**, 395–405.

Loitsyanskii, L. G. (1966) *Mechanics of liquids and gases*, Pergamon.

Lomholt S. (1996) Boundary Layer Dynamics in Two-dimensional Flows, *Risø-I-1063(EN)*, Risø National Laboratory, Roskilde, Denmark, 112 pp.

Lorenz, E.N. (1953) The interaction between mean flow and random disturbances, *Tellus*, **5**, 238–250.

Lorenz, E.N. (1963) Deterministic nonperiodic flow, *J.Amosph. Sci.*, **20**, 130–141.

Lorenz, E. N. (1985) The growth of errors in prediction, in M. Ghil, R. Benzi and G. Parisi, (eds.), *Turbulence and predicatbility in Geophysical fluid dynamics and climate dynamics*, North-Holland, pp.243-265.

Lu, S.S. and Willmarth, W.W. (1973) Measurements of the structure of the Reynolds stress in a turbulent boundary layer, *J. Fluid Mech.*, **60**, 481–511.

Lumley, J.L. (1970), *Stochastic tools in turbulence*, Academic Press.

Lumley, J.L. (1978) Computational modelling of turbulent flows, *Adv. Appl. Mech.*, **18**, 123-176.

Lumley, J.L. (1981) Coherent Structures in Turbulence, in R. Meyer, editor, *Transition and Turbulence*, AP, pp. 215–242.

Lumley, J.L. (1987) Review of 'Turbulence and random processes in fluid mechanics' by M.T. Landahl and E. Mollo-Christensen, *J. Fluid Mech.*, **183**, 566–567.

Lumley, J.L. (1989) The State of Turbulence Research, in W. K. George and R. Arndt, editors, *Advances in Turbulence*, Hemisphere & Springer, pp. 1–10.

Lumley J.L. (1992) Some comments on turbulence, *Phys. Fluids*, **A4**, 203–211.

Lumley J.L. (1996) Turbulence and turbulence modelling, in J.L. Lumley, A.Acrivos, L.G. Leal and S. Leibovich, editors, *Research trends in Fluid Dynamics*, pp. 167–177, AIP.

Lumley, J.L. and Yaglom, A.M. (2000) A century of turbulence, *Advances in turbulence*, **8**, 3–12.

Lundgren, T.S. (1982) Strained spiral vortex model for turbulent fine structure, *Phys. Fluids* **25**, 2193–2203.

Lüthi, B., Burr, U., Kinzelbach, W. and Tsinober, A. (2001) Velocity derivatives in turbulent flow from 3D-PTV measurements, in Proceedings of the Second International Symposium on *Turbulence and Shear Flow Phenomena*, Stockholm, 27-29 June 2001, in press.

Machiels, L. and Deville, M.O. (1998) Numerical simulation of randomly forced turbulent flows, *J. Comp. Phys.*, **145**, 246–279.

Mahrt, L. and Howell, J.F. (1994) The influence of coherent structures and microfronts on scaling laws using global and local transforms, *J. Fluid Mech.*, **260**, 247–270.

Majda, A.J. and Kramer, P.R. (1999) Simplified models for turbulent diffusion: Theory, numerical modelling, and physical phenomena, *Phys. Rep.*, **314**, 237–574.

Malecot, Y. (1998) Intermittence en turbulence 3D: statistiques de la vitesse et de la vorticité, Ph. D. IMG, Grenoble.

Malkus, W.V.R. (1996) Statistical stability criteria for turbulent flow, *Phys. Fluids*, **8**, 1582–1587.

Mankbadi, R.R. (1994) *Transition, Turbulence and Noise: Theory and applications for scientists and engineers*, Kluwer.

Mann, J., Ott, S., and Andersen, J.S. (1999) Experimental study of relative, turbulent diffusion, *RISOE-R-1036(EN)*, RISOE Natl. Lab., Roskilde, Denmark; see also Ott, S. and Mann, J. An experimental investigation of the relative diffusion of particle pairs in three-dimensional turbulent flow (2000) *J. Fluid Mech.*, **422**, 207–223.

Mansour, N.N. and Wray, A.A. (1994) Decay of isotropic turbulence at low Reynolds number, *Phys. Fluids*, **6**, 808–814.

Marasli, B., Champagne, F. and Wygnanski, I.J. (1991) On linear evolution of unstable disturbances in a plane turbulent wake, *Phys. Fluids*, **A4**, 665–674.

Martin, J.N., Ooi, A., Chong, M.S. and Soria, J. (1998) Dynamics of the velocity gradient tensor invariants in isotropic turbulence, *Phys. Fluids*, **10**, 2336–2346.

Mathieu, J. and Scott, J.F. (2000) *Turbulent flows*, vol. I, An introduction, Cambr. Univ. Press., Cambridge, UK.

Maxworthy, T., Caperan, Ph. and Spedding, G.R. (1987) Two dimensional turbulence and vortex dynamics in a stratified fluid, *Third International Symposium on Density-Stratified Flows*, Caltech, Pasadena.

McComb, W.D. (1990) *The physics of Fluid Turbulence*, Clarendon Press, Oxford.

McComb, W.D. and Watt, A.G. (1992) Two-field theory of incompressible-fluid turbulence, *Phys. Rev.*, **A46**, 4797–4812.

McEwan, A.D. (1970) Inertial oscillations in a rotating fluid cylinder, *J. Fluid Mech.*, **40**, 603–640.

McEwan, A.D. (1976) Angular momentum diffusion and the initiation of cyclones, *Nature*, **260**, 126–128.

McIntyre, M.E. (1993) On the role of wave propagation and wave breaking in atmosphere-ocean dynamics, in S. Bodner, J. Singer, A. Solan and Z. Hashin, editors, *Theoretical and applied mechanics 1992*, pp. 281–304, Elsevier.

Melville, W.K. (1993) The role of wave breaking in air-sea interaction, in S. Bodner, J. Singer, A. Solan and Z. Hashin, editors, *Theoretical and applied mechanics 1992*, pp. 99–117, Elsevier.

Meneveau, C., (1991), Analysis of turbulence in the otthonormal wavelet representation, *J. Fluid Mech.*, **232**, 469–520.

Meneveau, C. and Katz, J. (2000) Scale-invariance and turbulence models for large-eddy simulation, *Annu. Rev. Fluid Mech.*, **32**, 1–32.

Metzger, M.M., Klewicki, J.C., Bradshaw, K.L. and Sadr, R (2001) Scaling the near-wall axial turbulent stress in the zero pressure gradient boundary layer, *Phys. Fluids*, **13,** 1819–1821.

Meunier, P., Leweke, T. (2000) Merging and three-dimensional instability in a co-rotating vortex pair, in A. Maurel & P. Petitjeans, editors *Vortex Structure and Dynamics*, *Lecture Notes in Physics*, pp. 235–245, Springer-Verlag, Berlin.

Miles, J. (1984) Resonant motion of a spherical pendulum, *Physica*, **D11**, 309-323.

Min, I.A., Mezic, I. and Leonard, A. (1996) Levy stable distributions for velocity and velocity difference in systems of vortex elements, *Phys. Fluids*, **8**, 1169–1180.

Moffatt, H. K. (1988) Topological approach to problems of vortex dynamics and turbulence, in *Progress in Astronautics and Aeronautics*, **112**, 141–152.

Moffatt, H.K. (1990) Fixed points of turbulent dynamical systems and suppression of nonlinearity, in J. L. Lumley, editor,*Whither turbulence? Turbulence at the crossroads*, Springer, Berlin, pp. 250–257.

Moffatt, H. K. (1993) Spiral structures in turbulent flow, in: T. Dracos and A. Tsinober, editors, *New Approaches and Concepts in Turbulence*, pp. 121-129, Birkäuser, Basel.

Moffatt, H.K. and Tsinober, A., editors (1990) *Topological fluid dynamics*, Cambr. Univ. Press.

Moffatt, H.K. and A. Tsinober, A. (1992) Helicity in laminar and turbulent flows, *Ann. Rev. Fluid Mech.*, **24**, 281–312.

Moffatt, H.K., Zaslavsky, G.M., Comte, P. and Tabor, M., editors (1992) *Topological aspects of the dynamics of fluids and plasmas*, Kluwer.

Moin, P. and Kim, J. (1985) The structure of the vorticity field in turbulent channel flow. Part 1. Analysis of instantaneous fields and statistical, correlations, *J. Fluid Mech.*, **155**, 441–464.

Moin, P. and Manesh, K. (1998) Direct numerical simulation: a tool in turbulence research, *Annu. Rev. Fluid Mech.*, **30**, 539–78.

Moisy, F., Tabeling, P. and Williame, H., (1999) Kolmogorov equation in fully developed turbulence, *Phys. Rev. Lett.*, **82**, 3994–3997.

Mollo-Christensen, E., (1973) Intermittency in large-scale turbulent flows, *Annu. Rev. Fluid Mech.*, **5**, 101–118.

Monin, A.S. (1986) Hydrodynamic instability, *Soviet Phys. Uspekhi*, **29**, 843-868.

Monin, A.S. (1987) On the negative viscosity in global circulations, *Doklady AN SSSR*, **293**, 70–73.

Monin, A.S. and Yaglom, A.M. *Statistical fluid mechanics, vol. 1*, (MIT Press, 1971), 2nd Russian edition, (Gidrometeoizdat, St. Petersburg, 1992); vol. 2, (MIT Press, 1975), 2nd Russian edition, (Gidrometeoizdat, St. Petersburg, 1996).

Monin, A.S. and Yaglom, A.M. (1997, 1998) *Statistical fluid mechanics, The mechanics of turbulence*, New English edition vol. 1, Chapters 2 and 3, CTR Monographs, NASA Ames - Stanford University.

Moreau, R. (1990) *Magnetohydrodynamics*, Kluwer.

Moser, R.D., Kim, J. and Mansour, N.N. (1999) Direct numerical simulation of turbulent channel flow up to $Re_\tau = 590$, *Phys. Fluids,* **11**, 943–945.

Moser, R.D., Rogers, M.M., and Ewing, D.W. (1998) Self-similarity of time-evolving plane wakes, *J. Fluid Mech.*, **367**, 255–289.

Mowbray, D.E. and Rarity, B.S.H. (1967), A theoretical and experimental investigation of the phase configuration of internal waves of small amplitude in a density stratified fluid, *J. Fluid Mech.*, **28**, 1–16.

Mueller, T.J., Nelson, R.C., Kegelman, J.T. and Morkovin, M.V. (1981) Smoke visualization of boundary-layer transition on a spinning axisymmetric body, *AIAA J.*, **19**, 1607-1608.

Mullin, T., editor, (1993) *The nature of chaos*, Clarendon press, Oxford.

Muttray, H. (1932) Die experimentellen Tatsachen des Wiederstandes ohne Auftrieb, *Handbuch der Experimentalphysik*, **4**, teil 2, pp. 235–332, Leipzig.

Mydlarski, L. and Warhaft, Z. (1996) On the onset of high Reynolds number grid generated wind tunnel turbulence, *J. Fluid Mech.*, **320**, 331–368.

Mydlarski, L. and Warhaft, Z. (1998) Three-point statistics and the anisotropy of a turbulent passive scalar, *Phys. Fluids,* **10**, 2885–2894.

Nagib, H.M., Lavan, Z, and Fejer, A.A. (1971) Stability of pipe flow with superposed solid body rotation, *Phys. Fluids,* **14**, 766–68.

Nelkin, M. (1995) Inertial range scaling of intense events in turbulence, *Phys. Rev.*, **E52**, R4610–R4611.

von Neumann, J., (1949) Recent theories of turbulence, A report to the Office of Naval Research. *Collected works,* **6** (1963), 437–472, ed. A.H. Taub, Pergamon.

Nie, Q. and S. Tanveer, S. (1999) A note on third order structure functions in turbulence, *Proc. Roy. Soc.*, **A455**, 1615–1635.

Niemela, J.J., Skrbek, L., Sreenivasan, K.R. and Donnelly, R.D. (2000) Turbulent convection at very high Rayleigh numbers, *Nature,* **404**, 837–840.

Nieuwstadt, F.T.M. and Brethouwer, G. (2000) Turbulent transport and mixing, *Advances in turbulence,* **8**, 133–140.

Nikitin, N.B. (1995) Spatial approach to numerical modelling of turbulence in pire flows, *Physics-Doklady,* **40**767–770.

Nikitin, N.B. (2000) Numerical investigation of laminar-turbulent transition in a circular pipe under periodic disturbances at the entrance, *Fluid Dynamics,* in press.

Nikitin, N.B. and Chernyshenko, S.I. (1997) On the nature of the organized structures in turbulent near-wall flows, *Fluid Dynamics,* **32**(1), 18–23; also 3rd European Fluid Mechanics Conference, 15-18 September, 1997, Göttingen.

Nomura, K.K. and Post, G.K. (1998) The structure and dynamics of vorticity and rate-of-strain in incompressible homogeneous turbulence, *J. Fluid Mech.*, **377**, 65–98.

Nore, C., Abid, M. and Brachet, M.E. (1997) Decaying Kolmogorov turbulence in a model superflow, *Phys. Fluids.*, **9**, 2644-2669.

Normal R.E. (1993) Strange spatio-temporal patterns, self-organized universal crises and politico-dynamics via symmetric chaos, neural networks and the fractal wavelet transform of the magic of names. *Nonlinear Science Today,* **3**, 13 - 14.

Novikov, E. A. (1959) K voprosy o predskazuemosty sinopticheskich protsessov, Isvest. Akad. Nauk SSSR, Ser. Geof., **11**, 1721-1724; English translation: (1960) Contribution to the Problem of the Predictability of Synoptical Processes, *Proc. Acad. Sci. USSR, Geoph. Ser.*, **11**, 1209-1211.

Novikov, E.A. (1963) Variation in the dissipation of energy in a turbulent flow and the spectral distribution of energy, *Prikl. Math. Mech.*, **27**, 944–946; English translation (1964) *J. Appl. Math. Mech.*, **27**, 1445–1446.

Novikov, E.A. (1967) Kinetic equations for a vortex field, *Doklady Akad. Nauk SSSR,* **177**(2), 299–301; English translation (1968) *Soviet Physics-Doklady,* **12**(11), 1006–1008.

Novikov E.A. (1971) Intermittency and scale similarity of the structure of turbulent flow, *Appl. Math. Mech.*, **35**, 231–253.

Novikov, E.A. (1974) Statistical irreversibility of turbulence, *Archives of Mechanics (Archiwum Mechaniki Stosowanej)*, **4**, 741–745.

Novikov E.A. (1990a) The effects of intermittency on statistical characteristics of turbulence and scale similarity of breakdown coefficients, *Phys. Fluids*, **A2**, 814–820.

Novikov E.A. (1990b) The internal dynamics of flows and formation of singularities, *Fluid Dyn. Res.*, **6**, 79–89.

Novikov E.A. (1993a) Statistical balance of vorticity and a new scale for vortical structures in turbulence, *Phys. Rev. Lett.*, **71**, 2718–2720.

Novikov E.A. (1993b) A new approach to the problem of turbulence, based on the conditionally averaged Navier-Stokes equations, *Fluid Dyn. Res.*, **12**, 107–126.

Novikov E.A. (1997) Lagrangian infinitesimal increments of vorticity in 2D turbulence, *Phys. Lett*, **A236**, 65–68 .

Novikov E.A. (2000) Vortical structure and modelling of turbulence, in Hunt, J.C.R. and Vassilicos, J.C., editors, *Turbulence Structure and Vortex Dynamics*, Cambridge University Press, 140–143.

Noullez, A., Wallace, G., Lempert, W., Miles, R.B. and Frisch U. (1997) Transverse velocity increments in turbulent flows using the RELIEF technique, *J. Fluid Mech.*, **339**, 287–307.

O'Neil, J. and Meneveau, C. (1997) Subgrid-scale stresses and their modelling in a turbulent plane wake, *J. Fluid Mech.*, **349**, 253–293.

Oberlack, M. (1999) Symmetries of the Navier-Stokes equations and their implications for subgrid models in large-eddy simulation of turbulence, in A. Gyr, W. Kinzelbach and A. Tsinober, editors, *Fundamental Problematic Issues in Turbulence*, pp. 247–255, Birkhäuser.

Oberlack, M. (1999) Symmetries, invariance and scaling laws in inhomogeneous turbulent shear flows, *Flow, Turbulence, and Combustion*, **62**, 111-135.

Ohkitani, K. (1994) Kinematics of vorticity: Vorticity-strain conjugation in incompressible fluid flows, *Phys. Rev.*, **E 50**, 5107–5110.

Ohkitani, K. (1998) Stretching of vorticity and passive vectors in isotropic turbulence, *J. Phys. Soc. Japan*, **67**, 3668–3671.

Ohkitani, K. and Kishiba, S. (1995) Nonlocal nature of vortex stretching in an inviscid fluid, *Phys. Fluids*, **7**, 411–421.

Omurtag, A. and Sirovich, L. (1999) On low-dimensional modelling of channel turbulence, *Theor. Comput. Fluid Dynamics*, **13**, 115–127.

Onsager, L. (1945) The distribution of energy in turbulence, *Phys. Rev*, **68**, 286.

Onsager, L. (1949) Statistical Hydrodynamics, *Suppl., Nuovo Cimento*, **VI**, ser. IX, 279-287.

Ooi, A., Martin, J., Soria, J. and Chong, M.S. (1999) A study of the evolution and characteristics of invariants of the velocity-gradient tensor in isotropic turbulence, *J. Fluid Mech.*, **381**, 141–174.

Orlandi, P. (1997) Helicity fluctuations and turbulent energy production in rotating and non-rotating pipes, *Phys. Fluids*, **9**, 1–12.

Orlanski, J. (1975) A rational subdivision of scales for atmospheric problems, *Bull. Amer. Meteorol. Soc.*, **56**, 527–530.

Orszag, S. (1977), Lectures on the statistical theory of turbulence, in: R. Balian and J.-L.Peube, editors, *Fluid Dynamics*, pp. 235–374, Gordon and Breach.

Oster, D. and Wygnanski, I. (1982) The forced mixing layer between parallel streams, *J. Fluid Mech.*, **123**, 91–130.

Ott, E. (1999) The role of Lagrangian chaos in the creation of multifractal measures, in A. Gyr, W. Kinzelbach and A. Tsinober, editors, *Fundamental Problematic Issues in Turbulence*, pp. 381–403, Birkhäuser.

Overholt, M.R. and Pope, S.B. (1996) Direct numerical simulation of a passive scalar with imposed mean gradient in isotropic turbulence, *Phys. Fluids*, **8**, 3128–3148.

Overholt. M.R. and Pope, S.B. (1998) A deterministic forcing scheme for direct numerical simulation of turbulence, *Computers and fluids*, **27**, 11–28.

Österlund, J.M., Johanson, A.V., Nagib, H.M. and Hites, M.H. (2000) A note on the overlap region in turbulent boundary layers, *Phys. Fluids*, **12**, 1–4 and 2360-2363.

Panton, R.L., (ed.), (1997) *Self-Sustaining mechanisms of wall turbulence*, Comp. Mech. Publ.

Pao, Y.-H. (1969) Origin and structure of turbulence in stably stratified media, in Pao, Y.-H. and Goldburg, A., editors, *Clear air turbulence and its detection*, pp. 73–99, Plenum.

Pao, Y.-H. and Goldburg, A. (1969) *Clear air turbulence and its detection*, Plenum.

Paret, J. and Tabeling, P. (1998) Intermittency in the two-dimensional inverse cascade of energy: experimental observations, *Phys. Fluids*, **10**, 3126–3136.

Pedretti, C. (1982) Leonardo da Vinci Nature studies from the Royal Library at Windsor Castle, Johnson Reprint Corp.

Perry A.E. and Fairlie, B.D. (1974) Critical points in flow patterns, *Adv. Geophys.*, **18B**, 299–315.

Phillips, O.M. (1966) *The dynamics of the upper ocean*, Cambr. Univ. Press.

Phillips, W.R.C., Wu, Z. and Lumley, J.L (1996) On the formation of longitudinal vortices in a turbulent boundary layer over a wavy terrain, *J. Fluid Mech.*, **326**, 321–341.

Pierrehumbert, R.T. and Widnall, S.E. (1982) 'The two- and three-dimensional instabilities of a spatially periodic shear layer, *J. Fluid Mech.*, **114**, 59–82.

Pinton, J.-F., Holdsworth, P.C.W. and Labbé, R. (1999) Power fluctuations in a closed turbulent shear flow, *Phys. Rev.*, **E60**, R2452–2455.

Pinsky, M., Khain, A. and Tsinober A. (2000) Accelerations in isotropic and homogeneous turbulence and Taylor's hypothesis, *Phys. Fluids*, **12**, 3195–3204.

Piomelli, U. (1999) Large-eddy simulation: achievements and chalenges, *Progress in Aerospace Sci.*, **35**, 335-362.

Piquet, J. (1999) *Turbulent flows. Models and Physics*, Springer.

Podvigina, and Pouquet, A. (1994) On the non-linear stability of the 1:1:1 ABC flow, *Physica*, **D75**, 471–508.

Pomeau, Y. (1995) Statistical approach to 2D turbulence, in P. Tabeling and O. Cardoso, editors, *Turbulence - A tentative Dictionary*, pp. 117–123, Plenum.

Pope, S.B. (2000) *Turbulent flows*, Cambr. Univ. Press., Cambridge, UK.

Popham, A.E. (1994) *The drawings of Leonardo da Vinci*, Pimlico.

Porthérat, A., Sommeria, J. and Moreau, R. (2000) An effective two-dimensional model for MHD flows with transverse magnetic field, *J. Fluid Mech.*, **424**, 75–100.

Porter, D.H., Woodward, P.R. and Pouquet, A. (1998) Inertial range structures in decaying compressible turbulent flows, *Phys. Fluids*, **10**, 237–245.

Prandtl, L. and Tietjens, O.G. (1934) *Applied Hydro- and Aeromechanics*, Dover.

Prasad, J.K., Rasheed, A., Kumar, S. and Sturtevant, (2000) The late-time development of the Richtmyer-Meshkov instability, *Phys. Fluids*, **12**, 2108–2151.

Praskovsky, A. (1998) Longitudinal and transverse structure functions in high Reynolds number turbulent flows, unpublished manuscript.

Praskovsky, A., Gledzer, E.B., Karyakin, M. Yu. and Zhou, Y. (1993) The sweeping decorrelation hypothesis and energy-inertial scale interaction in high Reynolds number flows, *J. Fluid Mech.*, **248**, 493–511.

Proudman, I. and Reid, W.H. (1954) On the decay of a normally distributed and homogeneous turbulent velocity field, *Phil. Trans. Roy. Soc.*, **A247**, 163–189.

Pullin, D.I. and Saffman P.S. (1997) Vortex dynamics and turbulence, *Annu. Rev. Fluid Mech.*, **30**, 31–51.

Pumir, A., Shraiman, B.I. and Siggia, E.D. (1997) Perturbation theory for the δ-correlated model of passive scalar advection near the Batchelor limit, *Phys. Rev.*, **E55**, R1263–R1266.

Rempfer, D. (1999) On dynamical systems obtained via Galerkin projections onto low-dimensional bases of eigenfunctions, in: A. Gyr, W. Kinzelbach and A. Tsinober, editors, in: A. Gyr, W. Kinzelbach and A. Tsinober, editors, *Fundamental Problematic Issues in Turbulence*, pp. 233–245, Birkhäuser.

Rempfer, D. (2000) On low-dimensional Galerkin models for fluid flow, *Theor. Comput. Fluid Dynamics,* **14**, 75–88.

Renner, C., Peinke, J. and Friedrich, R. (2001) Experimental indications for Markov properties of small-scale turbulence, *J. Fluid Mech.* ,**433**, 383–409.

Reynolds O. (1883) An experimental investigation of the circumstances which deteremine whether the motion of water shall be direct or sinuous, and the law of resistance in parallel channels, *Phil. Trans. Roy. Soc.* **174**, 935–982.

Reynolds. O. (1894) Study of fluid motion by means of coloured bands, *Nature,* **50**, 161–164.

Reynolds. O. (1895) On the dynamical theory of turbulent incompressible viscous fluids and the determination of the criterion, *Phil. Trans. Roy. Soc.*, **186**, 123–161.

Ricca, R. and Berger, M. (1996) Topological ideas and fluid mechanics, *Physics Today,* **49**, 28–34.

Ricca, R.L., Samuels, D., Barenghi, C.F. (1999) Evolution of vortex knots, *J. Fluid Mech.* ,**391**, 29–44.

Richardson, L.F. (1922) Weather prediction by numerical process, Cambridge University Press.

Richter, J.P. (1970) The Notebooks of Leonardo da Vinci, Vol. 1 & 2, Dover.

Ricou, F.P. and Spalding, D.P. (1961) Measurements of entrainment by axisymmetric turbulent jets, *J. Fluid Mech.*, **11**, 21–32.

Riley, J.J. and Lelong, M.-P. (2000) Fluid motions in the presence of strong stable stratification, *Annu. Rev. Fluid Mech.*, **32**, 613–657.

Rivera, M., Vorobieff, P. and Ecke, R.E. (1998) Turbulence in flowing soap films: velocity, vorticity, and thickness fields, *Phys. Rev. Lett.*, **81**, 1417-1420.

Roche, P.M., Castaing, B., Chabaud, B. and Herbal, B. (2001) Observation of 1/2 power law in Rayleigh-Bénard cell, *Phys. Rev.*, **E63**, 045303—1-4, in press.

Rogachevskii, I. and Kleorin, N. (1997) Intermittency and anomalous scaling for magnetic fluctuations, *Phys. Rev.*, **E56,** 417-426.

Rohr, J., Allen, J., Losee, J. and Latz, M.I. (1997) The use of bioluminescence a flow diagnostic, *Phys. Lett.*, **A228**, 408–416.

Roshko, A. (1976) Structure of turbulent shear flows: a new look, *AIAA J.*, **14**, 1349–1357.

Roshko, A. (1993) Instability and turbulence in shear flows, in S. Bodner, J. Singer, A. Solan and Z. Hashin, editors, *Theoretical and applied mechanics 1992*, pp. 1–24, Elsevier.

Rossby, C.G. (1947) On the distribution of angular velocityin gaseous envelopes under the influence of large-scale horizontal mixing processes, *Bull. Amer. Met. Soc.*, **28**, 53-68

Roux, S., Muzy, J.F. and Arneodo, A., (1999) Detecting vorticity filaments using wavelet analysis: about the statistical contribution of vorticity filaments to intermittency in swirling turbulent flows, *Eur. Phys. J.*, **B8**, (*Condensed Matter Physics*), 301–322.

Ruelle, D. (1990) The turbulent fluid as a dynamical system, in: L. Sirovich, editor, *New perspectives in turbulence*, Springer, pp. 123–138.

Ruetsch, G.R. and Maxey, M.R. (1991) Small-scale features of vorticity and passive scalar fields in homogeneous turbulence, *Phys. Fluids*, **A3**, 1587–1597.

Ruetsch, G.R. and Maxey, M.R. (1992) The evolution of small-scale structures in homogeneous turbulence, *Phys. Fluids*, *A4,* 2747–2760.

Ruiz-Chavaria, G., Baudet, C. and Ciliberto, S. (1996) Scaling laws and dissipation scale of a passive scalar in fully developed turbulence, *Physica*, **D99**, 369-380.

Saddoughi, S.G. (1997) Local isotropy in complex turbulent boundary layers at high Reynolds number, *J. Fluid Mech.*, **348**, 201–245.

Saffman, P.G. (1960) Lectures on homogeneous turbulence, in N.J. Zabusky, editor, *Topics in nonlinear physics*, pp. 485–614, Springer.

Saffman, P.G. (1978) Problems and progress in the theory of turbulence, in *Structure and Mechanics of turbulence, II*, ed. Fiedler, H., *Lect. Notes Phys.*, Springer, **76**, pp. 274–306.

Saffman, P.G. (1991) in *The Global Geometry of Turbulence*, NATO ASI Ser. **B 268**, ed. J. Jimenez, Plenum, p. 348.

Sain, A. (1998) Multiscaling in three-dimensional fluid turbulence, Ph.D. Thesis, Department of Physics, Indian Institute of Science, Bangalore.

Sain, A., Manu and Pandit, R. (1998) Turbulence and multiscaling in the randomly forced Navier-Stokes equation, *Phys. Rev. Lett.*, **81**, 4377–4380.

Sandham, N. and Tsinober A. (2000) Kinetic energy, enstrophy and strain rate in near-wall turbulence, *Advances in turbulence*, **8**, 407–410.

Sanford, T.B., Carlson, J.A., Dunlap, J.H., Prater, M.D. and Lien, R.-C. (1999) An electromagnetic vorticity and velocity sensor for observing fine scale kinetic fluctuations in the ocean, *J. Atm. Ocean. Techn.*, **16**, 1647–67.

Sarkar, S. (1995) The stabilizing effect of compressibility in turbulent shear layers, *J. Fluid Mech.*, **282**, 163–186.

Savill, A.M. (1987) Recent progress in rapid distortion theory, *Annu. Rev. Fluid Mech.*, **19**, 531–575.

Schiller, L. (1932) Fallversuche mit Kugeln und Scheiben, in L. Schiller, editor, *Handbuch der Experimentalphysik*, vol. **4**. part, I, pp. 339–387.

Schlichting, H. (1979) *Boundary layer theory*. McGraw-Hill.

Schultz-Grunov, F. (1980) Sudden transition to turbulence, in R. Eppler and H. Fasel, editors, *Laminar-turbulent transition, Springer*, pp. 389–395.

Schwarz, K.W. (1990) Evidence for organized small-scale structure in fully developed turbulence, *Phys. Rev. Lett.* **64**, 415–418.

Scorer, R.S. (1978) *Environmental aerodynamics*, Wiley.

Serrin, J. (1959) Mathematical principles of classical fluid mechanics, in S. Flugge and C. Truesdell, editors, *Handbuch der Physik*, Band **VIII/1**, Stromungsmechanik, pp. 125–263, Springer.

She, Z.-S. (1991) Intermittency and non-Gaussian statistics in turbulence, *Fluid Dyn. Res.*, **8**, 143–158.

She, Z.-S., Jackson, E. and Orszag, S.A. (1990) Intermittent vortex structures in homogeneous isotropic turbulence, *Nature*, **344**, 226–229.

She, Z.-S., Jackson, E. and Orszag, S.A. (1991) Structure and dynamics of homogeneous turbulence: models and simulations, *Proc. Roy. Soc. Lond.*, **A 434**, 101–124.

Shebalin, J.V. and Woodruff, S.L. (1997) Kolmogorov flow in three dimensions, *Phys. Fluids*, **9**, 164–170.

Shen, X. and Warhaft, Z. (2000) The anisotropy of the small-scale structure in high Reynolds number ($Re_\lambda = 1,000$) turbulent shear flow, Phys. Fluids, **12**, 2976–2989.

Sherman, F.S. (1990) *Viscous flow*. McGraw-Hill.

Shlesinger, M.S. (2000) Exploring phase space, *Nature*, **405**, 135, 137.

Shraiman, B. and E. Siggia, E. (1999) Fluctuations and mixing of a passive scalar in turbulent flow, in Chaté, H., Villermaux, E. and Chomaz, J.M., editors, *Mixing: Chaos and Turbulence*, NATO ASI Series, **B373**, pp. 361-383, Plenum Press.

Shraiman, B. and Siggia, E. (2000) Scalar turbulence, *Nature*, **405**, 639-646.

Shtilman, L., Spector, M. and Tsinober, A. (1993) On some kinematic versus dynamic properties of homogeneous turbulence, *J. Fluid Mech.*, **247**, 65–77.

Sieber, A. (1987) Experiments on attractor-dimension for turbulent pipe flow, *Phys. Lett.*, **A122**, 467–470.

Siggia, E.D. (1981) Numerical study of small-scale intermittency in three-dimensional turbulence, *J. Fluid Mech.*, **107**, 375–406.

Siggia, E. (1994) High Rayleigh number convection, *Annu. Rev. Fluid Mech.*, **26**, 137–168.

Sigurdson, L.W. (1997) Flow visualization in turbulent large-scale structure research, in *Atlas of Visualization III*, edited by The Visualization Society of Japan, pp. 99–113, CRC Press.

Sinai, Ya.G. (1999) Mathematical problems of turbulence, *Physica*, **A263**, 565–566.

Sirovich, L. (1997) Dynamics of coherent structures in wall bounded turbulence, in Panton, R.L., editor, *Self-Sustaining Mechanisms of wall turbulence*, Comp. Mech. Publ., pp.333–364.

Sirovich, L., Ball, K.S. and Handler, R.A. (1991) Propagating structures in wall bounded turbulent flows, *Theor. Comput. Fluid Dyn.*, **2**, 307–317.

Smale, S., (1998) Mathematical problems for the next century, *The Mathematical Intelligencer*, **20**, 7–15.

Smith, G.M. and Wei, T. (1994) Small-scale structure in colliding off-axis vortex rings, *J. Fluid Mech.*, **259**, 281–290.

Smyth, W.D. and Moum, J.N. (2000) Length scales in stably stratified mixing layers, *Phys. Fluids*, **12**, 1327–1342.

Smyth, W.D. and Moun, J.N. (2000) Anisotropy of turbulence in stably stratified mixing layers, *Phys. Fluids*, **12**, 1343–1362.

Soldati, A., editor, (2000) Turbulence modulation and control, Springer.

Sommeria, J. (2001) Two-dimensional turbulence, in M. Lesieur, A. Yaglom and F.David, editors, *New trends in turbulence*, *Les Houches 2000*, EDP/Springer, in press.

Soria, J., Sondergaard, R., Cantwell, B.J., Chong, M.S., and Perry, A.E. (1994) A study of the fine-scale motions of incompressible time-developing mixing layers, *Phys. Fluids*, **6**, 871–884.

Sparrow, E.M., Husar, R.B. and Goldstein, R.J. (1970) Observations and other characteristics of thermals, *J. Fluid Mech.*, **41**, 793–800.

Spedding, G.R., Browand, F.K. and Fincham, A.M., (1996) Turbulence, similarity scaling and vortex geometry in the wake of a towed sphere in a stratified fluid, *J. Fluid Mech.*, **314**, 53–103.

Sreenivasan, K.R. and Antonia, R. (1997) The phenomenology of small-scale turbulence, *Annu. Rev. Fluid Mech.*, **29**, 435–472.

Sreenivasan, K.R. and Dhruva, B. (1998) Is there scaling in high-Reynolds number turbulence?, *Prog. Theor. Phys.*, *Suppl.*, No. 130, 103–120.

Sreenivasan, K.R. and White, C.M. (2000) The onset of drag reduction by dilute polymer additives, and the maximum drag reduction asymptote, *J. Fluid Mech.*, **409**, 149–164.

Staquet, C. and Sommeria, J. (1996) Internal waves, turbuence and stratified flows: a report on Euromech Colloquium 339, *J. Fluid Mech.*, **314**, 349–371.

Starr, V. (1968) *Physics of negative viscosity phenomena*, McGraw Hill.

Stewart, R.W. (1959) The Problem of Diffusion in a stratified fluid, in Frenkiel, F.N. and Sheppard, P.A., editors, *Atmospheric Diffusion and Air Pollution*, *Adv. Geophys.*, **6**, pp. 303–311, AP.

Stewart, R. W. (1969) Turbulence and Waves in Stratified Atmosphere, *Radio Sci*, **4**, 1269–1278.

Stoyan, D. and Stoyan, H. (1992) *Fractals, random shapes and point fields*, Wiley.

Su, L.K. and Dahm, W.J.A. (1996b) Scalar imaging velocimetry measurements of the velocity gradient tensor field in turbulent flows. II. Experimental results, *Phys. Fluids*, **8**, 1883–1906.

Tabor, M. and Klapper, I. (1994), Stretching and alignment in chaotic and turbulent flows, *Chaos, Solitons & Fractals*, **4**, 1031–1055.

Taguchi, Y-H., 1995 Numerical study of granular turbulence and the appearance of the $k^{-5/3}$ spectrum without flow, *Physica*, **D80**, 61–71.

Tam, C.K.W. (1972) On the noise of a nearly expanded supersonic jet, *J. Fluid Mech.*, **51**, 69–95.

Tan-Attichat, J., Nagib, H.M., Loehrke, R.I. (1982) Interaction of free-stream turbulence with screens and grids: a balance between turbulence scales, *J. Fluid Mech.*, **114**, 501–528.

Tanaka, M. and Kida, S. (1993) Characterization of vortex tubes and sheets, *Phys. Fluids* **A 5**, 2079–2082 (1993).

Tao, B., Katz, J. and Meneveau, C. (2000) Geometry and scale relationships in high Reynolds number turbulence determined from three-dimenisonal holographic velocimetry, *Phys. Fluids*, **12**, 941-4.

Taylor G.I. (1935) The statistical theory of turbulence, *Proc. Roy. Soc. London*, **A151**, 421–478.

Taylor G.I. (1937) The statistical theory of isotropic turbulence, *J. Aeronaut. Sci.*, **4**, 311–315.

Taylor G. I. (1938a) Production and dissipation of vorticity in a turbulent fluid, *Proc. Roy. Soc. London*, **A164**, 15–23.

Taylor G. I. (1938b) The spectrum of turbulence, *Proc. Roy. Soc. London*, **A164**, 476–490.

Tennekes, H. (1973) Intermittency of the small-scale structure of atmospheric turbulence, *Boundary layer meteorology*, **4** , 241–250.

Tennekes, H. (1975) Eulerian and Lagrangian time microscales in isotropic turbulence, *J. Fluid Mech*, **67**, 561–567.

Tennekes, H. (1976) Fourier-Transform Ambiguity in Turbulence Dynamics, *J. Atmosp. Sci*, **33**, 1660-1163.

Tennekes, H. (1989) Two- and three-dimensional turbulence, in J.R. Herring and J.C. McWilliams, editors, *Lecture notes on turbulence, NCAR-GTP 1987*, pp. 1 – 73, World Scientific, Singapore.

Tennekes, H. (1993) Karl Popper and the accountability of numerical weather forecasting, Workshop *Predictability and chaos in the geosciences*, Boulder, 7-10 September 1993, 343–346.

Tennekes, H. and Lumley, J.L. (1972) *A first course of turbulence*, MIT Press.

Thorpe, S.A. (1971) Experiments on instability of stratified shear flows: miscible fluids, *J. Fluid Mech.*, **46**, 299–319.

Thorpe, S.A. (1987) Transitional phenomena and the development of turbulence in stratified fluids: a review, *J. Geophys. Res.*, **92**, C5, 5231-48.

Tikhomirov, V. M., editor, (1991) *Selected works of A. N. Kolmogorov*, **I**, Kluwer.

Tong, P., Goldburg, W.I. Huang, J.S. and Witten, T.A. (1990) Anisotropy in turbulent drag reduction, *Phys. Rev. Lett.*, **65**, 2780–3.

Tollmien, W. (1931) Turbulente strömungen, in L. Schiller, editor, *Handbuch der Experimentalphysik*, vol. **4**. part, I, pp. 291–342.

Townsend, A.A., (1948) Local isotropy in the turbulent wake of cylinder, *Austr. J. Sci. Res.*, **1**, 161 - 174.

Townsend, A.A. (1951) On the fine scale structure of turbulence, *Proc. Roy. Soc.*, **A208**, 534–542.

Townsend, A.A. (1954) The uniform distortion of homogeneous turbulence, *Quart. J. Mech. Appl. Math,*, **7**, 104–127.

Townsend, A.A., (1976) *The structure of turbulent shear flow*, Cambr. Univ, Press.

Tritton, D. J. (1985) Experiments on turbulence in geophysical fluid dynamics, I. Turbulence in rotating fluids, in M. Ghill, R. Benzi and G. Parisi *Turbulence and predictability in geophysical fluid dynamics*, pp. 172–192, North-Holland.

Tritton, D. J. (1988) *Physical fluid dynamics*, 2nd ed., Clarendon press, Oxford.

Tsinober, A. (1990a) On one property of Lamb vector in isotropic turbulent flow, *Phys. Fluids,* **A2**, 484–486.

Tsinober, A. (1990b) Turbulent drag reduction versus structure of turbulence, in A. Gyr, editor, *Structure of turbuence and drag reduction*, pp. 313–334, Springer.

Tsinober, A. (1990c) MHD flow drag reduction, in D.M. Bushnell and J.N. Hefner, editors, *Viscous drag reduction in boundary layers*, pp. 327–349.

Tsinober, A. (1993a) Some properties of velocity derivatives in turbulent flows as obtained from experiments (laboratory and numerical), in R. Benzi, C. Basdevant and S. Ciliberto, editors, *Turbulence in Spatially Extended Systems*, pp. 117–127, Nova Science.

Tsinober, A. (1993b) How important are direct interactions between large and small scales in turbulent flows?, in: T. Dracos and A. Tsinober, editors, *New Approaches and Concepts in Turbulence*, pp.141–151, Birkäuser, Basel.

Tsinober, A. (1995a) Variability of anomalous transport exponents versus different physical situations in geophysical and laboratory turbulence, in: *Levy Flights and Related Topics in Physics*, *Lecture notes in physics*, **450**, editors, M. Schlesinger, G. Zaslavsky and U. Frisch, Springer, pp. 3–33.

Tsinober, A. (1995b) On geometrical invariants of the field of velocity derivatives in turbulent flows, in *Actes Congrès Francais de Mécanique* , **3**, 409–412.

Tsinober, A. (1996a) Geometrical statistics in turbulence, *Advances in Turbulence*, **6,** 263 - 266.

Tsinober, A. (1996b) Review of *Turbulence: The legacy of A. N. Kolmogorov* by U. Frisch, Cambridge University Press (1995), *J. Fluid Mech.*, **317**, 407–411.

Tsinober, A. (1998a) Is concentrated vorticity that important?, *Eur. J. Mech.*, **B**/*Fluids*, **17**, 421–449.

Tsinober, A. (1998b) Turbulence - Beyond Phenomenology, in S. Benkadda and G.M. Zaslavsky, editors, *Chaos, Kinetics and Nonlinear Dynamics in Fluids and Plasmas*, *Lecture Notes in Physics,* **511**, 85-143, Springer.

Tsinober, A. (1999) On statistics and structure(s) in turbulence, Isaac Newton Institute for Mathematical Sciences, Preprint NI99010-TRB. A synopsis of lectures given at the IUTAM/IUGG Symposium, *Developments in Geophysical Turbulence*, NCAR, Boulder, Colorado, USA, June 16-19, 1998 and at the Workshop on *Perspectives in Understanding of Turbulent Systems*, Isaac Newton Institute, Cambridge, January 13-22, 1999.

Tsinober, A. (2000) Vortex stretching versus production of strain/dissipation, in Hunt, J.C.R. and Vassilicos, J.C., editors, *Turbulence Structure and Vortex Dynamics*, Cambridge University Press, 164–191.

Tsinober, A. (2001) Nonlocality in turbulence, in R.J. Donnelly, W.F. Winen and Barenghi, C., editors, *Quantized vortex dynamics and superfluid turbulence,* Springer, in press.

Tsinober, A. and Galanti, B. (2001) Geometrical statistics of passive vectors, submitted.

Tsinober, A., Eggels, J.G.M. and Nieuwstadt, F.T.M. (1995) On alignments and small scale structure in turbulent pipe flow, *Fluid Dyn. Res.*, **16**, 297–310.

Tsinober, A. Kit, E. and Dracos, T. (1992) Experimental investigation of the field of velocity gradients in turbulent flows, *J. Fluid Mech.*, **242**, 169–192.

Tsinober, A. and Ortenberg, M. (2001) The dynamics of errors in numerical turbulence, submitted.

Tsinober, A., Ortenberg, M. and Shtilman, L. (1999) On depression of nonlinearity in turbulence, *Phys. Fluids*, **11**, 2291 - 2297.

Tsinober, L. Shtilman and H. Vaisburd (1997) A study of vortex stretching and enstrophy generation in numerical and laboratory turbulence, *Fluid Dyn. Res.*, **21**, 477–494.

Tsinober, A., Shtilman, L., Sinyavskii, A. and Vaisburd, H. (1995) Vortex stretching and enstrophy generation in numerical and laboratory tubulence, in *Small-scale structures in three-dimensional hydro- and magnetohydrodynamic turbulence, editors, M. Meneguzzi, A. Pouquet and P.-L. Sulem, Springer, Lect. Notes Phys.*, **462**, pp. 9–16.

Tsinober A., Vedula, P. and Yeung, P.K. (2000) Random Taylor hypothesis and the behaviour of local and convective accelerations in isotropic turbulence, *Phys. Fluids*, **13**, in press.

Tsui, Y. (1991) Review: Turbulence modification in fluid-solid flows. In D.E. Stock et al., editors, *Gas-Solid Flows*, ASME-FED, **110**, 1–6.

Tsuji, Y. and Dhruva, B. (1999) Intermittency feature of shear stress in high-Reynolds-number turbulence, *Phys. Fluids*, **11**, 3017 – 3025.

Turner, J.S. (1985) Multicomponent convection, *Annu. Rev. Fluid Mech.*, **17**, 11–44.

Turner, J.S. and Chen, C.F. Two-dimensional effects in double-diffusive convection, *J. Fluid Mech*, **63**, 577–592.

Vaillancourt, P.A. and Yau, M.K. (2000) Review of particle-turbulence interactions and consequences for cloud physics, *Bull. Amer. Meteorol. Soc.*, **81**, 285–298.

Van Atta, C.W. (1974) Sampling techniques in turbulence measurements, *Annu. Rev. Fluid. Mech.*, **6**, 75–91.

Van Dyke, M. (1982) *An Album of Fluid Motion*, The Parabolic Press.

Van Zandt, T.E. (1982) A universal spectrum of buoyancy waves in the atmosphere, *Geophys. Res. Lett.*, **9**, 575-578.

Vasil'kov, A.P., Yefimov, S.V. and Kondranin, T.V. (1992) Remote measurement of bioluminescence in the sea, *Oceanology, 32*, 530–533.

Vassilicos, J.C. (2001), editor, *Intermittency in Turbulent Flows*, Cambridge University Press.

Vassilicos, J.C. and Brasseur, J. (1996), Self-similar spiral flow structure in low Reynolds number isotropic and decaying turbulence, *Phys. Rev.*, **E54**, 467–485.

Vedula, P. and Yeung, P.K. (1999) Similarity scaling of acceleration and pressure statistics in numerical simulations of isotropic turbulence, *Phys. Fluids*, **11**, 1208–1220.

Vergassola, M. (1996) Anomalous scaling for passively advected magnetic fields, *Phys. Rev.*, **E53**, R3021–R3024.

Villermaux, E., Innocenti, C. and Duplat., J. (2001) Short circuits in the Corrsin–Obulhov cascade, *Phys. Fluids*, **13**, 284–289.

Villermaux, E., Sixou, B. and Gagne, Y. (1995) Intense vortical structures in grid-generated turbulence, *Phys. Fluids*, **7**, 2008–2013.

Vincent, A. and Meneguzzi, M. (1991) The spatial structure and statistical properties of homogeneous turbulence, *J. Fluid Mech.*, **225**, 1–20.

Vincent, A. and Meneguzzi, M. (1994) The dynamics of vorticity tubes in homogeneous turbulence, *J. Fluid Mech.*, **258**, 245–254.

Vincent, A.P. and Yuen, D.A. (2000) Transition to turbulent thermal convection beyond $Ra = 10^{10}$ detected in numerical simulations, *Phys. Rev.*, **E61**, 5241–5246.

Virk, P.S. (1971) Drag reduction in rough pipes, *J. Fluid Mech*, **45**, 225–246.

Waleffe, F. (1990) Proposal for a self-sustaining mechanism in shear flows, *CTR, Stanford University and NASA Ames*, March 1990, unpublished manuscript.

Waleffe, F. and J. Kim (1998) How streamwise rolls and streaks self-sustain in a shear flow. Part 2, *AIAA 98-2997*, 1–8.

Warhaft, Z. (2000) Passive scalars in turbulent flows, *Annu. Rev. Fluid Mech.*, **32**, 203–240.

Warholic, M.D., Massah, H., Hanratty, T.J. (1999) Influence of drag-reducing polymers on turbulence: effects of Reynolds number, concentration and mixing, *Experiments in Fluids*, **27**, 461-472.

Webber, G.A., Handler, R.A. and Sirovich, L. (1997) The Karhunen-Loeve decomposition of the minimal channel, *Phys.Fluids*, **9**, 1054–1066.

Wei, T. and , Willmarth, W.W. (1989) Reynolds-number effects on the structure of a turbulent channel flow. *J. Fluid Mech.*, **204**, 57–95.

Wei, T. and , Willmarth, W.W. (1992) Modifying turbulent structure with drag-reducing polymer additives in turbulent channel flow. *J. Fluid Mech.*, **245**, 619–641.

Weisbrot, I. and Wygnanski, I. (1988) On coherent structures in a highly excited mixing layer, *J. Fluid Mech.*, **195**, 137–159.

Weiss, J. (1981) The dynamics of enstrophy transfer in two-dimensional hydrodynamics, La Jolla Inst. Tech. Rep., LJI-TN-81-121, La Jolla California. See also (1991), *Physica*, **D48**, 273–294.

Werlé, H. (1987) *Transition et Turbulence*, ONERA, Note Technique 1987-7.

Wijngaarden, L. van (2000) Future trends in fluid mechanics, in *Mechanics at the turn of the century*, IUTAM, 85-89.

Wiltse, J.M. and Gledzer, A. (1998) Direct excitation of small-scale motions in free shear layers, *Phys. Fluids*, **10**, 2026–2036.

Wosnik, M., Castillo, L. and George, W.K. (2000) A theory of turbulent pipe and channel flows, *J. Fluid Mech.*, **421**, 115–145.

Wygnanski, I., Champagne, F. and Marasli, B. (1986) On the large-scale structures in two-dimensional, small-deficit, turbulent wakes, *J. Fluid Mech.*, **168**, 31–71.

Xu, X., Bajaj, K.M.S. and Ahlers, G. (2000) Heat transport in turbulent Rayleigh-Benard convection, *Phys. Rev. Lett.*, **84**, 4357–4360.

Xin, Y.-B., Xia, K.-Q. and Tong, P. (1996) Measured velocity boundary layers in turbulent convection, *Phys. Rev. Lett.*, **77**, 1266–9.

Yanitskii, V.E. (1982) Transport Equation for The Deformation- Rate Tensor and Description of an Ideal Incompressible Fluid by a System of Equations of the Dynamical Type, *Sov. Phys. Dokl*, **27**, 701-703.

Yeung, P.K., Brasseur, J. and Wang, Q. (1995) Dynamics of direct large-small-scale coupling in coherently forced turbulence: concurrent physical- and Fourier-space views, *J. Fluid Mech*, **283**, 743–95.

Zakharov, V.E., editor (1990) *What is integrability?*, Springer.

Zakharov, V.E., L'vov, V.S. and Falkovich, G. (1993) *Kolmogorov spectra of turbulence*, **I**, Springer.

Zaslavsky, G.M., 1999, Chaotic dynamics and the origin of statistical laws, *Physics Today*, **51**, 39–45.

Zel'dovich, Ya.B., Ruzmaikin, A.A. and Sokoloff, D.D. (1983) *Magnetic fields in astrophysics*, Gordon and Breach.

Zel'dovich, Ya.B., Molchanov, S.A., Ruzmaikin, A.A. and Sokolov, D.D. (1988) Intermittency, diffusion and generation in a nonstationary random medium, *Sov. Sci. Rev.*, **C7**, 1–110.

Zel'dovich, Ya.B., Ruzmaikin, A.A. and Sokoloff, D.D. (1990) *The almighty chance*, World Scientific.

Zhou, T. and Antonia, R. A., (2000) Reynolds number dependence of small-scale structure of grid turbulence, *J. Fluid Mech.*, **406**, 81–107.

Zimin, V.D. and Frik, P.G. (1988) *Turbulent convection*, Nauka, Moscow, in Russian.

Mechanics

FLUID **MECHANICS AND ITS APPLICATIONS**

Series Editor: R. Moreau

Aims and Scope of the Series

The purpose of this series is to focus on subjects in which fluid mechanics plays a fundamental role. As well as the more traditional applications of aeronautics, hydraulics, heat and mass transfer etc., books will be published dealing with topics which are currently in a state of rapid development, such as turbulence, suspensions and multiphase fluids, super and hypersonic flows and numerical modelling techniques. It is a widely held view that it is the interdisciplinary subjects that will receive intense scientific attention, bringing them to the forefront of technological advancement. Fluids have the ability to transport matter and its properties as well as transmit force, therefore fluid mechanics is a subject that is particularly open to cross fertilisation with other sciences and disciplines of engineering. The subject of fluid mechanics will be highly relevant in domains such as chemical, metallurgical, biological and ecological engineering. This series is particularly open to such new multidisciplinary domains.

1. M. Lesieur: *Turbulence in Fluids.* 2nd rev. ed., 1990   ISBN 0-7923-0645-7
2. O. Métais and M. Lesieur (eds.): *Turbulence and Coherent Structures.* 1991
    ISBN 0-7923-0646-5
3. R. Moreau: *Magnetohydrodynamics.* 1990   ISBN 0-7923-0937-5
4. E. Coustols (ed.): *Turbulence Control by Passive Means.* 1990   ISBN 0-7923-1020-9
5. A.A. Borissov (ed.): *Dynamic Structure of Detonation in Gaseous and Dispersed Media.* 1991
    ISBN 0-7923-1340-2
6. K.-S. Choi (ed.): *Recent Developments in Turbulence Management.* 1991 ISBN 0-7923-1477-8
7. E.P. Evans and B. Coulbeck (eds.): *Pipeline Systems.* 1992   ISBN 0-7923-1668-1
8. B. Nau (ed.): *Fluid Sealing.* 1992   ISBN 0-7923-1669-X
9. T.K.S. Murthy (ed.): *Computational Methods in Hypersonic Aerodynamics.* 1992
    ISBN 0-7923-1673-8
10. R. King (ed.): *Fluid Mechanics of Mixing.* Modelling, Operations and Experimental Techniques. 1992   ISBN 0-7923-1720-3
11. Z. Han and X. Yin: *Shock Dynamics.* 1993   ISBN 0-7923-1746-7
12. L. Svarovsky and M.T. Thew (eds.): *Hydroclones.* Analysis and Applications. 1992
    ISBN 0-7923-1876-5
13. A. Lichtarowicz (ed.): *Jet Cutting Technology.* 1992   ISBN 0-7923-1979-6
14. F.T.M. Nieuwstadt (ed.): *Flow Visualization and Image Analysis.* 1993 ISBN 0-7923-1994-X
15. A.J. Saul (ed.): *Floods and Flood Management.* 1992   ISBN 0-7923-2078-6
16. D.E. Ashpis, T.B. Gatski and R. Hirsh (eds.): *Instabilities and Turbulence in Engineering Flows.* 1993   ISBN 0-7923-2161-8
17. R.S. Azad: *The Atmospheric Boundary Layer for Engineers.* 1993   ISBN 0-7923-2187-1
18. F.T.M. Nieuwstadt (ed.): *Advances in Turbulence IV.* 1993   ISBN 0-7923-2282-7
19. K.K. Prasad (ed.): *Further Developments in Turbulence Management.* 1993
    ISBN 0-7923-2291-6
20. Y.A. Tatarchenko: *Shaped Crystal Growth.* 1993   ISBN 0-7923-2419-6
21. J.P. Bonnet and M.N. Glauser (eds.): *Eddy Structure Identification in Free Turbulent Shear Flows.* 1993   ISBN 0-7923-2449-8
22. R.S. Srivastava: *Interaction of Shock Waves.* 1994   ISBN 0-7923-2920-1
23. J.R. Blake, J.M. Boulton-Stone and N.H. Thomas (eds.): *Bubble Dynamics and Interface Phenomena.* 1994   ISBN 0-7923-3008-0

Mechanics

FLUID MECHANICS AND ITS APPLICATIONS
Series Editor: R. Moreau

24. R. Benzi (ed.): *Advances in Turbulence V.* 1995 ISBN 0-7923-3032-3
25. B.I. Rabinovich, V.G. Lebedev and A.I. Mytarev: *Vortex Processes and Solid Body Dynamics.* The Dynamic Problems of Spacecrafts and Magnetic Levitation Systems. 1994
ISBN 0-7923-3092-7
26. P.R. Voke, L. Kleiser and J.-P. Chollet (eds.): *Direct and Large-Eddy Simulation I.* Selected papers from the First ERCOFTAC Workshop on Direct and Large-Eddy Simulation. 1994
ISBN 0-7923-3106-0
27. J.A. Sparenberg: *Hydrodynamic Propulsion and its Optimization.* Analytic Theory. 1995
ISBN 0-7923-3201-6
28. J.F. Dijksman and G.D.C. Kuiken (eds.): *IUTAM Symposium on Numerical Simulation of Non-Isothermal Flow of Viscoelastic Liquids.* Proceedings of an IUTAM Symposium held in Kerkrade, The Netherlands. 1995 ISBN 0-7923-3262-8
29. B.M. Boubnov and G.S. Golitsyn: *Convection in Rotating Fluids.* 1995 ISBN 0-7923-3371-3
30. S.I. Green (ed.): *Fluid Vortices.* 1995 ISBN 0-7923-3376-4
31. S. Morioka and L. van Wijngaarden (eds.): *IUTAM Symposium on Waves in Liquid/Gas and Liquid/Vapour Two-Phase Systems.* 1995 ISBN 0-7923-3424-8
32. A. Gyr and H.-W. Bewersdorff: *Drag Reduction of Turbulent Flows by Additives.* 1995
ISBN 0-7923-3485-X
33. Y.P. Golovachov: *Numerical Simulation of Viscous Shock Layer Flows.* 1995
ISBN 0-7923-3626-7
34. J. Grue, B. Gjevik and J.E. Weber (eds.): *Waves and Nonlinear Processes in Hydrodynamics.* 1996 ISBN 0-7923-4031-0
35. P.W. Duck and P. Hall (eds.): *IUTAM Symposium on Nonlinear Instability and Transition in Three-Dimensional Boundary Layers.* 1996 ISBN 0-7923-4079-5
36. S. Gavrilakis, L. Machiels and P.A. Monkewitz (eds.): *Advances in Turbulence VI.* Proceedings of the 6th European Turbulence Conference. 1996 ISBN 0-7923-4132-5
37. K. Gersten (ed.): *IUTAM Symposium on Asymptotic Methods for Turbulent Shear Flows at High Reynolds Numbers.* Proceedings of the IUTAM Symposium held in Bochum, Germany. 1996 ISBN 0-7923-4138-4
38. J. Verhás: *Thermodynamics and Rheology.* 1997 ISBN 0-7923-4251-8
39. M. Champion and B. Deshaies (eds.): *IUTAM Symposium on Combustion in Supersonic Flows.* Proceedings of the IUTAM Symposium held in Poitiers, France. 1997 ISBN 0-7923-4313-1
40. M. Lesieur: *Turbulence in Fluids.* Third Revised and Enlarged Edition. 1997
ISBN 0-7923-4415-4; Pb: 0-7923-4416-2
41. L. Fulachier, J.L. Lumley and F. Anselmet (eds.): *IUTAM Symposium on Variable Density Low-Speed Turbulent Flows.* Proceedings of the IUTAM Symposium held in Marseille, France. 1997
ISBN 0-7923-4602-5
42. B.K. Shivamoggi: *Nonlinear Dynamics and Chaotic Phenomena.* An Introduction. 1997
ISBN 0-7923-4772-2
43. H. Ramkissoon, *IUTAM Symposium on Lubricated Transport of Viscous Materials.* Proceedings of the IUTAM Symposium held in Tobago, West Indies. 1998 ISBN 0-7923-4897-4
44. E. Krause and K. Gersten, *IUTAM Symposium on Dynamics of Slender Vortices.* Proceedings of the IUTAM Symposium held in Aachen, Germany. 1998 ISBN 0-7923-5041-3
45. A. Biesheuvel and G.J.F. van Heyst (eds.): *In Fascination of Fluid Dynamics.* A Symposium in honour of Leen van Wijngaarden. 1998 ISBN 0-7923-5078-2

Mechanics

FLUID MECHANICS AND ITS APPLICATIONS

Series Editor: R. Moreau

46. U. Frisch (ed.): *Advances in Turbulence VII*. Proceedings of the Seventh European Turbulence Conference, held in Saint-Jean Cap Ferrat, 30 June–3 July 1998. 1998   ISBN 0-7923-5115-0
47. E.F. Toro and J.F. Clarke: *Numerical Methods for Wave Propagation*. Selected Contributions from the Workshop held in Manchester, UK. 1998   ISBN 0-7923-5125-8
48. A. Yoshizawa: *Hydrodynamic and Magnetohydrodynamic Turbulent Flows*. Modelling and Statistical Theory. 1998   ISBN 0-7923-5225-4
49. T.L. Geers (ed.): *IUTAM Symposium on Computational Methods for Unbounded Domains*. 1998   ISBN 0-7923-5266-1
50. Z. Zapryanov and S. Tabakova: *Dynamics of Bubbles, Drops and Rigid Particles*. 1999   ISBN 0-7923-5347-1
51. A. Alemany, Ph. Marty and J.P. Thibault (eds.): *Transfer Phenomena in Magnetohydrodynamic and Electroconducting Flows*. 1999   ISBN 0-7923-5532-6
52. J.N. Sørensen, E.J. Hopfinger and N. Aubry (eds.): *IUTAM Symposium on Simulation and Identification of Organized Structures in Flows*. 1999   ISBN 0-7923-5603-9
53. G.E.A. Meier and P.R. Viswanath (eds.): *IUTAM Symposium on Mechanics of Passive and Active Flow Control*. 1999   ISBN 0-7923-5928-3
54. D. Knight and L. Sakell (eds.): *Recent Advances in DNS and LES*. 1999 ISBN 0-7923-6004-4
55. P. Orlandi: *Fluid Flow Phenomena*. A Numerical Toolkit. 2000   ISBN 0-7923-6095-8
56. M. Stanislas, J. Kompenhans and J. Westerveel (eds.): *Particle Image Velocimetry*. Progress towards Industrial Application. 2000   ISBN 0-7923-6160-1
57. H.-C. Chang (ed.): *IUTAM Symposium on Nonlinear Waves in Multi-Phase Flow*. 2000   ISBN 0-7923-6454-6
58. R.M. Kerr and Y. Kimura (eds.): *IUTAM Symposium on Developments in Geophysical Turbulence* held at the National Center for Atmospheric Research, (Boulder, CO, June 16–19, 1998) 2000   ISBN 0-7923-6673-5
59. T. Kambe, T. Nakano and T. Miyauchi (eds.): *IUTAM Symposium on Geometry and Statistics of Turbulence*. Proceedings of the IUTAM Symposium held at the Shonan International Village Center, Hayama (Kanagawa-ken, Japan November 2–5, 1999). 2001   ISBN 0-7923-6711-1
60. V.V. Aristov: *Direct Methods for Solving the Boltzmann Equation and Study of Nonequilibrium Flows*. 2001   ISBN 0-7923-6831-2
61. P.F. Hodnett (ed.): *IUTAM Symposium on Advances in Mathematical Modelling of Atmosphere and Ocean Dynamics*. Proceedings of the IUTAM Symposium held in Limerick, Ireland, 2–7 July 2000. 2001   ISBN 0-7923-7075-9
62. A.C. King and Y.D. Shikhmurzaev (eds.): *IUTAM Symposium on Free Surface Flows*. Proceedings of the IUTAM Symposium held in Birmingham, United Kingdom, 10–14 July 2000. 2001   ISBN 0-7923-7085-6
63. A. Tsinober: *An Informal Introduction to Turbulence*. 2001   ISBN 1-4020-0110-X; Pb: 1-4020-0166-5

Kluwer Academic Publishers – Dordrecht / Boston / London

Mechanics

***SOLID* MECHANICS AND ITS APPLICATIONS**
Series Editor: G.M.L. Gladwell

90. Y. Ivanov, V. Cheshkov and M. Natova: *Polymer Composite Materials – Interface Phenomena & Processes*. 2001   ISBN 0-7923-7008-2
91. R.C. McPhedran, L.C. Botten and N.A. Nicorovici (eds.): *IUTAM Symposium on Mechanical and Electromagnetic Waves in Structured Media* held in Sydney, NSW, Australia, 18-22 Januari 1999. 2001   ISBN 0-7923-7038-4
92. D.A. Sotiropoulos (ed.): *IUTAM Symposium on Mechanical Waves for Composite Structures Characterization*. Proceedings of the IUTAM Symposium held in Chania, Crete, Greece, June 14-17, 2000. 2001   ISBN 0-7923-7164-X
93. V.M. Alexandrov and D.A. Pozharskii: *Three-Dimensional Contact Problems*. 2001   ISBN 0-7923-7165-8
94. J.P. Dempsey and H.H. Shen (eds.): *IUTAM Symposium on Scaling Laws in Ice Mechanics and Ice Dynamics*. held in Fairbanks, Alaska, U.S.A., 13-16 June 2000. 2001   ISBN 1-4020-0171-1

Kluwer Academic Publishers – Dordrecht / Boston / London